Biological
MOLECULES

Molecular and Cell
BIOCHEMISTRY

Biological
MOLECULES

SMITH AND WOOD

CHAPMAN & HALL

University and Professional Division

London · New York · Tokyo · Melbourne · Madras

UK	Chapman & Hall, 2–6 Boundary Row, London SE1 8HN
USA	Chapman & Hall, 29 West 35th Street, New York NY100001
JAPAN	Chapman & Hall Japan, Thomson Publishing Japan, Hirakawacho Nemoto Building, 7F, 1-7-11 Hirakawa-cho, Chiyoda-ku, Tokyo 102
AUSTRALIA	Chapman & Hall Australia, Thomas Nelson Australia, 102 Dodds Street, South Melbourne, Victoria 3205
INDIA	Chapman & Hall India, R. Seshadri, 32 Second Main Road, CIT East, Madras 600 035

First edition 1991

© 1991 Chapman & Hall

Typeset in 10/11½pt Palatino by EJS Chemical Composition,
Midsomer Norton, Bath, Avon
Printed in Hong Kong

ISBN 0 412 40780 9

British Library Cataloguing in Publication Data

Biological molecules.
 1. Biochemistry
 I. Wood, E.J. (Edward J.) II. Smith, C.A. III. Series
574.192

 ISBN 0–412–40780–9

Library of Congress Cataloging-in-Publication Data

Smith, C.A. (Chris A.)
 Biological molecules/C.A. Smith and E. Wood.
    p.   cm. — (Molecules and cell biochemistry)
 Includes bibliographical references and index.
 ISBN 0–412–40780–9
  1. Biomolecules. 2. Biopolymers. 3. Biochemistry.
  I. Wood, Edward J., 1941– . II. Title. III. Series.
QP514.2.S575   1991
574.19'2—dc20

91–10094
CIP

Copy Editors: Sarah Firman and Judith Ockenden
Sub-editor: Simon Armstrong
Production Controller: Marian Saville
Layout Designer: Geoffrey Wadsley (after an original design by Julia Denny)
Illustrators: Capricorn Graphics

Contents

Editors' foreword

This book is one of a series of brief fundamental texts for junior under-graduates and diploma students in biological science. The series, Molecular and Cell Biochemistry, covers the whole of modern biochemistry, integrating animal, plant and microbial topics. The intention is to give the series special appeal to the many students who read biochemistry for only part of their course and who are looking for an all-encompassing and stimulating approach. Although all books in the series bear a distinct family likeness, each stands on its own as an independent text.

Many students, particularly those with less numerate backgrounds, find elements of their biochemistry courses daunting, and one of our principal concerns is to offer books which present the facts in a palatable style. Each chapter is prefaced by a list of learning objectives, with short summaries and revision aids at the ends of chapters. The text itself is informal, and the incorporation of marginal notes and information boxes to accompany the main text give a tutorial flavour, complementing and supporting the main narrative. The marginal notes and boxes relate facts in the text to applicable examples in everyday life, in industry, in other life sciences and in medicine, and provide a variety of other educational devices to assist, support, and reinforce learning. References are annotated to guide students towards effective and relevant additional reading.

Although students must start by learning the basic vocabulary of a subject, it is more important subsequently to promote understanding and the ability to solve problems than to present the facts alone. The provision of imaginative problems, examples, short-answer questions and other exercises are designed to encourage such a problem-solving attitude.

A major challenge to both teacher and student is the pace at which biochemistry and molecular biology are advancing at the present time. For the teacher and textbook writer the challenge is to select, distill, highlight and exemplify, tasks which require a broad base of knowledge and indefatigable reading of the literature. For the student the challenge is not to be over-whelmed, to understand and ultimately to pass the examination! It is hoped that the present series will help by offering major aspects of biochemistry in digestible portions.

This vast corpus of accumulated knowledge is essentially valueless unless it can be used. Thus these texts have frequent, simple exercises and problems. It is expected that students will be able to test their acquisition of knowledge but also be able to use this knowledge to solve problems. We believe that only in this way can students become familiar and comfortable with their knowledge. The fact that it is useful to them will mean that it is retained, beyond the last examination, into their future careers.

The present series was written by lecturers in universities and polytechnics who have many years of experience in teaching, and who are also familiar with current developments through their research interests. They are, in addition, familiar with the difficulties and pressures faced by present-day students in the biological sciences area. The editors are grateful for the co-operation of all their authors in undergoing criticism and in meeting requests to re-write (and sometimes re-write again), shorten or extend what they originally wrote. They are also happy to record their grateful thanks to those many individuals who very willingly supplied illustrative material promptly and generously. These include many colleagues as well as total strangers whose response was positive and unstinting. Special thanks must go to the assessors who very carefully read the chapters and made valuable suggestions which gave rise to a more readable text. Grateful thanks are also due to the team at Chapman & Hall who saw the project through with good grace in spite, sometimes, of everything. These include Dominic Recaldin, Commissioning Editor, Jacqueline Curthoys, formerly Development Editor, Simon Armstrong, Sub-editor, and Marian Saville, Production Controller.

Finally, though, it is the editors themselves who must take the responsibility for errors and omissions, and for areas where the text is still not as clear as students deserve.

Contributors

DR B. CATLEY   *Department of Biological Sciences, Heriot-Watt University, Edinburgh, UK.   Chapter 6.*

DR P. GACESA   *Department of Biochemistry, University of Wales College of Cardiff, Cardiff, UK.   Chapters 4 and 5.*

DR J.J. GAFFNEY   *Department of Biological Sciences, Manchester Polytechnic, Manchester, UK.   Chapters 7 and 8.*

DR J. HARLAND   *School of Health Sciences, The Liverpool Polytechnic, Liverpool, UK.   Chapter 3.*

DR C.A. SMITH   *Department of Biological Sciences, Manchester Polytechnic, Manchester, UK.   Chapters 2, 7 and 8.*

DR E.J. WOOD   *Department of Biochemistry, University of Leeds, Leeds, UK. Chapters 1 and 2.*

Preface

The book looks at the basis of biological molecules largely from biological point of view. It aims to put biochemistry in its biological context with examples from medicine, nutrition and agriculture. Inevitably the molecules described and discussed tend to be macromolecules, or molecules such as the lipids, that associate non-covalently to form large aggregates, because these are the types of molecules that characterize living organisms.

A background of chemistry and some physics is required, although these aspects are explained as gently as possible, recognizing that many students in the biological sciences regard these as difficult areas. This is not to imply that physics and chemistry are not important, for they certainly are. Nevertheless, many teachers in universities and polytechnics recognise that it is impractical and unrealistic to demand a rigorous study of chemistry, for example, before students can be allowed to start on the study of biological chemistry. This state of affairs may not be the most desirable, but it exists. Therefore, the present text assumes some chemical knowledge and introduces additional chemistry in a biological, functional context where appropriate. This is helped by the deliberate use of side-notes, glossaries and boxes of additional information of relevant interest.

The eight chapters in this volume examine aspects of biological molecules that are features of, and are essential to, any biological science course, including proteins, polysaccharides, nucleic acids and lipids, with special emphasis on enzymes. It is hoped that the treatment will kindle students' interest and follow them to see the vital relevance of molecular structure to biological function.

In the past many student have had difficulty in visualizing molecular structures, and this has to some extent inhibited their enjoyment and understanding of biochemistry. There is no excuse for this now. In the last 5–10 years a vast amount of research, coupled with explosive developments in computer-graphics technology, has enabled us to visualize clearly the structures of biological molecules, and especially of macromolecules, in atomic detail. Wherever appropriate such structural models have been included to enable readers to comprehend more easily how these macromolecules fulfil their biological roles.

Abbreviations

A	adenine (alanine)
ACP	acyl carrier protein
ACTH	adrenal corticotrophic hormone
ADP	adenosine diphosphate
Ala,A	alanine
AMP	adenosine monophosphate
cAMP	adenosine 3′,5′-cyclic monophosphate
Arg,R	arginine
Asn,N	asparagine
Asp,D	aspartic acid
ATP	adenosine triphosphate
ATPase	adenosine triphosphatase
C	cytosine (cysteine)
CDP	cytidine diphosphate
CMP	cytidine monophosphate
CTP	cytidine triphosphate
CoA(CoASH)	coenzyme A
CoQ(Q)	coenzyme Q, ubiquinone
Cys, C	cysteine
d-	2-deoxy-
D	aspartic acid
d-Rib	2-deoxyribose
DNA	deoxyribonucleic acid
cDNA	complementary DNA
e-	electron
E	glutamic acid
E	oxidation-reduction potential
F	phenylalanine
F	the Faraday (9.648×10^4 coulomb mol^{-1})
FAD	flavin adenine dinucleotide
Fd	ferredoxin
fMet	N-formyl methionine
FMN	flavin mononucleotide
Fru	fructose
g	gram
g	acceleration due to gravity
G	guanine (glycine)
G	free energy
Gal	galactose
Glc	glucose

Gln, Q	glutamine
Glu, E	glutamic acid
Gly, G	glycine
GDP	guanosine diphosphate
GMP	guanosine monophosphate
GTP	guanosine triphosphate
H	histidine
H	enthalpy
Hb	haemoglobin
His, H	histidine
Hyp	hydroxyproline (HOPro)
I	isoleucine
Ig G	immunoglobulin G
Ig M	immunoglobulin M
Ile, I	isoleucine
ITP	inosine triphosphate
J	Joule
K	degrees absolute (Kelvin)
K	lysine
L	leucine
Leu, L	leucine
lu x	natural logarithm of $x = 2.303 \log_{10} x$
Lys, K	lysine
M	methionine
M_n	relative molecular mass, molecular weight
Man	manose
Mb	myoglobin
Met, M	methionine
N	asparagine
N	Avogadro's number (6.022×10^{23})
N	any nucleotide base (e.g. in NTP for nucleotide triphosphate)
NAD^+	nicotinamide adenine dinucleotide
$NADP^+$	nicotinamide adenine dinucleotide phosphate
P	proline
Pi	inorganic phosphate
PPi	inorganic pyrophosphate
Phe, F	phenylalanine
Pro, P	proline
Q	coenzyme Q, ubiquinone
Q	glutamine
R	arginine
R	the gas constant ($8.314 \, J \, K^{-1} \, mol^{-1}$)
Rib	ribose
RNA	ribonucleic acid
mRNA	messenger RNA

rRNA	ribosomal RNA
tRNA	transfer RNA
s	second
s	sedimentation coefficient
S	svedberg unit (10^{-13} seconds)
S	serine
SDS	sodium dodecylsulphate
ser, S	serine
T	thymine
T	threonine
TPP	thiamine pyrophosphate
Trp, W	tryptophan
TTP	thymidine triphosphate (dTTP)
Tyr, Y	tyrosine
U	uracil
UDP	uridine diphosphate
UDP-Glc	uridine diphosphoglucose
UMP	uridine monophosphate
UTP	uridine triphosphate
V	valine
V	volt
Val, V	valine
W	tryptophan
Y	tyrosine

Greek alphabet

A	α	alpha	N	ν	nu
B	β	beta	Ξ	ξ	xi
Γ	γ	gamma	O	o	omicron
Δ	δ	delta	Π	π	pi
E	ε	epsilon	P	ϱ	rho
Z	ζ	zeta	Σ	σ	sigma
H	η	eta	T	τ	tau
Θ	θ	theta	Y	υ	upsilon
I	ι	iota	Φ	ϕ	phi
K	κ	kappa	X	χ	chi
Λ	λ	lambda	Ψ	w	psi
M	μ	mu	Ω	ω	omega

Objectives

After reading this chapter you should be able to:

☐ define biochemistry as one of the sciences of life and appreciate that all forms of life obey the laws of chemistry;

☐ outline the key properties of the molecules of life, especially the macromolecules – proteins, nucleic acids and polysaccharides;

☐ describe briefly the energy relationships by which organisms maintain themselves and drive their activities;

☐ explain how interactions account for many, if not all, biological phenomena such as highly specific and efficient catalysis and the formation of selectively permeable biological membranes.

1.1 Introduction: Life is based on carbon

Chemical analysis of practically any organism on earth, be it animal, plant or microorganism, shows that of all the elements in the periodic table, very few are present in living things. What is more, these are always the same few elements and they are almost always found in similar proportions. The elements carbon, hydrogen, nitrogen and oxygen, in combination, account for over 95% of living matter. If phosphorus and sulphur are added to the list, the figure rises to over 98% (Table 1.1).

Table 1.1 *Elemental composition (expressed as %)* of animals, plants and microorganisms*

Element	Humans	Green plant	Bacterium
Oxygen	62.81	77.80	73.68
Carbon	19.37	11.34	12.14
Hydrogen	9.31	8.72	9.94
Nitrogen	5.14	0.83	3.04
Phosphorus	0.63	0.71	0.60
Sulphur	0.64	0.10	0.32

* Note that these do not quite add up to 100% because small amounts of other elements are also present. An example is the iron in haemoglobin.

Table 1.2 *Chemical composition of a typical bacterial cell*

Compound	Percentage of total weight
Water	70
Macromolecules	
Proteins	15
Nucleic acids (DNA, RNA)	7*
Polysaccharides	3
Lipids	2
Small organic molecules	2†
Inorganic ions	1

* Made up of about 1% DNA and 6% RNA.
† This includes building blocks for making macromolecules and other molecules in the process of being synthesized or degraded.

Although these six elements are combined in many different ways, chemical analysis of any organism again shows that, at the molecular level, it consists of relatively few types of compounds. Water makes up about 70% of most organisms (Table 1.2). Of the remainder, the major compounds are **carbohydrates, fats (or lipids), amino acids**, mostly combined to form proteins, and **nucleotides**, many of them in combination as the nucleic acids RNA (ribonucleic acid) and DNA (deoxyribonucleic acid).

Although many other compounds are present, these four categories account for the vast bulk of **organic material** in organisms. These compounds are called 'organic' because they are based on the element carbon.

The carbon atom

The chemistry of compounds containing carbon is called *organic chemistry*. (The simplest carbon compounds such as CO, CO_2 and the carbonates are excluded from this definition.) This branch of chemistry was originally called 'organic' because it used to be thought that only living organisms could manufacture such compounds. This idea had to be abandoned after Wohler showed that an organic compound, urea, could be made from an inorganic one in the laboratory. Today, several million organic compounds are known and characterized, and the majority of them have been synthesized in the laboratory, not in organisms.

Organic compounds are based on carbon in combination with itself and with hydrogen, oxygen, nitrogen, and a few other elements. It is the structure of the carbon atom that holds the key to this diversity of chemical structure. A carbon atom has six protons and six electrons, two in the inner shell and four in the outer shell (Fig. 1.1). Given this arrangement, carbon has little tendency

☐ Wöhler in 1878 showed that ammonium cyanate could be turned into urea simply by the action of heat:

$$NH_4OCN \rightarrow H_2N.CO.NH_2$$

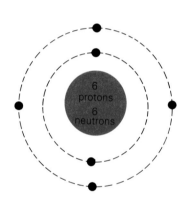

Fig. 1.1 The carbon atom has four electrons in the outer shell.

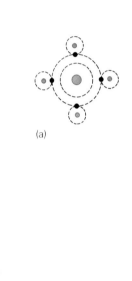

(a)

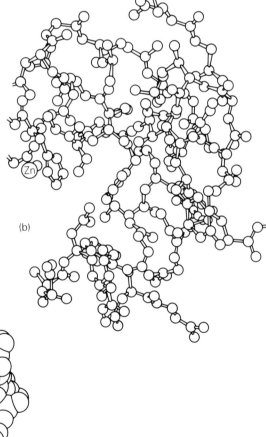

(b)

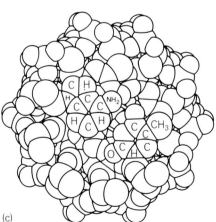

(c)

Fig. 1.2 (a) The structure of methane, CH_4, the simplest carbon compound and, in contrast, the structure of (b) a protein molecule, carboxypeptidase, and (c) a cross-section through the double helix of DNA. In (c) the part of the structure shown represents a tiny fraction of the whole molecule.

Organic chemistry: the chemistry of carbon compounds (excepting carbonates and CO_2, CO, etc.); all the rest is inorganic chemistry.

Reference Widom, J.M. and Edelstein, S.J. (1981) *Chemistry, An Introduction to General, Organic and Biological Chemistry*, W.H. Freeman, San Francisco, USA. An imaginative text that bridges the gap between chemistry and biochemistry.

to gain or lose electrons—that is, to become ionized. Instead, it forms stable covalent bonds with itself and other atoms. The forces that hold together the simplest organic compound, methane (CH_4), also hold together huge molecules with millions of atoms (Fig. 1.2). Because carbon atoms can link together to form chains, networks, and circles, there is virtually no limit to the molecular architecture of organic compounds (Fig. 1.3). Some of these, fashioned by organic chemists, are vital in our everyday lives as the products of the pharmaceutical and plastics industries.

☐ When atoms share electrons so that their outermost electron orbitals achieve the rare gas structure (2 or 8 electrons) they are said to be joined by a covalent bond.

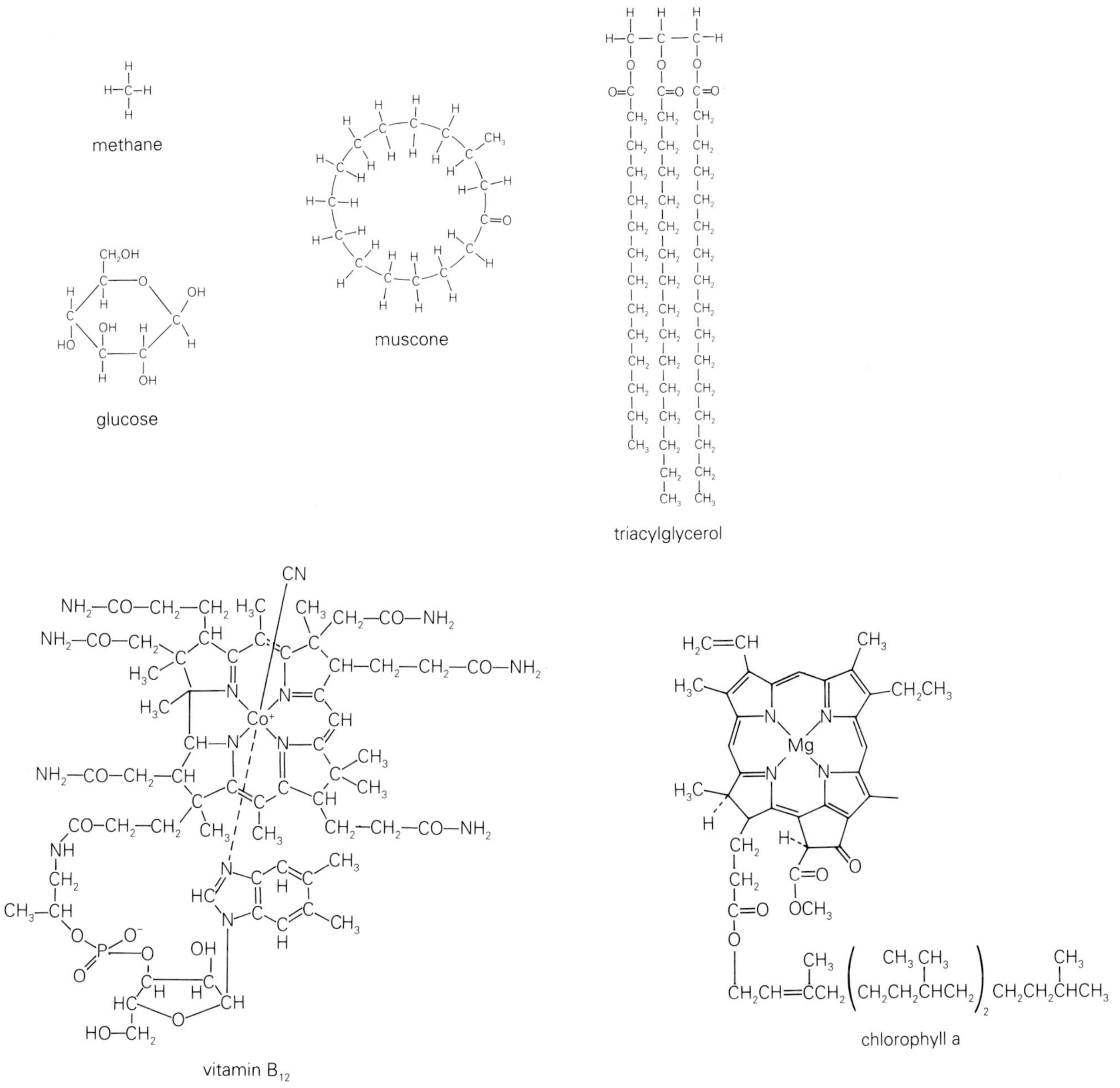

methane

glucose

muscone

triacylglycerol

vitamin B_{12}

chlorophyll a

Fig. 1.3 Carbon atoms can link with other carbon atoms and to many other atoms to form a vast range of organic compounds. The same forces that hold methane (CH_4) together, also enable extremely complicated but stable molecules to be formed. In many structures, such as triacylglycerol, the predominant motif is chains. In others, such as the sugar, glucose, it is rings. Muscone, the sex attractant of the Tibetan musk deer, is also based on a carbon ring. Other vital molecules, such as chlorophyll and vitamin B_{12}, incorporate chains, rings, and metal ions.

POLYMERS AND MACROMOLECULES. A polymer is formed by linking together hundreds or thousands of molecular units (monomers) to form a large molecule. Polymers may be relatively simple chains of identical monomer units or may be extremely complex. The chains may be branched or unbranched. If the units are identical, the result is a **homopolymer**. If the units differ, the product is a **heteropolymer**. Figure 1.4 shows two types of synthetic polymer.

Compounds in organisms do not differ fundamentally from those used in organic chemistry. Organisms also manipulate their molecules in a variety of ways to store or release energy, to act as catalysts, to store information, to construct the mechanical elements they need, and in many other ways. The majority of the **biomolecules** that enable organisms to function are polymers. Biological polymers, or **macromolecules**, are also built up from monomers consisting of small molecular building blocks. Some are homopolymers of the sugar, or **monosaccharide** unit, glucose. Others, such as proteins, are heteropolymers formed by linking together a selection chosen from 20 kinds of amino acids (Fig. 1.5).

Box 1.1
Nomenclature

Organic chemical compounds are named according to an unambiguous scheme settled by international agreement through the International Union of Pure and Applied Chemistry (IUPAC). The complete set of rules is a lengthy document that occupies a large book. This is absolutely essential for organic chemistry. For example, there are 802 possible compounds of molecular formation $C_{13}H_{28}$ and any of these may be encountered by an organic chemist. Thus CH_2=$CHCH_2CH_3$ is but-1-ene and CH_3CH=$CHCH_3$ is but-2-ene.

On the whole biochemists do not use this nomenclature, preferring the so-called 'trivial' names. Thus, $HOCH_2CH(OH)CH_2OH$ is propane-1,2,3-triol, but is usually referred to as glycerol, and ethanoate is referred to as acetate. It is not that biochemists disagree with the IUPAC nomenclature system, quite the contrary. Rather, it is a matter of habit, convenience and practicality. This may be illustrated as follows:

1. Biochemists deal with relatively few organic compounds. There may seem a lot to the novice, but compared with the millions of possible compounds known to organic chemistry, there are very few.
2. It is a great intellectual help to be able to identify any of these compounds instantly because this will immediately establish its pedigree: where it came from, what it usually does in the cellular machinery, and where it goes. Thus, the amino acid, methionine, is 2-amino-4-methylmercaptobutyric acid. The name *methionine* instantly identifies this compound as one of the 20 amino acids found in proteins, and a 'methionyl residue' is recognizable as a building block of a polypeptide chain.
3. It would be both impossible and stupid to try to give a systemic name to a protein. The molecular formula of human haemoglobin is $C_{2592}H_{4664}O_{832}N_{812}S_8Fe_4$ and the position in space of every atom in the molecule is known. The name 'haemoglobin' conveys over 100 years of haemoglobin and protein research. This is an extreme example, but there are many other compounds where a systemic name would be unhelpful.
4. Many trivial names carry information about the history and origin of compounds. Thus, citric acid from citrus fruits; even chemists use 'benzene' and 'glucose'. Such trivial names encapsulate a great deal. They may seem to present the novice with problems, but it is better to get down to learning them, rather than fighting against them. It will be many years before biochemistry texts deal with ethanoyl CoA rather than acetyl CoA, and the systematic name of the 'CoA' portion would be practically unrecognizable.
5. When necessary, full IUPAC names are used by biochemists in order to be quite specific and to avoid ambiguity. It appears not to be necessary very often. A survey of a recent issue of any biochemistry journal will confirm this.

Homopolymers: *formed by combining identical monomer building blocks, e.g. AAAAA…*
Heteropolymers: *combinations of more than one type of monomer. Structures may be simple as in –(A.B)$_n$– or highly complicated as in –ACBABDEG–.*

Biomolecules: *molecules, large and small, found in living organisms and manufactured by cells.*
Macromolecules: *means 'large molecules', but this is not very helpful as it does not define what 'large' is. In biological systems it means molecules with M$_r$ in the thousands to millions, and includes proteins, nucleic acids and polysaccharides.*

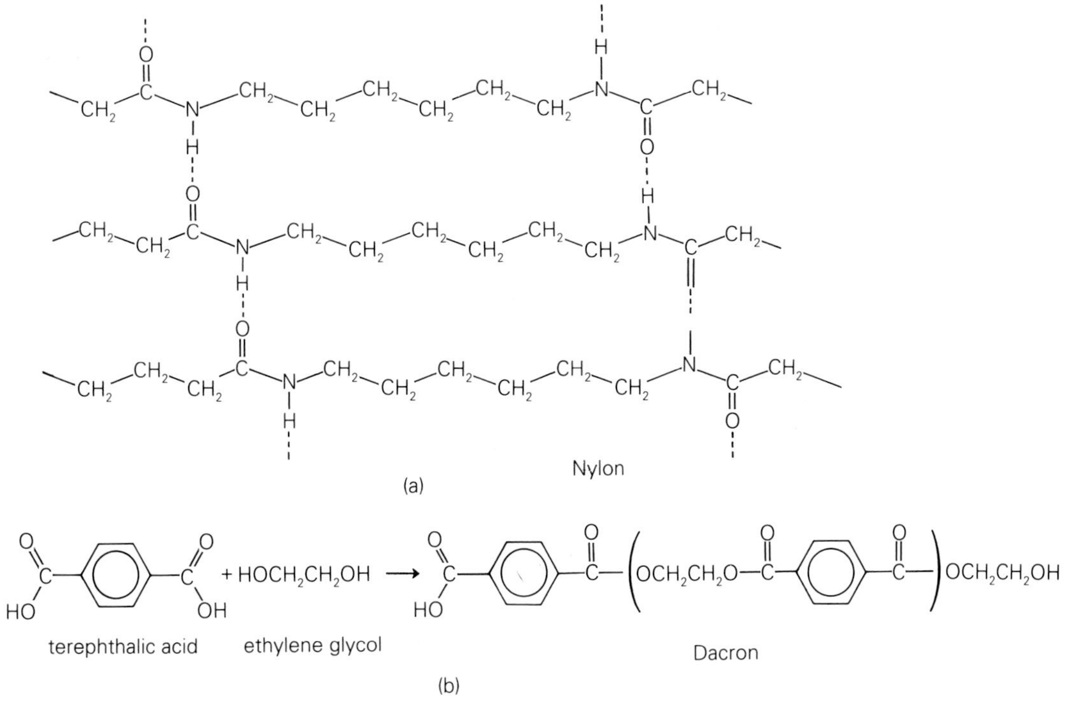

Nylon

(a)

terephthalic acid   ethylene glycol → Dacron

(b)

Fig. 1.4 Some synthetic polymers. (a) Nylon is a polyamide in which the bonds between monomers are similar to those in polypeptides. Hydrogen bonds between neighbouring polymer chains account for the high tensile strength of nylon. (b) Dacron is a polyester formed from terephthalic acid and ethylene glycol.

Box 1.2 Carbohydrates

Carbohydrates (see Chapter 6) are polyhydroxy compounds that contain in addition an aldehyde or a ketone. They are related structurally to glyceraldehyde or hydroxy-acetone, which may be regarded as the simplest members of the family.

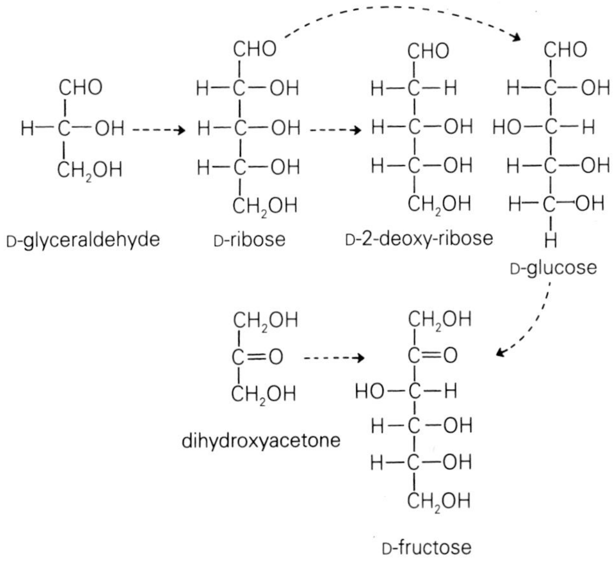

D-glyceraldehyde   D-ribose   D-2-deoxy-ribose   D-glucose

dihydroxyacetone   D-fructose

Some relationships between carbohydrates are shown: the dotted lines indicate relationships, *not* how one compound is converted to another. These units are called monosaccharides.

Monosaccharides may link together in a variety of ways, but always by condensation reactions, to form di-, tri-, oligo- and polysaccharides of wide biological importance. The chains may be branched or unbranched. Starch, glycogen and cellulose, for example, are all polymers of D-glucose.

Vitamin B_{12} (cobalamin), M_r 1355, would not be regarded as a macromolecule while insulin, M_r 5780, would.

Monosaccharides: *a very large range of carbohydrates of general formula $(CH_2O)_n$, where n is commonly 3–7. Such units link together to form disaccharides, oligosaccharides and polysaccharides.*

Reference Atkins, P.W. (1987) *Molecules* (Scientific American Library), W.H. Freeman, New York, USA. A beautiful picture book of molecules.

Analysis of proteins (see Chapters 2 and 3) from whatever source shows that they are all made up from a selection from a standard set of 20 amino acids.

In proteins, the selection taken from the 20 amino acids are linked together by peptide bonds formed by condensation reactions.

$$H_3\overset{+}{N}-\underset{\underset{\text{amino acid 1}}{}}{\overset{\overset{R'}{|}}{CH}}-CO_2^- + H_3\overset{+}{N}-\underset{\underset{\text{amino acid 2}}{}}{\overset{\overset{R''}{|}}{CH}}-CO_2^- \xrightarrow{-H_2O} H_3\overset{+}{N}-\overset{\overset{R'}{|}}{CH}-CO-NH-\underset{\underset{\text{dipeptide}}{}}{\overset{\overset{R''}{|}}{CH}}-CO_2^-$$

Proteins may contain hundreds or even thousands of amino acid units or 'residues'. A peptide is formed from a few amino acids, while a polypeptide is formed from many. The term protein is used to describe the biologically active form: this may have one or more polypeptides and each of these is folded up into its 'correct' form.

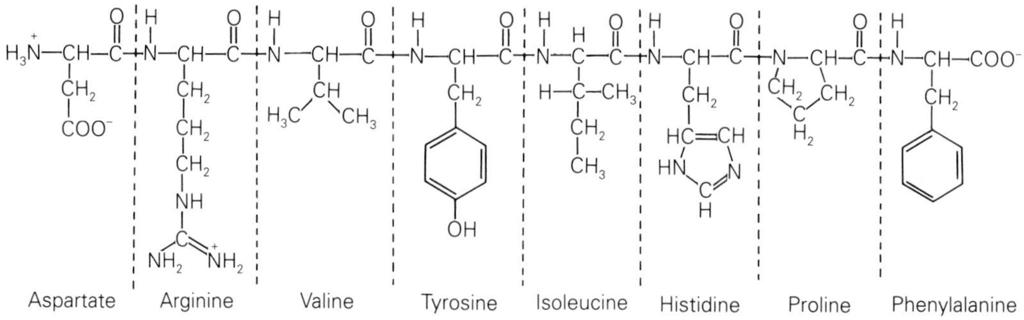

Fig. 1.5 Peptides, polypeptides and proteins are all made by linking together a selection of 20 amino acids. The molecule shown is angiotensin II, an octapeptide (composed of eight amino acid units). Angiotensin is a hormone with a role in controlling the blood pressure. Proteins may contain polypeptide chains hundreds or even thousands of amino acid units long.

Labels: Aspartate, Arginine, Valine, Tyrosine, Isoleucine, Histidine, Proline, Phenylalanine

Table 1.3 *Macromolecules are polymers*

Monomer	Polymer
Monosaccharides	Polysaccharides
Amino acids	Proteins
Nucleotides	Nucleic acids

☐ Nucleotides have the structure: nitrogenous base–sugar–phosphate, and in DNA or RNA four different nitrogenous bases occur. When nucleotides are linked together *via* the sugar–phosphate portions a 'nucleic acid' is produced.

Exercise 1

For a peptide of 20 different amino acids, each of which only occurs once, how many possible sequential arrangements are there?

Clearly, heteropolymers offer the possibility of a colossal number of different structures. Think of a necklace with 20 different types of beads: there are many, many possibilities for the order in which the beads can be strung. This analogy is useful because the finished article will still be recognizable as a necklace, regardless of the order of the beads. In a similar way, the polymers or macromolecules in organisms can all be recognized as belonging to one of only three classes of compounds. Those made of amino acids are **proteins**; those made of sugars are **polysaccharides**; and the third class, made by linking together nucleotides, are called **nucleic acids** (Table 1.3).

Analysis of any cell from any organism reveals that they all contain proteins, polysaccharides, and nucleic acids. Different organisms contain different *types* as proteins, polysaccharides and nucleic acids (Table 1.4), with very different chemical (and more importantly) different biological properties. Nevertheless, they are all easily recognizable as belonging to a particular class of macromolecule.

SHAPE AND CONFIGURATION. Macromolecules interact with each other and with a wide variety of other molecules found in the cell. The idea of interaction carries with it the implication of recognition. Two molecules come together and interact because they have the appropriate complementary shapes that fit closely together.

Molecules have three-dimensional shapes. Carbon may join with four other atoms in a tetrahedral arrangement (Fig. 1.6). If the four atoms are all different, there are two possible ways of joining them to the carbon atom. The

The nucleic acids are *ribonucleic acid* (RNA) and *deoxyribonucleic acid* (DNA). DNA in the nucleus of cells carries the genetic information and RNA is involved in the various processes that enable the genetic information in DNA to be 'translated' into proteins.

Analysis shows that nucleic acids contain three types of unit: phosphates, pentose sugars, and nitrogenous bases derived from the heterocyclic compounds purine and pyrimidine. Both DNA and RNA are long polymers consisting of a sugar–phosphate backbone with, on each of the sugars, a base. Because any one from a set of four different bases may be used on each sugar, the sequence of bases on the polymer forms a code, that is, *information* is stored in the sequence of bases.

GAGAAAATGGAGAGAGGCACGGACACAGAGACTTGGTCAT...

In RNA the pentose sugar is ribose: in DNA it is deoxyribose. Although this may seem a trivial difference, the nature of the resulting polymers is very different and each has its own specific biological role. The enzymes that deal with DNA do not, in general, recognize RNA, for example.

The links in the sugar–phosphate backbone are formed by condensation reactions. They are to the 3 and the 5 positions of the pentose sugar rings. Thus, the chains of DNA and RNA have 'direction'.

The unit base–sugar (which may also exist free) is called a *nucleoside*. Confusingly, the unit base–sugar–phosphate (which may also exist free) is called a *nucleotide*.

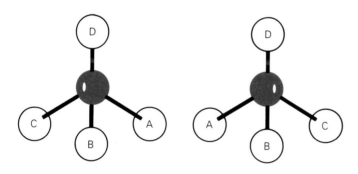

Fig. 1.6 Because the valencies on the carbon atom have a tetrahedral arrangement, the three-dimensional molecule shown here can be built in two ways which are mirror images of each other. This applies to any carbon atom that carries four different groups. Although the chemical properties of such a pair of compounds are similar to each other, the biological properties may be very different. This is because biological recognition works by the interactions between molecules with complementary shapes. Typically one molecule will be biologically active and the other totally inactive.

Table 1.4 *Biological functions of some macromolecules*

Proteins	
Enzymes	Catalysis
Hormones	Messengers
Collagen	Structural
Polysaccharides	
Cellulose	Structural
Starch, glycogen	Storage
Chondroitin sulphate	Connective tissue matrix
Nucleic acids	
DNA	Information storage
RNA	Information transfer, structural

two compounds are in fact mirror images of one another (Fig. 1.6). Although two such compounds have practically identical chemical properties and closely similar physical properties, their behaviour in biological systems is usually quite distinct. One compound is readily accepted by a biological system, the other is ignored, or may even be toxic! This is because biological systems work by recognizing shapes, or molecular **configurations**—the precise positions of atoms and groups of atoms relative to each other.

When a macromolecule is built from monomer units that can exist as mirror-image pairs, the 'correct' member of the pair is always chosen. This ensures that the resulting macromolecule will have the shape and configuration required. The amino acids provide a good example. Each amino acid can exist in one of two possible configurations that are mirror images of each other. To make a protein whose shape is the 'correct' one for biological activity, only one configuration of the amino acids is utilized.

Box 1.5
Asymmetry of carbon compounds

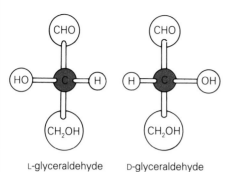

L-glyceraldehyde   D-glyceraldehyde

Fischer projections

$$\overset{1}{C}HO$$
$$HO-\overset{2}{C}-H$$
$$\overset{3}{C}H_2OH$$

$$\overset{1}{C}HO$$
$$H-\overset{2}{C}-OH$$
$$\overset{3}{C}H_2OH$$

$$CHO$$
$$HO-\!\!\!|\!\!\!-H$$
$$CH_2OH$$

$$CHO$$
$$H-\!\!\!|\!\!\!-OH$$
$$CH_2OH$$

$$\overset{1}{C}O_2H$$
$$H-\overset{2}{C}-NH_2$$
$$\overset{3}{C}H_3$$

Fischer projection formula: $-NH_2$ on right, therefore related to D-glyceraldehyde (Note that D-alanine does not occur in proteins)

$$CO_2H$$
$$H-\!\!\!-\overset{|}{NH_2} \quad \longleftarrow \quad Observer$$
$$CH_3$$

Priority rule $NH_2 > COOH > CH_3 > H$
(this has to be looked up)

Chirality of L-isoleucine = (2S), (3S)-isoleucine

$$\overset{1}{C}O_2H$$
$$H_2N-\overset{2}{C}-H$$
$$H_3C-\overset{3}{C}-H$$
$$\overset{4}{C}H_2$$
$$\overset{5}{C}H_3$$

both C2 and C3 are asymmetric

The four valencies of the carbon atom point to the corners of an equilateral tetrahedron. When a carbon atom has four different substituents, two distinct spatial arrangements are possible, one of which is the mirror image of the other. These are called *enantiomers* and the two molecules are not superimposable.

Glyceraldehyde, like many biomolecules, possesses at least one asymmetric carbon and can therefore exist in isomeric forms. Because biological reactions require specific three-dimensional structures to be recognized, usually only one of the possible isomers is biologically active. The two mirror image isomers may have closely similar chemical and physical properties making them difficult to distinguish in the chemical laboratory. To a biological organism, however, they are as different as chalk and cheese.

Fischer projection formulae are a way of writing such structures unambiguously in two dimensions, provided the rules are kept. It is understood, in the projection, that bonds pointing horizontally are coming out of the plane of the paper whereas those pointing vertically go down below the plane of the paper (see figure). Carbon atom number 1 must always be written at the top: for the glyceraldehyde shown in the figure, this is the aldehyde carbon. Such representations may not be removed from the two-dimensional world of the paper, although they may slide around on the paper.

One physical property that allows enantiomers to be distinguished is that solutions of such compounds rotate the plane of polarized light characteristically: clockwise, to the right $(+)$; or anticlockwise, to the left $(-)$. (Polarized light is produced by passing light through a crystal of iceland spar (calcite) or through certain polaroid materials, as used in sunglasses). The isomers that rotate the plane of light to the right are called *dextrorotatory* and the ones that rotated it to the left, *laevorotatory*.

In the Fischer convention, a particular asymmetric structure is designated by placing a D- or an L- before the name. In the D- isomer (the functional group in question in glyceraldehyde), an $-OH$ group (see figure) is written on the right, and in the L-isomer, on the left. This D- and L-designation has nothing to do with the direction in which the plane of polarized light is rotated. It was assigned quite arbitrarily by Fischer, D- to dextrorotatory glyceraldehyde, and L- *vice versa*; Fischer knew he had a 50 : 50 chance of being wrong. It was not until 50 years later as a result of X-ray crystallographic studies on a derivative of this compound, that he was shown to have made the correct choice for glyceraldehyde.

The carbohydrates are designated by the D- and L- convention. Unfortunately, monosaccharides may have a number of asymmetric carbon atoms. The Fischer convention applies to the asymmetric carbon atoms with the highest number. (This will be C5 in a hexose, for example, because C6 is not asymmetric). This convention *relates* monosaccharides structurally to the two glyceraldehydes, but it says nothing about their optical rotation. Thus, both D-glucose and D-fructose are *related* to D-glyceraldehyde: in fact, their optical rotations are dextrorotatory for glucose but laevorotatory for fructose. This is because each asymmetric carbon makes its own contribution to the total observed optical rotation.

The modern convention for dealing with this problem of specifying exactly which isomer is being referred to is the R,S convention (R for *rectus*, right; S for *sinister*, left.) Asymmetric molecules are said to be **chiral** which comes from a Greek word for handedness. In the R,S notation the four groups surrounding the carbon atom in question are ranked according to a priority sequence. This is decided by a number of rules, of which the most important is that higher atomic number precedes lower. The convention requires that the carbon atom is viewed down the axis connecting this central carbon atom to the group having lowest priority according to the rules. It is then asked whether the sequence of groups, according to their priority, goes clockwise (R) or anticlockwise (S). Each chiral carbon atom in turn can be examined thus. In the example shown, the amino acid, L-isoleucine, has the R,S designation (2S), (3S)-isoleucine.

1.2 Chemical bonds

Chemical reactions are the making and breaking of the bonds that link atoms together in a molecule. Life processes consist of building up and breaking down molecules; in other words, they are chemical reactions. Consequently, it is important to know something about chemical bonds in order to be able to understand biochemical processes.

Covalent bonds

The majority of bonds in organic compounds are covalent bonds. A covalent bond is one in which electrons are shared between atoms. Such bonds are strong. The **bond energy**—that is, the energy required to break the bond, is comparatively high: that for a carbon–carbon bond is around $350 \, \text{kJ mol}^{-1}$. It is necessary to use comparative terms here: some of the bonds dealt with later on are 'weak' in comparison with a covalent bond (Table 1.5). Making and breaking covalent bonds is the stuff of organic chemistry. In organisms the processes and reactions by which covalent bonds are made and broken are identical to those in organic chemistry, but the *means* by which this is achieved are quite different. Unlike the organic chemist, the cell operates between 0 and 40°C, at normal atmospheric pressure, and in aqueous solution.

Table 1.5 *Bonds and forces other than covalent bonds*

Type of bond	Example	Comments
Ionic bond	$-CO_2^-$ on glutamyl residue and $-NH_3^+$ on lysyl residue	Potentially very strong, depending upon the dielectric constant
Hydrogen bond	$C{=}O$ and NH in adjacent peptide bonds	Up to $20 \, \text{kJ mol}^{-1}$: form because oxygen is electronegative creating dipoles
Hydrophobic effects	Tendency of hydrophobic (hydrocarbon-like) molecules to keep away from aqueous phase	Placing hydrocarbons in an aqueous medium requires energy to change water structure
van der Waals' forces	All atoms have a weak tendency to stick together over short ranges	Results from electrostatic attraction of positively charged nucleus of one atom and negatively charged electrons of the other. Weak and fluctuating $4 \, \text{kJ mol}^{-1}$

Ionic bonds

Ionic bonds form as a result of the attraction between ions of opposite charge. Some atoms have a tendency to gain negatively charged electrons, becoming negatively charged themselves. Other atoms tend to lose electrons and become positively charged. Thus, a negatively charged ion will attract a positively charged one (and *vice versa*), and an ionic bond is said to join them. Such bonds are formed in many inorganic compounds such as sodium chloride, NaCl, which should properly be represented as Na^+Cl^-.

Similar bonds may also form between charged organic molecules. For example, an ionic bond will form between a negatively charged carboxyl ion and a positively charged amino group: $R-CO_2^- \; ^+H_3N-R'$. Such ionic bonds may be quite strong, but their precise strength depends on the environment of the charged pair.

□ Atoms have a tendency to lose or gain electrons to achieve the rare gas structure, but in doing so become positively or negatively charged. For example, sodium tends to lose an electron to form the positively charged sodium ion (Na^+).

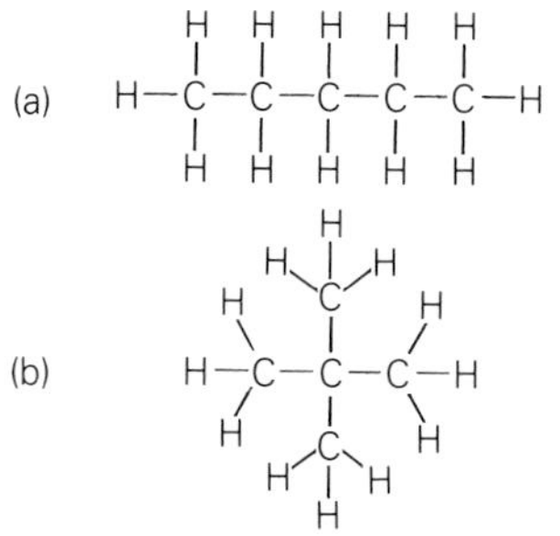

Van der Waals' forces

Covalent bonds between atoms are strong and so too may be ionic forces. However, there are in addition a number of much weaker forces between atoms and molecules. These forces arise because, for certain atoms at any one instant, the distribution of electric charge on them may be non-uniform even though, over time, the electronic charge is distributed uniformly. This fluctuating distribution of charge gives the atom **polarity**: one part of it is slightly negatively charged with respect to the other parts. This polarity induces an opposite polarity in neighbouring atoms and molecules such that a temporary negative charge in one atom may be attracted to a temporary positive charge in another.

Such forces, called van der Waals' forces, are about 100 times weaker than a covalent bond. Nevertheless, they tend to pull molecules together. For example, the boiling point of n-pentane is 36°C whereas that of neopentane is 9.5°C (Fig. 1.7). Both are non-polar; it is *only* the van der Waals' forces that hold the molecules together in the liquid. Nevertheless, although each has the same number of carbon and hydrogen atoms, the molecule of neopentane is more compactly arranged and has less of its surface exposed to neighbouring molecules. Neopentane, therefore, has the lower boiling point because its molecules can separate more easily.

When two molecules have complementary shapes hundreds or even thousands of van der Waals' forces may form between them. Although each individual interaction is very weak, several hundred such bonds can 'stick' molecules together very tightly. Many biological interactions are governed by complementary shapes and van der Waals' forces.

Fig. 1.7 Structures of (a) n-pentane and (b) neopentane. Remember that these molecules are three-dimensional because the carbons have a tetrahedral arrangement of their valencies (see Fig. 1.6).

Box 1.6
Attractions between charged groups (salt linkages)

Consider the attraction between a carboxylate group ($-CO_2^-$) in contact with a protonated amino group ($-NH_3^+$) (see figure). The approximate distance between the centres of positive and negative charges is about 0.25 nm.

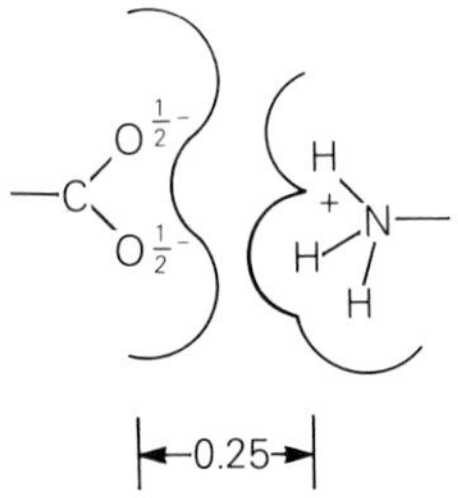

Applying Coulomb's Law to compute the force, F, between the two particles:

$$F = \frac{qq'}{\varepsilon r^2}$$

$F = 8.9875 \times 10^9 \times qq'/\varepsilon r^2$ newtons [where r is the distance in metres, q and q' are the charges in coulombs (1 electronic charge = 1.602×10^{-19} coulombs), ε is the dielectric constant, and F is the force in newtons. The dielectric constant is 1 for a vacuum, about 2 for a hydrocarbon solvent, and 78.5 for water at 25°C.

In a hydrocarbon solvent ($\varepsilon = 2$), the force F for $r = 0.25$ is 7.7×10^{14} Nmol^{-1} (i.e. $F \times$ Avogadro's number to obtain the force per mole). To move the charges further apart by 0.01 nm would require 7.7 kJ mol^{-1}, a considerable amount of energy.

In contrast, in water with its much higher dielectric constant, this force would be reduced by 2/78.5 or almost 40-fold.

The majority of organisms live in temperatures of 0–40°C. Life below 0°C is impossible unless an organism maintains its temperature. Most organisms regulate their intracellular pH to around 7.0 although many live in conditions one or two pH units on either side of this, and some bacteria survive in acid springs. Yet other bacteria survive in hot springs at over 90°C. Many organisms live at the bottom of the ocean under high pressure, and some microorganisms can live at high salinity (e.g. *Halobacterium*).

It intrigues biochemists to know how bacteria can survive high temperatures, not only because of basic scientific curiosity but also because of possible uses for heat-stable proteins extracted from such organisms. One example that has suddenly become extremely important is an enzyme called DNA polymerase, from *Thermus aquaticus*. This enzyme is used in the polymerase chain reaction (see *Cell Biology*) which is set to revolutionize the techniques of molecular biology. Because it is stable at over 95°C, solutions may be heated to this temperature to separate the two strands of a DNA double helix without destroying the enzyme that will later act upon them.

Hydrogen bonds

If a hydrogen atom is bonded to a highly electronegative atom, such as oxygen or nitrogen, the electronegative atom draws the bonding electron closer to it. As a result, the hydrogen atom carries a small positive charge and will be attracted to other negatively charged atoms and groups. This attraction is referred to as a hydrogen bond. Such bonds are about ten times stronger than van der Waals' forces but about ten times weaker than covalent bonds. Hydrogen bonds account for the unexpectedly high boiling point of water (H_2O) compared with that of hydrogen sulphide (H_2S) and of ammonia (NH_3) compared with phosphine (PH_3).

1.3 The properties of water

All life exists in a watery environment and all the reactions and interactions inside a cell occur in the presence of water. Not surprisingly, therefore, the properties of water are a key factor in our understanding of biochemistry.

☐ The energies of different bonds may be compared thus: that of the C–C bond approaches 400 kJ mol⁻¹, that of a hydrogen bond is up to 40 kJ mol⁻¹, and that of a van der Waals' interaction is almost 4 kJ mol⁻¹. These figures are only approximate but the factor of ten difference between each is easy to remember.

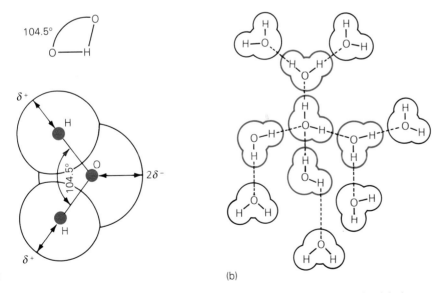

(a)   (b)

Fig. 1.8 (a) Structure of the water molecule. (b) Water molecules in solution interact with each other via hydrogen bonds.

The water molecule consists of two hydrogen atoms covalently bonded to an electronegative oxygen atom (Fig. 1.8a). The oxygen carries a small negative charge while the two hydrogens carry small positive charges. Consequently, water molecules tend to cohere through a network of hydrogen bonds (Fig. 1.8b). Bulk water has considerable 'structure' because of these interactions, although it is changing all the time as the relatively weak hydrogen bonds break and re-form.

Many of the properties of water can be explained by this extensive hydrogen bonding between water molecules. Water is liquid over a wide range of temperatures and the bulk of the water on the earth's surface is liquid, providing a major habitat for living things. Ice floats, and a body of water freezes from the top down. A top layer of ice on the body of water acts as an insulator and frequently the water lower down does not freeze, allowing aquatic life to continue. The heat of evaporation of water is much higher than for many other liquids: on average three hydrogen bonds must be broken before a water molecule can escape. Many organisms make use of this property to cool themselves.

Water acts as a solvent for polar molecules, mainly those with which it can form hydrogen bonds. Polar molecules include both charged and uncharged molecules or parts of molecules. Thus, water will form hydrogen bonds with compounds containing $-CO_2^-$, $-NH_2$ or $-NH_3^+$, $-OH$ groups, and many others. Additionally, the ions of a pair (such as Na^+Cl^-) will tend to separate in water (Fig. 1.9) each becoming surrounded by a 'shell' of water molecules the so-called water of ionization.

□ It takes more than 2000 joules to change one gram of liquid water into vapour. This is five times as much as is needed for a gram of ether and twice as much as for a gram of ammonia.

Exercise 3

List three properties of water essential to life.

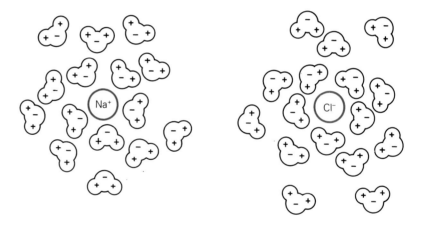

Fig. 1.9 Water molecules accommodate the separation of ions in solution. Each ion has 'shells' of water molecules.

In contrast, non-polar molecules (for example hydrocarbons) are insoluble in water. Such molecules interrupt the hydrogen-bonded structure of water without forming favourable interactions with water molecules. Fatty material (lipid) tends to coalesce into oily droplets in water because then the molecules cause less disruption to the hydrogen-bonded network of water molecules around them. To make non-polar molecules mix with water it is necessary to use energy to rearrange the water molecules, by breaking hydrogen bonds, to create a cavity in the water.

The hydrophobic effect

The apparent reluctance of non-polar molecules to dissolve in water has led to such compounds being referred to as **hydrophobic** or 'water-hating'

Hydrophobic: means water-hating, in contrast to hydrophilic, which means water-loving. These terms may be used to describe whole molecules or groups that form part of molecules. Thus, a molecule may be said to have hydrophobic regions.

molecules. This is in contrast to compounds that are water-soluble and are called hydrophilic or 'water-loving'. This terminology is useful, although the idea of molecules hating water is misleading. As we now know, non-polar compounds do not mix freely with water because energy is needed to make them do so.

Certain molecules are hydrophilic in some parts of their structure and hydrophobic in others. Water-soluble protein molecules, for example, tend to have hydrophobic groups on the inside of a roughly spherical molecule and hydrophilic groups on the outside in contact with water (Fig. 1.10). Such an arrangement stabilizes the macromolecule in its biologically active form. A consequence of this arrangement is that any ionic bonds that form in the interior of the molecule will be much more stable, given the hydrophobic environment there.

The fact that some molecules have both polar and non-polar regions has a much wider significance because it provides the basis for biological *organization*. First, however, we need to look at the ways organisms manage energy.

--- Exercise 4 ---

Explain why methane (CH_4) would not be expected for form hydrogen bonds with water.

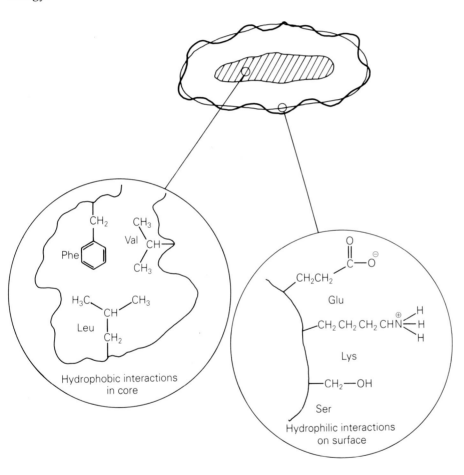

Fig. 1.10 Globular proteins typically have a hydrophobic core with hydrophilic amino acid residues on the surface that interact with the surrounding aqueous solvent.

--- Exercise 5 ---

To make French dressing, why do you need to shake the oil and vinegar together vigorously?

1.4 Energy

Life activities such as muscular contraction, the orientating of leaves towards light, the beating of flagella, the conduction of nerve impulses, and building new living matter, are all energy-requiring processes (Table 1.6). Organisms need not only to obtain energy to drive these processes, but must also have the necessary mechanisms to transform it.

Table 1.6 *Energy-requiring life activities. All of the processes listed require an input of energy. The source of this energy must be either photosynthesis or the oxidation of food materials*

MOVEMENT:Mechanical
Muscular contraction, locomotion and protoplasmic streaming: from whole organism to part of cell

GROWTH: Chemical
Making more of an organism's constituents by chemical synthesis from simpler precursors. Duplicating the genetic material. Growth or cell division

CONCENTRATION: Osmotic
Selectively accumulating or transporting compound or ions: maintaining a concentration gradient across a biological membrane

CHARGE SEPARATION: Electrical
Making and maintaining a gradient change (ions) across a membrane. Nervous conduction, electric eels

PHOTOCHEMICAL REACTIONS: Light
Carrying out chemical reactions that generate light (glow worms, fireflies)

□ Different forms of energy can be interconverted. Light energy may be absorbed by a sugarcane plant and used to convert CO_2 into sucrose, which represents a potential store of chemical energy. Subsequently, an athlete may consume the sucrose and release the energy from it as muscular movement (kinetic energy).

THE SUN IS THE PRIMARY SOURCE OF ENERGY. Practically all life derives its energy from sunlight, and plants and certain bacteria have evolved mechanisms for trapping this light energy in a process called *photosynthesis*. Photosynthesis may be looked upon as a reduction of CO_2 to organic compounds, the energy required to drive this reduction being supplied by sunlight. All other organisms—animals, and non-photosynthetic micro-organisms, and fungi (which are similar to plants but lack a photosynthetic apparatus)—survive by taking in molecules synthesized by plants and breaking them down to release the energy contained in them. It matters little whether these organisms feed on plant matter directly or obtain preformed organic molecules by consuming animals that have eaten plants.

The process of breaking down the organic molecules is called *catabolism*. Catabolic processes are mostly oxidations and they release energy. They may go to completion and oxidize the organic material to CO_2 and H_2O, or occur incompletely, the products being used as building blocks for biosynthetic activities.

Organisms also store energy. The sun does not always shine and animals

---

*Box. 1.8
Gasohol*

Gasohol is produced by fermenting the sugar (sucrose) in sugarcane and then distilling the ethanol produced. The fermentation is carried out by yeast under anaerobic conditions. This releases only part of the energy of the sugar (which is used by the yeast). The resulting dilute solution of ethanol needs to be concentrated in order to use it to run a car engine. The heat required for this is generated by burning the bagasse or spent sugarcane stems which consists mostly of cellulose. The origin of both the sucrose and the cellulose is the photosynthetic activity of the sugarcane plant. The ultimate source of the energy is sunlight.

Although the process is successful and is used widely in Brazil, the technology is fairly expensive. Economically, the process is entirely at the mercy of world oil prices. Early in 1990 gasohol production had almost stopped in Brazil because of this, creating problems for the many motorists whose cars have been specially converted to run on ethanol.

Photosynthesis: *the process by which plants and some bacteria take in CO_2 and use light energy to convert it to highly reduced carbon compounds such as carbohydrates, $(CH_2O)_6$.*

Catabolism: *the breaking down, usually by a process of oxidation, of food materials or stored molecules to release biologically useful energy to drive life processes.*

do not eat all the time. Therefore, it is necessary to put by fuel molecules, in some convenient form, to be used to maintain life on 'rainy days'.

The details of energy transformation in living matter are dealt with elsewhere. However, it is appropriate to consider a few basic principles here.

Thermodynamics

The study of energy and energy changes was originally based on measuring heat changes, and the branch of science that deals with energy is called thermodynamics because of this. Energy can be thought of as either stored or **potential energy** (glucose in a chocolate bar), or as **kinetic energy**, such as motion in the form of muscular movement (using the energy derived from this glucose in the chocolate bar).

When energy transformation takes place, some of it is turned into heat and is said to be 'lost'. What this means is that it is no longer available to do work because the heat is evenly distributed. In a petrol engine, only about 20% of the energy in the petrol (potential energy) released upon oxidation is actually turned into energy of motion. The remaining 80% is dissipated to the surroundings as heat.

The concept of useful energy (that can be made to do work) as opposed to that lost as heat, is encapsulated in the idea of **free energy**. This is the energy available to do work under conditions of constant temperature and pressure. It is the free energy released in the breakdown of organic molecules that is of importance to biochemical systems. Many of the mechanical devices we use (steam engines, petrol engines) are heat engines which work by making heat move from a region of high temperature to one of low temperature. However, organisms are uniform in temperature and cannot use temperature gradients to drive their reactions.

ENTROPY. Another important thermodynamic maxim states that all natural processes tend to proceed such that the extent of disorder or randomness of the system increases. This disorder is given the name **entropy**. Molecules, through the process of random motion or diffusion, tend to distribute themselves more randomly. Reversing this spontaneous trend requires the input of energy.

□ Because of this spontaneous tendency towards disorder, organisms need to spend energy just to maintain the *status quo*.

When the concept of entropy was formulated some people believed that living systems violated thermodynamic principles because they tended to increase in order—that is, their entropy decreased. However, if one looks at the living system *and* its environment, the total entropy increases. The living system becomes more organized but the environment becomes more random. Thus, living systems take in useful energy from external sources and release it into their environment in a less useful form. Their degree of order increases by means of the many reactions they perform.

It will be seen shortly that the idea of a living system and its environment requires that these be distinguished. In other words, there has to be a boundary dividing a living system from its surroundings. Before that, however, it is helpful to look briefly at the types of reactions that go on in living systems, noting especially whether they release energy or require energy.

Reactions in living systems

The sum total of the chemical processes in an organism is called ***metabolism***. This overall process is made up of reactions that produce energy (**catabolism**) and those that require energy (**anabolism**).

Although the total number of reactions that occur in a given cell of a given

Metabolism: the sum of reactions going on in a living cell. Anabolic reactions are the building-up reactions, and the catabolic reactions are the breaking-down (to provide energy) reactions.

□ Vertebrate voluntary muscles can work aerobically or non-aerobically. Aerobically, much more energy can be obtained per mole of glucose. However, in the flight or fight response, the large muscles need to contract strongly and rapidly. In this situation most of the blood is squeezed out of the muscle bed and there is no time between contractions for blood to flow. Consequently both oxygen and glucose supplies are cut off and energetic contraction can only be continued by using stored muscle glycogen anaerobically. This mode of action can only continue for a limited time. Eventually, blood must be allowed to flow again, to replenish the stores and to wash out the accumulated lactate.

organism can run into thousands or tens of thousands, all these reactions can be classified into a few fundamental types. A knowledge of these reaction types forms the basis of our understanding of biochemistry.

OXIDATION AND REDUCTION. In general, energy-releasing reactions are oxidations, while synthetic processes, such as building up plant material by photosynthesis, are reductions. It is useful to consider oxidation and reduction at the fundamental level of electron flow. Reduction implies that an element or compound gains electrons; oxidation suggests that it loses them. Losing electrons is a process that releases energy, and consequently oxidations yield energy.

Electron flow from one compound to another is a process uniquely important in living systems, and a wide variety of carrier molecules have evolved to participate in it. These compounds play key roles both in the release of biologically usable energy from fuel molecules and in tapping biologically useful energy during photosynthesis.

CONDENSATION AND HYDROLYSIS. The joining together of monomers to manufacture macromolecules almost always takes place by a process called **condensation**. The two molecules that condense lose the elements of a molecule of water and become joined through an oxygen atom (Fig. 1.11). This is true of amino acids linking to make proteins, monosaccharides joining to make polysaccharides, and nucleotides combining to make nucleic acids. As would be expected, making complex macromolecules from simple monomers requires energy. Exactly how organisms obtain it will be seen in later chapters.

The reverse process, the addition of water to release the monomers from a polymer, is called **hydrolysis**. In principle, such a process is spontaneous and proceeds with the release of energy. In practice, when macromolecules in living systems are broken down to their building blocks, this released energy

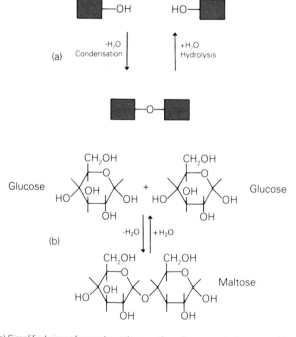

Fig. 1.11 (a) Simplified view of a condensation reaction: the removal of water and the formation of a link. The opposite process is called hydrolysis. (b) An example is the formation of the disaccharide, maltose, from two glucose molecules.

Reference Gayford, C.G. (1986) *Energy and Cells*, Macmillan, London. More than required here, but look at the introductory chapters.

is almost never used, and is lost as heat. Most often the breakdown of macromolecules to monomers takes place outside the cell (for example, in the intestines of vetebrates) before the building blocks are absorbed into the cells. There is no mechanism for capturing this released energy in a biologically useful form.

PHOSPHORYLATION. There are many other reactions that go on in living systems and they have two major aims. One is to re-tailor molecules according to the needs of the organism for building its own types of macromolecules. The other is to get the partially broken down molecules into a form such that energy can be released: usually this means into a form from which electrons can be extracted.

Although many of these reactions are considered in detail in later chapters it is appropriate here to mention the important reactions of phosphorylation and dephosphorylation. In the early part of the twentieth century it was discovered that many biomolecules carried phosphate groups. The first to be isolated and characterized were the sugar–phosphates (Figs 1.12 and 1.13).

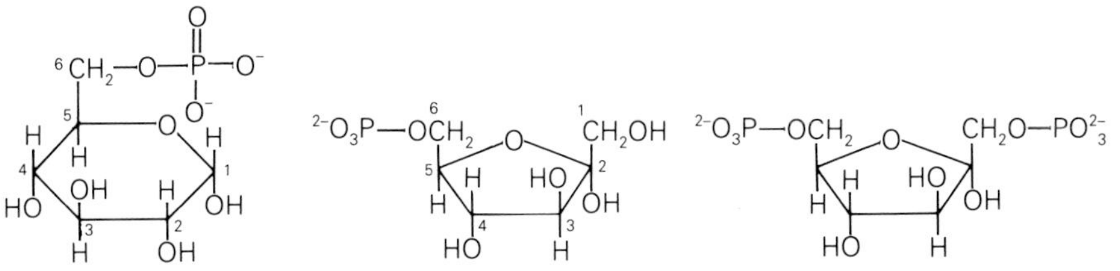

Glucose 6-phosphate   Fructose 6-phosphate   Fructose 1,6-bisphosphate

Fig. 1.12 Structures of some sugar–phosphate esters that occur in practically all cells. The number indicates the position of the phosphate group on the sugar ring. As with many other ring structures, individual carbon atoms are omitted for ease of drawing.

In the 1930s it was found that all living systems contained small amounts of **pyrophosphate** compounds. The vital one, present in all organisms, was adenosine triphosphate (ATP). It gradually became clear that this molecule is almost universally used as a form of biological energy. In a variety of ways the release of energy through the hydrolysis of the pyrophosphates of ATP is linked to other energy-requiring reactions and processes.

As a simple example, it is easy to see that the ready hydrolysis of the pyrophosphate bonds of ATP means that this compound can act as a dehydrating agent. In principle, therefore, it has the potential for driving the reactions by which macromolecules are synthesized from their monomers through condensation.

Glucose   Phosphate   Glucose 6-phosphate

Fig. 1.13 Formation of a sugar–phosphate ester between glucose and phosphate. This reaction tends to go in the opposite direction to that shown, that is, the equilibrium favours the *hydrolysis* of glucose–glucose.

□ If glycogen is being synthesized in liver, there is no need to remove phosphate from the glucose 6-phosphate because this is the starting material for glycogen synthesis. However, if it is *muscle* glycogen that is being replenished, it is necessary for the liver to make free glucose which enters the blood and passes to the muscles where it is rephosphorylated before glycogen synthesis. Sugar–phosphates do not pass through cell membranes and are not transported in the bloodstream.

Exercise 6

Explain the meaning of the term 'nucleoside triphosphate'.

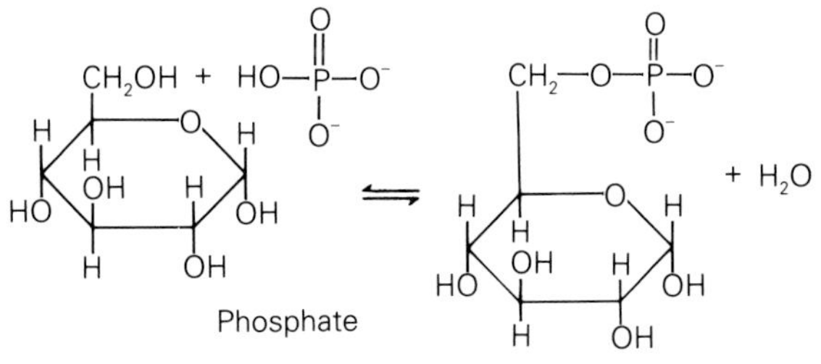

Pyrophosphates: *formed by the combination of two (or more) phosphates with the removal of a molecule of water.*

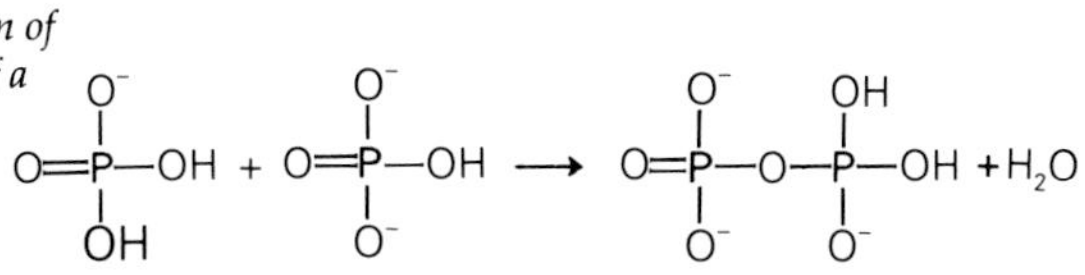

Box. 1.9
Adenosine triphosphate (ATP) and phosphorylations

Among the phosphate compounds found in cells, the one with the universal key role in energy metabolism is adenosine triphosphate (ATP) to which there will be repeated references in subsequent chapters. When muscle contracts, for example, the immediate source of energy is the hydrolysis of phosphate groups in ATP. Most other energy-requiring reactions in the cell are coupled to ATP hydrolysis.

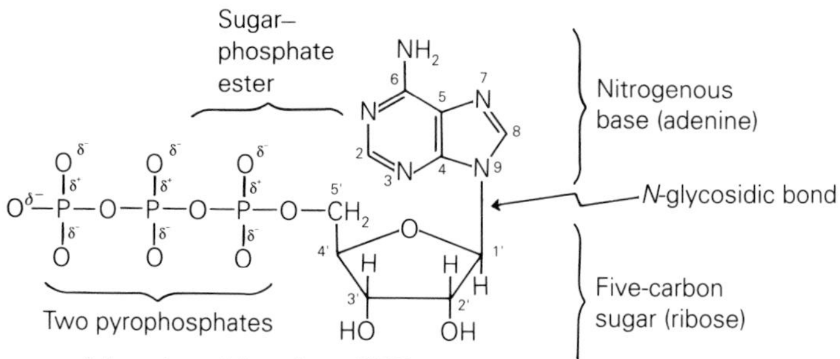

Adenosine triphosphate (ATP)

ATP is a **nucleotide**. Nucleotides are base–sugar–phosphate units and are closely related to monomers of nucleic acids. In ATP the base is **adenine**, the sugar is **ribose**, and to this are attached three phosphate groups. The first of these is attached directly to the ribose in a sugar–phosphate ester linkage. However, the next two phosphates are then linked on to the ester as pyrophosphates. Pyrophosphates are made by removing water between two phosphoric acid units. The reverse reaction, the hydrolysis of a pyrophosphate is thermodynamically favourable and requires water. So ATP hydrolysis can be used, formally, to remove water, in other words, it can act as a dehydrating reagent. However, direct dehydration is not possible in the 'wet' environment of the cell. Rather, a series of coupled reactions ensure that *the elements of water* are consumed during ATP hydrolysis.

ATP is also used when compounds become phosphorylated. ATP may be seen as donating a phosphate group to glucose in the formation of glucose 6-phosphate. The sugar–phosphate ester formed is more stable than the pyrophosphate in ATP that is hydrolysed.

Another important reaction in which ATP participates is the formation of cyclic adenosine monophosphate, cAMP (see figure). cAMP is a key compound in a number of aspects of hormonal control mechanisms involving the sending of signals through cell membranes to the interior of the cell.

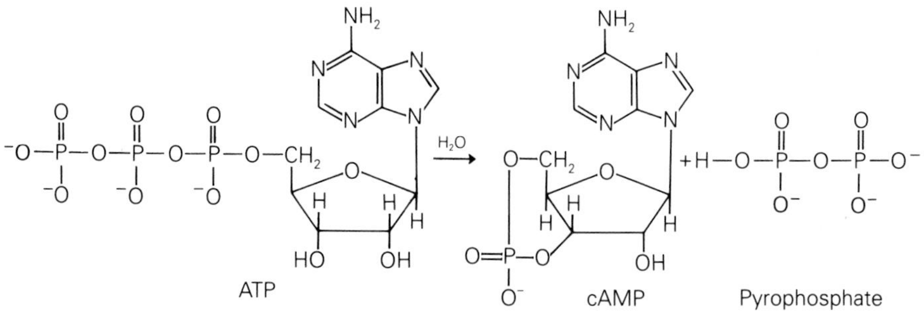

ATP → cAMP + Pyrophosphate

1.5 Biological organization

It was recognized as long ago as 1838 that all organisms are made up of **cells**. Cells are membrane-bounded entities, separated from one another and from the environment. Similar cells group together to form tissues, groups of tissues form organs, organs work together in systems, and systems combine to form whole organisms. Cells form the basis for biological organization. The cell theory is a major unifying concept in biology and may be summarized as follows:

- all living matter is composed of cells
- all cells arise from other cells
- all the metabolic reactions of a living organism take place within cells

Cellular organization

The human body contains about 100 different types of cells. Examination of plant tissues will reveal further, distinct types, and microbial cells are different again. Nevertheless, although they may differ widely in size and shape, all cells show common features. The most striking of these is that all cells, apart from bacteria, contain structures called **organelles**, such as a nucleus and mitochondria (Fig. 1.14). A characteristic feature of these organelles is that they, like the cell itself, are bounded by biological membranes. Cells and organelles form compartments and subcompartments in which different biological and metabolic activities may be kept separate. Why is compartmentalization vital and how is it achieved?

☐ Bacteria are unicellular organisms that fall into the category **prokaryotes**, lacking a distinguishable nucleus with nuclear membrane. In fact, they lack most of the typical membrane-bounded organelles found in eukaryotic cells.

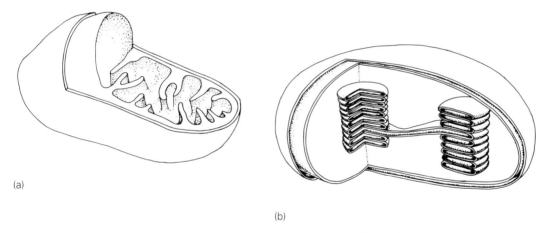

(a)

(b)

Fig. 1.14 Very simplified representation of two cellular energy-transforming organelles: (a) a mitochondrion, and (b) a chloroplast. Note that each has an inner and an outer membrane.

ORGANELLES HAVE DIFFERENT FUNCTIONS. One reason for having separate compartments inside a cell is that different organelles have different functions (Table 1.7). In a plant cell, for example, the chloroplasts are concerned with building up sugars from carbon dioxide and water. Mitochondria, on the other hand, are concerned with the oxidative degradation of compounds derived from sugars, with the accompanying production of ATP. It is not surprising that these two very different undertakings operate more efficiently within their own environments.

There are much more fundamental reasons for having compartments, however. The membranes surrounding these compartments (both the cell and its organelles) are not simply inert dividing walls. They are actively

Reference de Duve, C. (1984) *A Guided Tour of the Living Cell* (Scientific American Books), New York, USA. A well-illustrated and readable tour.

Reference Carroll, M. (1989) *Organelles*, Macmillan, London. A short introduction to organelles.

Box 1.10
The structure of cells

The function of the various cell organelles are now well understood. The nucleus contains the DNA which stores the cell's complement of information. The mitochondria and chloroplasts produce energy, or rather, transform energy. The endoplasmic reticulum is the site of synthesis of proteins that are destined to be secreted.

Bacterial cells differ from all other cells because they lack a defined nucleus and most of the organelles found in other eukaryotic cells. Bacteria might be looked on as being 'primitive' organisms but in fact they have survived from the earliest days of the evolution of life on earth, and have developed to be able to live in and exploit almost any habitat, from the human bloodstream to hot, sulphur-laden springs.

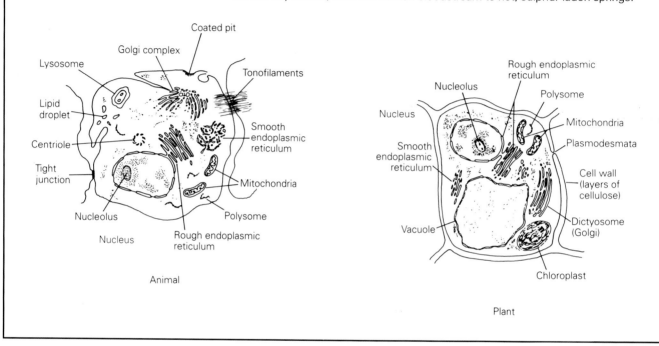

Animal

Plant

Table 1.7 *Organelles found in eukaryotic cells.* Organelles are typically bounded by a double or a single membrane. Not all cells contain all of the organelles listed here. For example, only photosynthetic cells would contain chloroplasts*

Organelle	Function
Nucleus	Carrying genetic information
Endoplasmic reticulum	Processing proteins
Golgi apparatus or 'complex'	Secretion
Secretory vesicles	
Lysosomes	Intracellular breakdown
Mitochondria	Energy production
Chloroplasts	Photosynthesis
Peroxisomes	Generate H_2O_2
Vacuoles	Storage?
Ribosomes†	Protein synthesis

* Cells other than bacterial cells are called eukaryotic cells and the organisms are called eukaryotes. Bacterial cells are prokaryotes, and were originally named thus because they lack a formed nucleus. In fact, they have none of the membrane-bounded organelles found in eukaryotic cells.
† Ribosomes are not strictly organelles because they are not membrane-bounded. Bacteria possess ribosomes.

involved in a great variety of processes, including **selective permeability** (letting certain things in and certain things out) and participating in energy production. Without intact functional membranes, organelles cannot carry out their functions, and the cell cannot survive.

MEMBRANES SEPARATE A CELL FROM THE ENVIRONMENT.   Cells need to be separated from, but in communication with, the environment. They need to take in nutrients (but exclude noxious materials), to excrete waste matter, and secrete finished products manufactured in the cell but used elsewhere. Nerve conduction depends on ion gradients across neuronal membranes. All cells require a selectively permeable membrane.

Even more fundamentally, cells can operate efficiently only by concentrating needed materials within their boundaries and keeping them there. A cell is not simply a 'blob of protoplasm'. The substances in such a structure would diffuse away, becoming more and more dilute. Cell machinery will not work at infinite dilution. Conversely, if water is allowed free access into a cell it would burst.

The membranes around the intracellular organelles have similar roles. But they also participate in the energy-producing mechanisms and in the traffic of molecules in and out of the organelles. A cell soon dies if it is treated with compounds that make the mitochondrial membrane permeable.

Membrane structure

Cell membranes are not mechanically strong structures, in contrast to the rigid and tough cellulose cell walls of plants or the 'string bag' cell walls of bacteria. Their role is to provide a selective barrier, not a tough wall. This in itself makes it difficult to investigate their structure and chemical nature because it is not easy to isolate and purify the 'stuff' of which cell membranes are made.

☐ The smaller cells are, the greater is their surface area to volume ratio. Assume that a cell is a cube of side x, then for a fixed total volume:

length of side (x) (μm)	20	10	2
surface area (μm^2)	2400	4800	24000
total volume (μm^3)	8000	8000	8000
surface area : volume ratio	0.3	0.6	3.0
number of cells	1	8	1000

The greater the surface area : volume ratio, the more efficient the transfer of materials into and out of the cell.

☐ Agents such as 2,4-dinitrophenol make the mitochondrial membrane leaky and the mitochondria can no longer produce biologically usable energy. The energy appears as useless heat.

Box 1.11
Keeping water out, or why bacterial cells have 'corsets'

The contents of a bacterial cell are a very concentrated mixture of macromolecules and small molecules in aqueous solution. Such a solution exerts a high osmotic pressure of several atmospheres, and there is a constant tendency for water to enter the cells to dilute the solution and equalize the osmotic pressure. However, this would not suit the bacterial cells and, indeed, would kill them. This may be observed experimentally. Bacterial cells can survive in water, but if their cell walls are gently removed by treatment with the enzyme lysozyme (which cleaves bonds in the cell wall), the cells burst. Bacterial cells whose walls have been removed can still be kept intact by placing them in a concentrated sugar solution that equalizes the osmotic pressure.

This tendency of water to rush in must be counteracted. There are three possible ways of doing this:

1. Bacteria could have impermeable cell membranes. However, this would not be feasible because nutrients need to pass in, and waste products out.
2. They could constantly pump water out. However, this would consume energy. Bacteria live in a highly competitive environment and so have evolved ways of avoiding spending this energy. They have 'chosen' the third solution.
3. They have a tough but highly permeable 'corset' round the cell, very much like a continuous string bag. Bacterial cells live at a high osmotic pressure. Water cannot get in because the cells are not allowed to increase in size. A selectively permeable membrane just inside the 'string bag' completes the structure. This arrangement costs no energy!

Reference Harowitz, H.J. (1987) *Mayonnaise and the Origin of Life*, Berkeley Publications, NY.

MEMBRANES ARE BASED ON LIPIDS. A class of compounds present in all cells (but so far mentioned only in passing) are the lipids or fatty materials. The simple **fats** and **oils** are lipids, which act as energy stores in many types of cells. Such molecules represent highly reduced forms of carbon. Consequently, their oxidation releases a great deal of energy. Simple fats and oils are hydrophobic, and tend to form droplets or globules in the aqueous interior of the cell. In this way they isolate themselves from other cellular activities, and are not directly involved in membrane structure.

☐ Globules of lipids such as triacylglycerols are of relatively low density and make cells buoyant.

The compounds that are involved in membrane structure are the complex lipids, especially the **phospholipids**. Phospholipids (whose structures are dealt with in detail in Chapter 00) are long molecules composed of hydrophobic chains of $-CH_2-$ units with a hydrophilic region at the end. The hydrophilic end arises from charged phosphate and nitrogenous bases, both of which can interact with water molecules (Fig. 1.15).

If some phospholipid is placed on water, it will tend to form a mono-

---

Box. 1.12 Lipids

Lipids (see Chapters 00 and 00) are the fats and oils whose character is predominantly hydrophobic. A number of different compounds as well as their constituent parts are referred to as lipids. The most characteristic and common feature of lipids is a long hydrocarbon chain, typically derived from a so-called long-chain fatty acid:

$$CH_3CH_2CH_2CH_2CH_2CH_2CH_2CH_2CH_2CH_2CH_2CH_2CH_2CH_2CH_2COO^-$$

Molecular model of palmitic acid

Such compounds may be saturated or unsaturated, i.e. having one or more double bonds. The double bonds are typically *cis* which creates kinks in the long molecules.

Number of C atoms	Structure	Name
Saturated		
12	$CH_3(CH_2)_{10}COO^-$	lauric acid or laurate
14	$CH_3(CH_2)_{12}COO^-$	myristic acid or myristate
16	$CH_3(CH_2)_{14}COO^-$	palmitic acid or palmitate
18	$CH_3(CH_2)_{16}COO^-$	stearic acid or stearate
Unsaturated		
16	$CH_3(CH_2)_5\dot{C}H{=}\dot{C}H(CH_2)_7COO^-$	palmitoleic acid or palmitoleate
18	$CH_3(CH_2)_7\dot{C}H{=}\dot{C}H(CH_2)_7COO^-$	oleic acid or oleate
18	$CH_3(CH_2)_4\dot{C}H{=}\dot{C}HCH_2\dot{C}H{=}$ $\dot{C}H(CH_2)_7COO^-$	linoleic acid or linoleate
18	$CH_3CH_2\dot{C}H{=}\dot{C}HCH_2\dot{C}H{=}\dot{C}HCH_2\dot{C}H{=}$ $\dot{C}H(CH_2)_7COO^-$	linolenic acid or linolenate

Three molecules of fatty acid can combine with a molecule of the trihydric alcohol glycerol to form a neutral, hydrophobic triacylglycerol (formerly called a triglyceride).. Phospholipids may be formed by replacing one of the three fatty acids in a triacyglycerol with a phosphate and a nitrogen-containing aliphatic base (See Fig. 1.15). Phospholipids are amphipathic: the molecules are water-soluble at one end but water-insoluble at the other.

Oils and fats: *these are triglycerides or triacylglycerols. Fats are solid at room temperature whereas oils are liquid.*

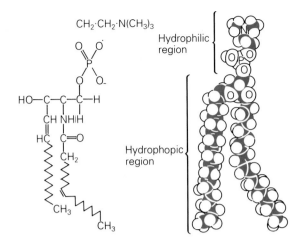

Fig. 1.15 Biological membranes are formed on the basis of a bilayer of phospholipids; (a) shows the chemical structure of a phospholipid, and (b) shows a space-filling model.

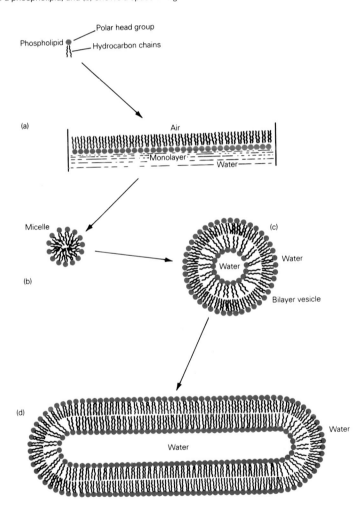

Fig. 1.16 Biological membranes are formed on the basis of a bilayer of phospholipid. (a) At an air–water interface, the phospholipid molecules form a monomolecular layer with their hydrophilic portions in the water. (b) When dispersed in water, the phospholipid molecules may form micelles with the hydrophobic portions inside the micelle. However, it is also possible to form a bilayer vesicle (c), where there is now water both inside and outside. (d) A large bilayer vesicle forms the basis of a cell membrane, separating two watery regions, inside and outside.

Exercise 7

What is the agent in mayonnaise that enables a stable emulsion to be formed?

Fig. 1.17 When mixtures of phospholipid and water are homogenized or sonicated to disperse the lipid in the water phase, liposomes may form. The photograph shows a liposome as viewed in the electron microscope after freeze-fracturing. The liposomes were frozen and then the frozen block of material was cleaved by a sudden shock with a knife. Fracturing took place at several planes, revealing the concentric layers of lipid forming the liposome. (Magnification × 18 900). Courtesy of Dr P.F. Knowles, University of Leeds, UK.

molecular layer on the surface, with the hydrophilic parts of the molecule in the water and the hydrophobic chains sticking out of it (Fig. 1.16a). If this arrangement is disturbed, by stirring for example, **micelles** form, loose 'compartments' with the hydrophobic parts of the molecules pointing towards the inside and the hydrophilic parts sticking outwards, interacting with water molecules (Fig. 1.16b). A very large, flattened micelle is, effectively, a **bilayer**. Here again the hydrophobic tails point inwards while the hydrophilic heads are in contact with the water. A large structure such as this can be arranged to form a compartment, separating one body of water from another (Fig. 1.16c, d). Sonicating a mixture of phospholipid and water produces multilayered vesicles called liposomes (Fig. 1.17).

MEMBRANES ARE MADE OF LIPID AND PROTEIN. A membrane constructed as just described will indeed isolate one body of water from another, but it will be almost impermeable. Neither hydrophilic nor hydrophobic substances will pass through it. Hydrophobic substances will tend not to approach the hydrophilic head groups. Such a structure is of no use to a cell that must selectively exchange materials with its environment.

Impermeable membranes can be modified by the addition of proteins. Cells manufacture a wide variety of proteins capable of being integrated into a lipid bilayer. Some can sit on one surface and some on the other. Yet others may occupy a position across both layers, providing a channel through them. Imagine a protein that has a hydrophobic belt around it (formed by hydrophobic amino acid residues) with, on either end, a hydrophilic region (Fig. 1.18). This is quite different from a typical globular protein in which most of the hydrophobic residues are internal (Fig. 1.10 and Chapter 2). Such **transmembrane proteins** can form controllable channels through which molecules of a certain size and shape can pass, while others are excluded (Fig. 1.19). Although the three-dimensional structures of only one or two membrane proteins are known in detail, there is no doubt that it is this sort of arrangement that enables membranes to carry out their biological functions. Membranes structure and function is discussed in greater detail in *Cell Biology*, Chapter 4.

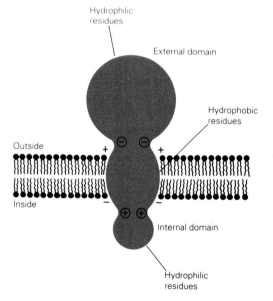

Fig. 1.18 Possible arrangement of a protein passing all the way through a membrane. One could envisage a 'belt' of hydrophobic amino acid residues around the molecule at the places where it interacts with the hydrophobic lipid of the lipid bilayer. The region on the 'inside' and 'outside' of the cell or organelle would have exposed hydrophilic residues.

Micelle: this term was first used in 1879 to describe a colloidal particle with a double layer of ions round it. With detergents and phospholipids in water it refers to the small vesicles which form, in which hydrophobic tails point inwards and hydrophilic heads point outwards.

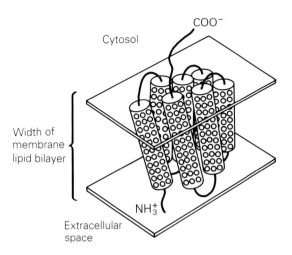

Fig. 1.19 A membrane protein constructed of a number of hydrophobic portions shown as cylinders, passing backwards and forwards across the membrane forming a channel through the centre of which ions or small molecules might pass in a controlled way.

INTERACTIONS. The assembly of lipid molecules into a membrane and their association with protein molecules is an example of **interaction**. The bonds formed between the various participants are non-covalent. Hydrophobic forces, hydrogen bonds, and van der Waals' attractions predominate.

This arrangement means that membranes are flexible and dynamic. All the molecules are in constant motion and interactions are constantly being broken and re-formed. Material may pass in or out by forming vesicles (Fig. 1.20) after which the membrane spontaneously reassembles.

☐ This dynamism can be observed experimentally. If specific protein molecules in the membrane are labelled with fluorescent tags, the areas of fluorescence can be seen to move around.

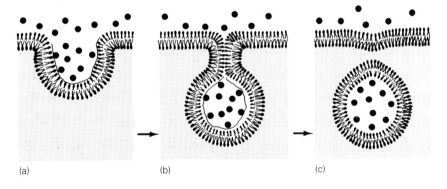

(a)   (b)   (c)

Fig. 1.20 Membranes are dynamic and flexible: they can form vesicles that bud off. The diagram shows the process of *endocytosis* by which particulate matter may be taken into a cell (a). A vesicle eventually pinches off (b). Afterwards the membrane reseals itself (c).

Exercise 8

Why are free fatty acids dangerous materials to have around in a cell or in an organism?

1.6 Interactions between protein and other molecules

It has been seen that lipids can interact with other lipids or with proteins. However, a great deal of what goes on in a cell, or indeed in an organism, is caused by proteins interacting with other proteins, or with small molecules (Table 1.8). The ideas are the same: the bonds involved are weak and non-covalent. They form easily and break easily. Just two examples will illustrate the fundamental importance of these concepts, although many more

Table 1.8 *Interactions between proteins and proteins, and proteins and other molecules*

Example	Function
Enzyme–substrate	Catalysis
Antibody–bacterial toxin	Protection
Protein–vitamin A	Transport
Haemoglobin–oxygen	Transport

examples will be encountered in the rest of this book. The two examples are enzyme catalysis and information transfer. They explain incidentally an enormous number of biological phenomena and show why macromolecules are so vital to biological activities.

Enzymes

Although the reactions in cells are often extremely complicated, they all proceed by very simple steps. But they would not take place at all, or at least would proceed only extremely slowly, if it were not for the ubiquitous presence of catalytic proteins called **enzymes**.

☐ By bringing the appropriate enzymes into play at appropriate times organisms can control which reactions they want to proceed and which they do not.

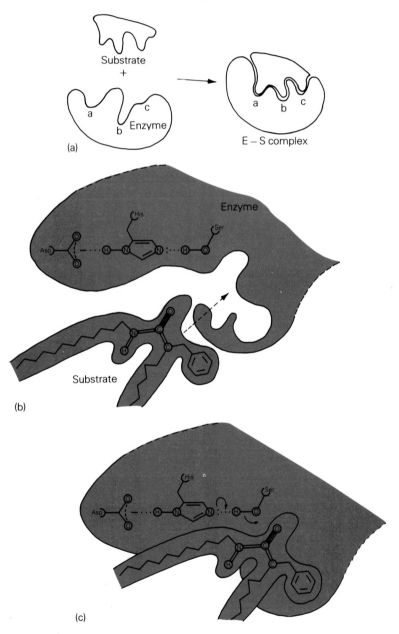

Fig. 1.21 An enzyme and its substrate come together (recognition) before catalysis takes place. (a) A general impression of how shapes interact specifically. At the chemical level, (b) and (c) show how this probably happens in the proteolytic enzyme, chymotrypsin.

Enzymes: macromolecular biological catalysts (see Chapters 4 and 5). The majority of enzymes are proteins, but recently certain RNA molecules have been discovered to possess catalytic activity.

Reference Ingle, M.R. (1986) *Enzymes, Energy and Metabolism*, Blackwell, Oxford. An inexpensive introduction: see the first two chapters.

Enzymes can best be understood by comparing how organic compounds are transformed in the chemical laboratory with what happens in the cell. In the laboratory, reactions may or may not proceed easily. If they do not, they are persuaded to by heating, or by applying pressure or by using strong acids and alkalis. At the end of an organic preparation it is usually necessary to purify the desired product from a number of unwanted by-products. Yields are frequently rather low.

None of these procedures is available in living cells. Cells operate at temperatures between 0 and 40°C, at about atmospheric pressure, and without strong acids and alkalis, which would destroy their structure. There are, furthermore, no ways of purifying the end product. Reactions must be made to go under mild conditions and with almost 100% yields.

Clearly enzymes are very specialized kinds of **catalysts**. They recognize the molecules they have to deal with (their **substrates**) with a very high degree of specificity. The region or cleft on the surface of an enzyme (called the **active site**), where the molecules react, has a shape complementary to that of its substrate (Fig. 1.21). Very close approaches are possible and very many van der Waals' interactions and hydrogen bonds form, orientating the substrate into its optimum location. Other chemical groups on the enzyme act in various ways so that the desired reaction is catalysed and the product rapidly released.

This will be dealt with in detail later (see Chapter 5). It is sufficient to understand at this stage that biological recognition takes place because molecules have complementary shapes and their interactions are stabilized by numerous weak bonds. Only macromolecules can carry out such functions.

Exercise 9

Why are enzymes so large compared with their substrates? (For example, hexokinase, M_r 52 000, has substrates glucose (M_r 180) and ATP (M_r 507).)

Information transfer

Biological organisms need to store and transfer information. The genetic instructions for building and running a cell need to be retained intact, used, and passed on to future cells at cell division. The information molecules of the

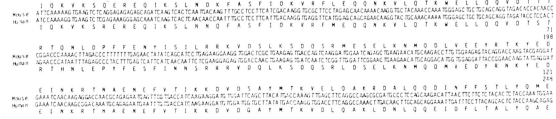

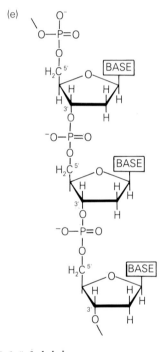

Fig. 1.22 Ways of storing information. (a) Using the binary system, numbers can be stored as a series of ones and noughts. There are many ways of doing this in a computer. (b) A bar code is machine-readable data of this type. (c) Musical notation is another way of storing information. (d) Shows part of the DNA sequences of mouse and human keratins: each triplet of bases 'translates' into one amino acid residue in the protein, and here the amino acids are shown as their single-letter abbreviations. (e) In biological systems, information is stored in DNA as a sequence of nitrogenous bases on a very long polymer.

Catalysts: *these speed up a chemical reaction by lowering the activation energy. Many chemical catalysts are known and used (e.g. plantinized asbestos) but biological catalysts, enzymes, are infinitely more specific in their reactions. In principle a catalyst should be unchanged at the end of the reaction.*

Substrates: *the substances that enzymes 'work' on. As a result of catalysis, substrates are transformed into products.*
Active site: *the region of an enzyme that is responsible for catalysis as well as specific recognition of a substrate.*

cell are the nucleic acids. A very brief outline will be given here of how nucleic acids carry out their functions: much more detail is given in Chapter 8 and *Molecular Biology and Biotechnology*.

DNA STORES INFORMATION. With the exception of certain viruses, all organisms store information coded in DNA. The information is stored in a linear fashion as a sequence of chemical groups on a very long polymer. There are many analogies with the field of computers. Data may be stored as a sequence of magnetic regions on a piece of tape or on a diskette, for example, and these may be 'translated' into the words of a book or the design of a car. Unlike computer data which is stored as binary numbers (a series of ones and zeros), the information in DNA is stored as a sequence of four chemical groups (Fig. 1.22), the nitrogenous bases. Obviously, the information needed to construct and operate an organism such as a human being is very extensive. Every cell of the body contains 46 molecules of DNA (one in each chromosome) which have a total length of over a metre.

REPLICATION OF INFORMATION. When a cell divides, its DNA has to be duplicated exactly—a process called **replication**. This is achieved largely by the use of hydrogen bonds and matching complementary shapes. The DNA occurs as a double-stranded molecule (the so-called double helix) in which the nitrogenous bases of one strand complement those of the other strand. Essentially the same information is present in each strand: if the sequence of bases in one strand is known then the sequence in the other strand can also be determined because the bases pair only in certain ways (Fig. 1.23).

To duplicate a double helix, it is necessary to pull apart the two strands, allow free bases to match up with the now exposed bases on the separated strands, and then 'zip up' the new complementary sequences.

□ Viruses are non-living. They consist of a piece of nucleic acid surrounded by a protein coat. However, when the nucleic acid enters a cell, the information contained in it takes over the machinery of the cell and diverts it into producing new viruses. These are subsequently released and go on to infect new cells. This is the 'life-cycle' of a virus, although it can only be said to be 'living' when it is in another cell.

Exercise 10

List three ways humans have devised for storing information and for each state how the information is coded.

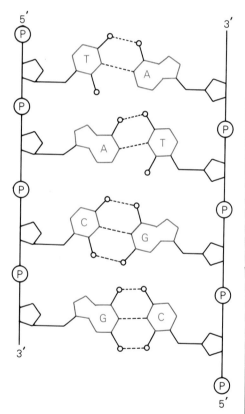

Fig. 1.23 DNA can replicate because the bases pair up across the double helix. The pairing is specific, so the sequence of bases in any new strand will have its order predetermined.

1.7 *Overview*

Life is based on the ability of carbon to form a myriad of molecules, some of which, the macromolecules, are extremely large. The macromolecules found in organisms are very diverse but practically all may be classified as being either proteins, polysaccharides or nucleic acids. They all have their specific biological roles. Life processes take place in water, and the properties of the water molecule are of crucial importance.

For life to continue, it is necessary for living organisms to obtain energy. They do so in a variety of ways and use this energy to drive their life processes. Although organisms operate according to the laws of thermodynamics, they do not function as heat engines. Instead, they obtain and use free energy and operate under conditions of low temperature (0–40°C), atmospheric pressure and neutral pH. To do so they must be separate from, but remain in communication with, their environment. They achieve this by the use of selectively permeable membranes consisting of lipid bilayers containing proteins.

The reactions that organisms carry out proceed because of the presence of highly effective macromolecular catalysts called enzymes, which specifically recognize the molecules with which they react.

Life also requires that information be stored and transmitted. These processes also require the participation of macromolecules (DNA) and involve weak interactions (hydrogen bonds) during molecular recognition in the duplication process.

Replication: *when a cell divides, its DNA needs to be duplicated so that each of the two daughter cells have exactly the same DNA complement as the mother cell. The process of producing a copy of this DNA is called replication.*

Reference Watson, J.D. (1968) *The Double Helix*, Weidenfeld & Nicholson, London, UK published in paperback by Penguin Books in 1970). One of the co-discoverers of the DNA double-helix gives his view of how it happened.

1. 20! That is, 20 × 19 × 18 × 17 ... which comes out to 2 × 10^{18}. The exclamation mark is a mathematical symbol meaning factorial 20.

2. All except the hydrocarbons, $-CH_3$ and $-CH_2CH_3$.

3. High latent heat of evaporation; less dense at 4°C then at 0°C so that ice floats; liquid at biological temperatures.

4. Methane is symmetrical (tetrahedral) with an even distribution of electrons.

5. Oil and water (the major component of the vinegar) do not mix. It takes energy to disrupt the hydrogen-bonded structure of the water so that the oil can get into the body of the water. Shaking supplies energy.

6. A nucleoside is a base–sugar unit: in ATP this part of the molecules is adenine–ribose. Adding a triphosphate unit gives ATP.

7. The oil is likely to be vegetable oil and the emulsifying agent egg yolk or lecithin. Lecithin is a phospholipid abundant in egg yolk.

8. Free fatty acids act as detergents. Potentially they could disrupt membranes and denature proteins.

9. The amino acid residues in the active site of an enzyme are held in extremely precise orientations with respect to each other, to create a three-dimensional shape into which substrate molecules fit with great specificity. A major role of that part of the protein which is not active site is probably scaffolding maintaining the positions of the active site residues, probably numbering 6–12 residues, and which frequently come from widely separated regions of the polypeptide chain. Of course not all substrates are small, but even when they are large (e.g. DNA, proteins) only a small part of them is in contact with the active site residues of the enzyme that is acting on them. **Reference:** Cornish-Bowden, A. (1986) Why are enzymes so big? *Trends in Biochemical Sciences*, **11**, 286.

10. (i) Magnetic tape, regions of magnetization;
    (ii) printed word, alphabet, words, etc.;
    (iii) compact disk, pits read by coherent light from laser.

QUESTIONS

FILL IN THE BLANKS

1. Biological membranes surround all cells as well as most cell _____ . They are constructed from a _____ made of _____ , in which the molecules point their _____ ends towards each other. However, membranes constructed solely in this way would be _____ to both hydrophilic and hydrophobic molecules. Therefore, the other major component of membranes is _____ which can regulate the _____ of materials into and out of cells. These molecules characteristically have _____ regions on their surfaces whereby they interact with the membrane _____ . This is in contrast to the majority of globular _____ whose _____ regions tend to be hidden inside the molecules.

Choose from: bilayer, hydrophobic (3 occurrences), impermeable, lipid (2 occurrences), movement, organelles, protein, proteins.

2. Biological macromolecules are typically formed from smaller molecules by linking them together in _____ reactions. Such reactions require an input of _____ to drive them, and the overall process involves removal of _____ . Biological macromolecules can _____ with other molecules, both large and small, in highly _____ ways. This is possible because macromolecules can form _____ shapes that recognize each other and _____ tightly together. The bonds involved in such _____ are almost always _____ , _____ bonds. Although such bonds are individually much _____ than C–C bonds, the fact that there are very _____ of them involved in an interaction makes the union _____ .

Choose from: bind, complementary, condensation, energy, interact, interactions, many, non-covalent, specific, strong, water, weak, weaker.

Reference *Molecules to Living Cells.* A collection of older articles taken from *Scientific American* gives the background to much of the work on DNA. Example: 'The genes of men and molds' by G.W. Beadle, first published 1948.

3. The protein haemoglobin is made up of _____ subunits, identical in _____ . Each subunit contains one _____ group which has a central _____ atom that functions in _____ transport. The subunits fit very tightly together and are joined by _____ bonds. The haem groups fit into _____ clefts in the polypeptides and are linked to the protein _____ . Because of the way in which the subunits fit tightly together, a change in _____ of one subunit leads to a _____ change in the other subunits. A consequence of this behaviour is that the _____ binding curve for haemoglobin is _____ in shape.

Choose from: conformational, four, haem, hydrophobic, iron, non-covalent, non-covalently, oxygen (2 occurrences), pairs, shape, sigmoidal.

MULTIPLE-CHOICE QUESTIONS

4. Which of the following types of molecule would you call macromolecules?
A. polysaccharides
B. phospholipids
C. triacylglycerols
D. proteins
E. nucleotides

5. Which are the reactions by which macromolecules are formed from monomeric building blocks?
A. reductions
B. oxidations
C. esterifications
D. phosphorylations
E. condensations

6. Which of the following are formed by linking together α-amino acids?
A. polypeptides
B. proteins
C. peptides
D. nucleic acids
E. biological membranes

7. Which of the following functions are fulfilled by proteins?
A. structure (e.g. in bone and tendon)
B. transport of oxygen in the blood
C. catalysis of metabolic reactions
D. acting as antibodies
E. carrying information

8. Which of the following carry genetic information that can be passed on to the daughter cells at division?
A. complex polysaccharides
B. RNA
C. complex lipids
D. polydeoxynucleotides
E. DNA

9. Which of the following molecules act as stores of biological energy?
A. phospholipids
B. starch
C. haemoglobin
D. glycogen
E. cellulose

10.   Which of the following life activities require an input of biological energy?
A. glycogen biosynthesis
B. muscular contraction
C. triacylglycerol hydrolysis
D. protoplasmic streaming
E. fermentation

11.   Which of the following are involved in the interaction and recognition of one molecule by another?
A. ionic forces
B. van der Waals' attractions
C. carbon–carbon bond formation
D. hydrogen bonds
E. hydrophobic interactions

12.   When two complementary DNA molecules join to form a double helix, the types of bond involved are:
A. peptide bonds
B. hydrogen bonds
C. salt linkages
D. hydrophobic interactions
E. sugar–phosphate links

SHORT-ANSWER QUESTIONS

13.   For each of the classes of macromolecule found in living cells, write down three biological functions of that type of macromolecule.

14.   Explain why it is *weak* forces and bonds (interactions) rather than strong covalent bonds, that give rise to biological specificity.

15.   Briefly describe the role in the living cell of each of the following: (a) DNA, (b) phospholipids, and (c) enzymes.

16.   Explain why a biological membrane composed only of phospholipid molecules would be ineffective.

17.   For the main types of macromolecules in cells describe briefly the features of the monomers which give the macromolecules their characteristic properties.

18.   From the properties of biological macromolecules you know about explain why it would not be feasible to use heat to drive metabolic reactions.

2

Globular proteins

Objectives

After reading this chapter you should be able to:

☐ describe the general roles of proteins in biological systems;

☐ explain how the various side groups of amino acids contribute to the structure and functions of proteins;

☐ describe the hierarchial levels of protein structure;

☐ evaluate the importance of forces that determine the structure of a given protein molecule;

☐ illustrate how the biological activities of a selected globular protein may be explained at the molecular level;

☐ explain what is meant by the term protein denaturation.

2.1 Introduction

If DNA is the *legislative* molecule of the cell, having the ability to direct the development of organisms and the activities of cells, then proteins are the *executive*, controlling and modulating cellular metabolism and integrating the physiological activities of multicellular organisms.

The importance of proteins was recognized early in the ninteenth century by Mulder (1838):

There is present in plants and animals a substance which ... is without doubt the most important of the known substances in living matter, and without it, life would be impossible on our planet. This material has been named protein.

The name *protein* testifies the importance attributed to this class of biomolecules.

All proteins are formed by joining together amino acids in a linear unbranched chain, although this arrangement may later be modified. The structures and types of amino acids found in proteins are described in Table 2.1. Most protein molecules exist in one or other of two general *conformations*: they can be extended, relatively rigid rods or compact and globular in shape. The former group are mainly **fibrous** proteins with mechanical, structural roles. This chapter concentrates largely on the structure of **globular** proteins, and how structure underpins activities and biological functions.

Legislative: *the branch or agency with the power to make laws and set limits.*
Executive: *the agency with the skill or abilities to perform functions, i.e. the 'doers'.*

Protein: *from the Greek* protos, *first. The Dutch chemist, Mulder, introduced this word in 1838 implying 'first rank', or the most important substance in living matter.*
Contormation: *the general structure or shape.*

2.2 General functions of proteins

Proteins are undoubtedly the most functionally diverse of biomolecules. In general, globular proteins function by 'recognizing' other molecules to which they specifically bind. Such precise binding is possible because the protein molecule has a site that is **complementary** to a site on the molecule recognized (Fig. 2.1). However, this binding is *not* fixed and rigid but, rather, exists in a

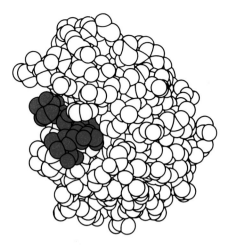

Fig. 2.1 Representation of a molecule of lysozyme and its substrate (highlighted) bound together at complementary surfaces. Redrawn from photograph originally supplied courtesy of Dr J.M. Burridge, IBM, UK.

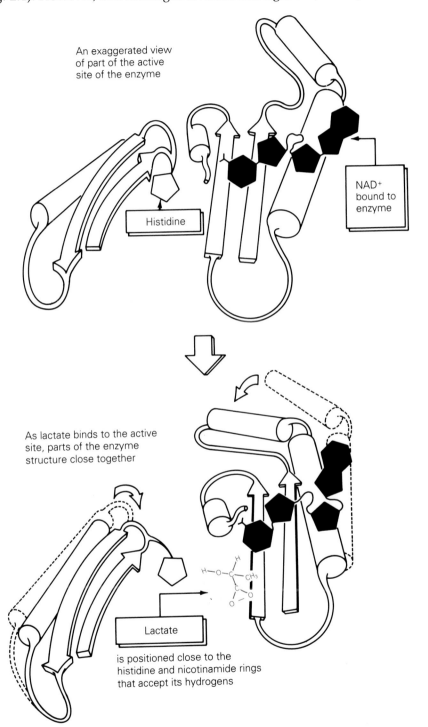

An exaggerated view of part of the active site of the enzyme

Histidine

NAD⁺ bound to enzyme

As lactate binds to the active site, parts of the enzyme structure close together

Lactate

is positioned close to the histidine and nicotinamide rings that accept its hydrogens

Fig. 2.2 Changes in the structure of the protein lactate dehydrogenase on the binding of lactate and NAD⁺ Redrawn from Wynn, C. (1989) *Biological Science Review*, **1**, 16–21.

Reference Creighton, T.E. (1984) *Proteins: Structure and Molecular Properties*, W.H. Freeman & Co., New York. Excellent general reference text. Covers many aspects of the composition, structure and activities of proteins.

dynamic equilibrium with recognition, binding and release occurring continuously. At any moment the proportion of bound molecules depends upon: (a) the relative concentrations of the protein and the molecule to which it binds, and (b) the strength of association between them. The latter depends on how well the complementary sites fit together and the types of interactions involved, for example, hydrophobic, ionic, hydrogen bonding, which occur between the sites. Once binding does occur, there is a conformational change in the protein-bound molecule, or in the complex of protein and protein-bound molecule (Fig. 2.2). This change is the signal that initiates the biochemical activity associated with the protein and forms the basis for the remarkable range of biological roles exhibited by proteins.

The formation of proteins is under the *direct* control of DNA. The growth and differentiation of cells, organs and organisms results from the orderly expression of information contained in the DNA molecules. However, a 'chicken and egg' situation exists, since the formation of proteins, and indeed the replication of DNA, requires the activity of pre-existing proteins.

Much of biochemistry is concerned with the remarkable protein catalysts called **enzymes**. Many reactions that normally proceed at barely measurable rates are typically accelerated by a factor of 10^8–10^{11} by the presence of the appropriate enzyme. In comparison with chemical catalysts, enzymes are also amazingly specific; a given enzyme catalyses only a single transformation or a group of similar reactions. Their catalytic power, their specificity and the fact that their activity can be regulated, mean that enzymes ensure that metabolism proceeds in an orderly fashion.

Specific transport proteins are a feature of living systems. The well-known blood protein, haemoglobin, transports O_2 in the blood of vertebrates. Examples of transport proteins in the serum are **albumin**, which can transport fatty acids; **lipoproteins**, which carry cholesterol and other lipids; and **transferrin**, which transports iron. Invertebrates have copper-containing proteins called **haemocyanins** (Fig. 2.3), which have O_2-carrying roles parallel to those of the vertebrate haemoglobins.

Other transport proteins have a different function. They are situated in biological membranes and allow materials to be transported across the membrane. For example, the Na^+, K^+-ATPase is a protein that pumps Na^+, out of cells and K^+ into cells, at the expense of metabolic energy.

Proteins play a key role in the co-ordination of metabolism. For example, neurons (Fig. 2.4) respond to specific signals via protein receptors on their surfaces. Indeed, many of these 'signals' are chemical ones, consisting of peptides or small proteins. Co-ordination in multicellular types is often mediated by hormonal signals; and in animals the hormone receptors and, indeed, some hormones themselves are proteins.

The movement of organisms is achieved by a dynamic function of protein molecules. Some bacteria are motile using extended appendages called flagella. Eukaryotic cells use cilia and flagella in locomotion, but multicellular animals move using skeletal muscle (Fig. 2.5). All of these locomotory activities depend upon the co-ordinated movements of sets of fibrous proteins.

Protein molecules are responsible for the mechanisms by which organisms protect themselves against parasites and toxins. Scavenging white blood corpuscles, called leukocytes, recognize invading microorganisms by means of protein receptor molecules on their surfaces, and then engulf them. **Antibodies** are serum proteins that can combine with **antigens** such as bacterial toxins, leading to their neutralization. Other proteins, such as **fibrinogen**, circulate in the blood and are able to form fibrous mats to seal wounds.

Mechanical support is given by several types of fibrous proteins both inside

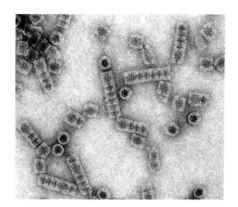

Fig. 2.3 Electron micrograph of haemocyanin molecules from the water snail *Lymnaea stagnalis*. The molecule is a cylinder of 30 nm diameter.

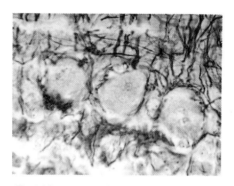

Fig. 2.4 Large neurons (Purkinje cells) of mammalian cerebellum.

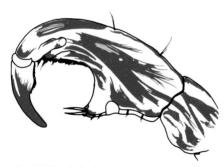

Fig. 2.5 Muscle (red) attachments in claw of louse. Redrawn from 14th prize-winning photograph from the 1984 Nikon Photomicrography Competition. Courtesy of Mrs H. Selena and Nikon.

and outside cells. **Tubulin** forms extended microtubules within the cytoplasm, which help determine the shape of the cell. Other proteins are found extracellularly and help organize the matrix that surrounds the cell. **Collagen** is a widely distributed extracellular protein, which imparts a high tensile strength to tissues such as cartilage, bone and the skin.

2.3 Structure of proteins

α-Amino acids (Fig. 2.6) are joined together by peptide bonds to give proteins. The addition of increasing numbers of amino acids gives peptides and then polypeptides. The distinction between a peptide and a polypeptide is rather arbitrary: if the M_r is less than 5000, the molecule is usually referred to as a peptide. Proteins show a variety of structures. Some consist of a single polypeptide chain, others of several, which fit together in beautiful and intricate ways (see later).

Amino acids

The amino acids of a protein are linked together by peptide bonds formed by condensation reactions (Fig. 2.7). Since the elements of water are lost in the condensation, the amino acids are more correctly called **residues**. Twenty different amino acids are found in proteins, although subsequent to their incorporation they may be subjected to a variety of chemical modifications, for example, phosphorylation, hydroxylation, methylation, acylation or glycosylation. Table 2.1 gives the structures and overall description of the 'standard' amino acids of proteins specified by the genetic code. These amino acids differ only in the nature of their 'R-groups' or side chains. Thus, the presence of 20 different types of amino acid residue side chains confers a variety of properties, which gives rise to the structure and therefore functions of proteins.

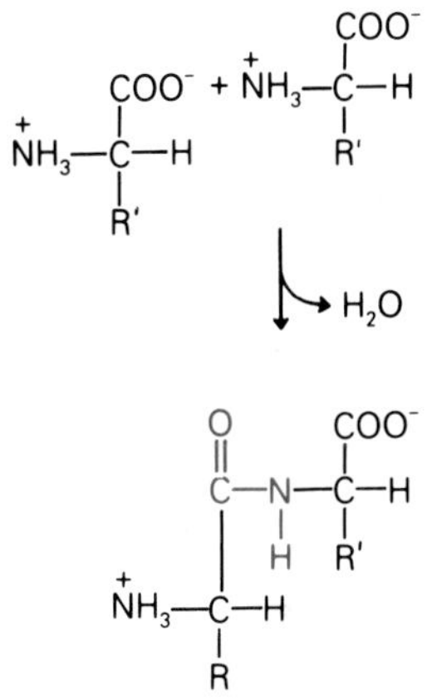

GLYCINE and ALANINE are chemically the simplest of the amino acids. Both are metabolically cheap to produce. Glycine may be regarded as lacking a side chain (i.e. R = –H). It occupies little space and is found in sites within protein molecules where space is restricted. Alanine seems to be the most abundant amino acid found in proteins. This is probably because the inclusion of a high portion of glycine into a protein would make the molecule

□ Collagen is the most common protein in bulk in vertebrate bodies, constituting about 25% of all protein (see Chapter 3).

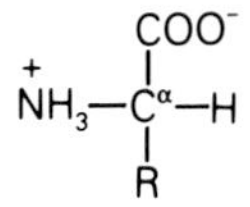

Fig. 2.6 General structure of an α-L-amino acid. This is an α-amino acid because the amino group is carried on the α-carbon atom (indicated).

Fig. 2.7 Peptide bond formation. This does not describe the biosynthesis of the bond but merely illustrates the reaction involved.

□ The 20 standard amino acids are those coded for by the genetic code (see also Chapter 8 and *Molecular Biology and Biotechnology*, Chapter 3). These 20 were first identified by Crick and Watson as long ago as 1953 as the fundamental units of all proteins.

Box 2.1
Aspartame

Aspartame is an 'artifical' sweetner 100–150 times sweeter than sucrose on a gram for gram basis. It is the methyl ester of the dipeptide aspartylphenylalanine (Asp–Phe).

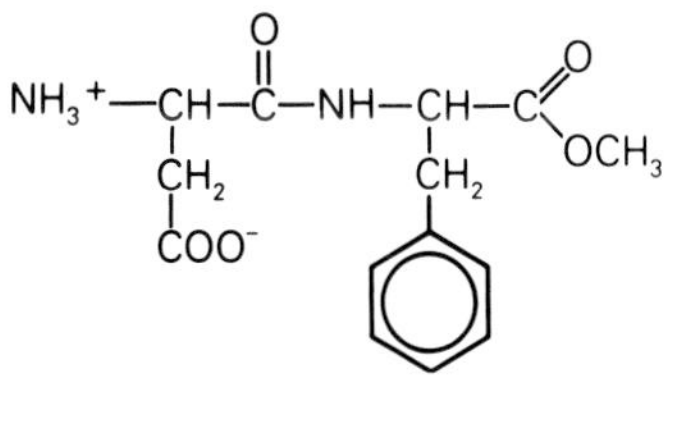

The sweetness of aspartame was discovered in 1969 and its application as a sweetening agent has been approved for a variety of culinary uses. Aspartame is synthesized from L-amino acids. Combinations of L- and D- or all D-amino acids are not sweet-tasting but bitter.

Reference Dickerson, R.E. and Geis, I. (1969) *The Structure and Action of Proteins*, Harper & Row, Philadelphia. Very old, but still excellent for starting to learn about protein structure.

Table 2.1 *Classification and structure of amino acids. Amino acids are often referred to by either a three letter or single letter abbreviation*

Group	Name	Abbreviation	Residue M_r* (pH 7.0)	Structure of side group
	Glycine	Gly, G	57	–H
Small polar	Threonine	Thr, T	101	
	Serine	Ser, S	87	
	Aspartate	Asp, D	114	
	Asparagine	Asn, N	114	
Small non-polar	Cysteine	Cys, C	103	
	Alanine	Ala, A	71	
	Proline	Pro, P	97	
Large non-polar	Valine	Val, V	99	

Table 2.1 Continued

Group	Name	Abbreviation	Residue M_r* (pH 7.0)	Structure of side group
	Isoleucine	Ile, I	113	
	Histidine	His, H	137	
	Leucine	Leu, L	113	
	Methionine	Met, M	131	
	Phenylalanine	Phe, F	147	
Large polar	Glutamate	Glu, E	128	
	Glutamine	Gln, Q	128	
	Lysine	Lys, K	129	

Table 2.1 Continued

Group	Name	Abbreviation	Residue M_r* (pH 7.0)	Structure of side group
	Arginine	Arg, R	157	
	Tryptophan	Trp, N	186	
Intermediate polarity	Tyrosine	Tyr, Y	163	

* Mean residue M_r = 118.75.

Exercise 1

What is the approximate M_r of a polypeptide chain of 300 amino acid residues?

excessively flexible. In contrast, alanine the next cheapest amino acids has a methyl group (R = –CH₃), which confers reduced mobility on the polypeptide compared with glycine.

VALINE, ISOLEUCINE and LEUCINE have individual branched aliphatic (that is non-cyclic) side chains that are relatively rigid. Such stiff side groups have little internal flexibility and facilitate the folding of protein molecules into specific shapes with appropriate biological properties.

PHENYLALANINE, TYROSINE and TRYPTOPHAN have planar aromatic rings that are separated from the polypeptide chain by single methylene groups (–CH₂–), which to some extent prevent the rigid rings from hindering the folding of the protein molecule.

Phenylalanine has the largest completely non-polar side group. It is unreactive, but can form non-covalent interactions, which are important in stabilizing the shape of protein molecules (see later).

Tyrosine, differs from phenylalanine by the presence of a phenolic hydroxyl group on the ring, making it relatively reactive chemically, although it is not charged at physiological pH. The hydroxyl group can participate in hydrogen bonding.

Tryptophan has the largest side group of all the 20 amino acids. It has an indole ring capable of participating in charge-transfer reactions and also in hydrogen bonding.

PROLINE has a non-polar and fully saturated five-membered ring. It is unique in that the side group is curled round so that the nitrogen and the α-carbon atoms form part of the ring. Proline lacks an amino group (NH₂–) and is described as an **imino** acid. This feature allows little flexibility and means the polypeptide must adopt a particular shape at sites of proline residues.

SERINE and THREONINE both contain alcoholic hydroxyl groups, which can participate in hydrogen bonding. The lone pairs of electrons on the oxygen atoms mean they can attack positvely charged chemical groups, and many enzymes have serine residues that function in catalysis. Often the immediate environment *within* the protein molecule ensures that a particular serine or threonine residue is especially reactive. Serine and threonine are also common sites for the chemical modifications of proteins. For example, phosphorylation and glycosylation often occur at specific serine or threonine residues.

CYSTEINE and METHIONINE both contain sulphur. In some proteins, specific pairs of cysteine residues are oxidized to form disulphide bridges (Fig. 2.8), which link different portions of a single polypeptide chain together (e.g. ribonuclease or α-lactalbumin, Fig. 2.9a), or different polypeptides together (e.g. insulin or immunoglobulins, Fig. 2.9b).

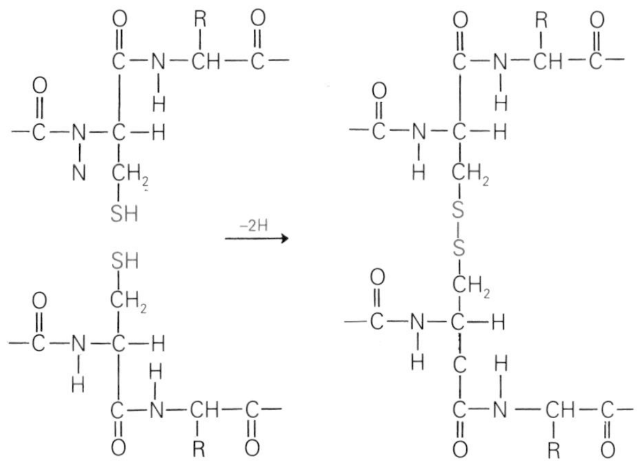

Exercise 2

The chemical reagent diisopropylphosphofluoridate reacts irreversibly with some serine residues in proteins. When an enzyme was treated with this reagent, only one of its 29 serine residues reacted. What is the most probable reason for this?

Fig. 2.8 Oxidization of two cysteine residues to form a disulphide bond.

Box 2.2
Alice's hatter and Japanese fishermen

Many enzymes are known which contain active sulphydryl groups of cysteine residues. These groups are easily oxidized by heavy metal ions such as Pb^{2+} or Hg^{2+}

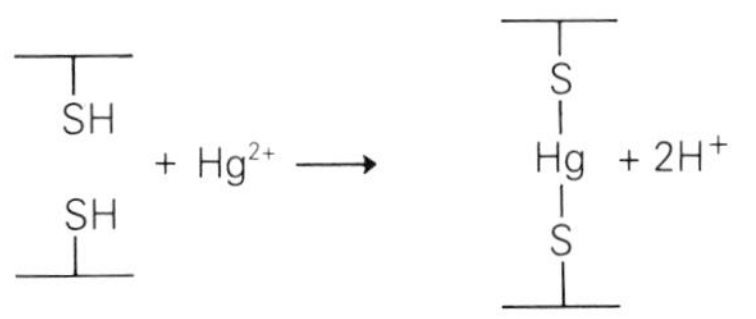

Such reactions can irreversibly inhibit enzymic activity and this is the biochemical basis of lead and mercury poisoning.

Industrial mercury poisoning was prevalent in the seventeenth century among workers producing felt hats. Such hats were made from the fine hairs of rabbits and hares. The felting process involved treating the hairs with Hg^{2+} salts, which disrupted their molecular structure making them softer and limp. Unfortunately, contamination of the air with mercury vapour led to nervous disorders in the workers, who developed symptomatic twitches. The idea for the Mad Hatter in *Alice in Wonderland* originated from this disease.

More recently, the dumping of mercury waste into the Bay of Minamata in Japan allowed mercury to enter the marine food chains. The high sea-food diet of local people led to over a 100 cases of Hg poisoning with 44 deaths.

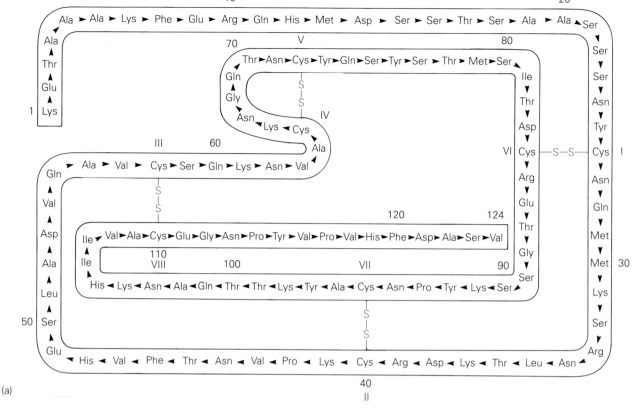

Fig. 2.9 Disulphide bonds in (a) ribonuclease and (b) insulin.

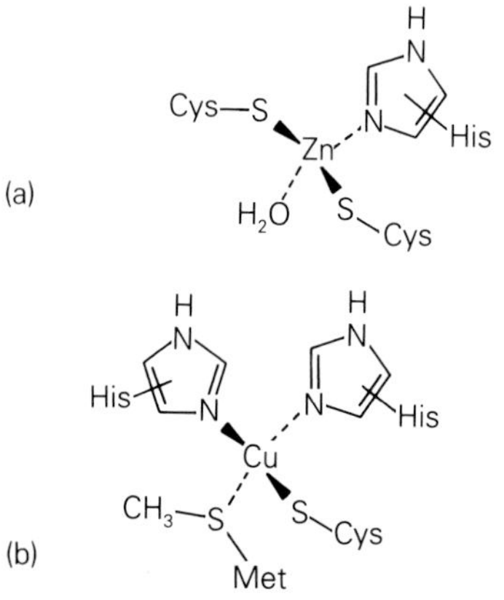

(b)

Gly.Ile.Val.Glu.Glu.Cys.Cys.Ala.Ser.Val.Cys.Ser.Leu.Tyr.Glu.Leu.Glu.Asp.Tyr.Cys.Asp
1 (NH₂—S—S) 10 20 (NH₂ NH₂ NH₂)

Phe.Val.Asp.Glu.His.Leu.Cys.Gly.Ser.His.Leu.Val.Glu.Ala.Leu.Tyr.Leu.Val.Cys.Gly.Glu.Arg.Gly.Phe.
1 (NH₂ NH₂) 10 20

Phe.Tyr.Thr.Pro.Lys.Ala
30

Fig. 2.10 Examples of involvement of cysteine, methionine and histidine residues in binding metal ions in the metalloproteins: (a) liver alcohol dehydrogenase, and (b) azurin and plastocyanin. Redrawn from Ibers, J.A. and Holm, R.H. (1980) *Science*, **209**, 227–35.

The lone pairs of electrons on the sulphur atom mean that the reactive sulphydryl group of cysteine may participate in catalysis, while both cysteine and methionine form links with metal ions in some metalloproteins (Fig. 2.10).

HISTIDINE has a relatively highly reactive heterocyclic aromatic side group. Within the physiological range of pH the ring can either be protonated or unprotonated (Fig. 2.11) and histidine is another residue that often plays a role in enzymic catalysis. The lone pairs of electrons on the ring nitrogens mean histidine may also form co-ordinated links with the metal ions that are often found in proteins (Fig. 2.10).

LYSINE and ARGININE both have long, flexible side chains containing basic amino groups. Both residues are generally found on the surface of protein molecules; their 'wobbliness' increasing the water solubility of the molecule. If arginine or lysine residues occur internally in protein molecules, they almost always participate in salt bridges (see later) or have a role in, for example, catalysis.

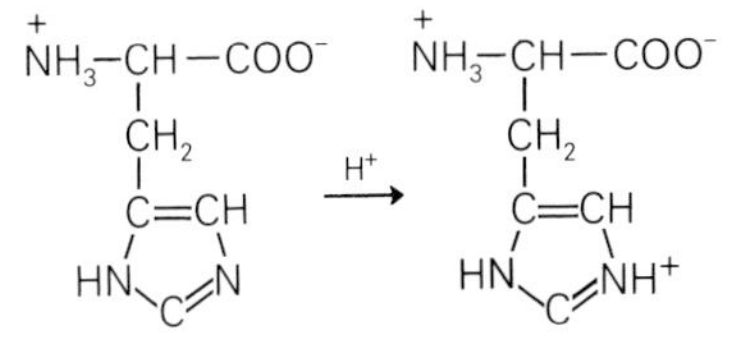

ASPARTATE and GLUTAMATE are capable of acting as hydrogen donors and acceptors in both catalysis and hydrogen bond formation.

ASPARAGINE and GLUTAMINE are also members of the standard 20 amino acids, and are capable of forming hydrogen bonds. The amide groups function as hydrogen donors, the carboxyls as acceptors.

Primary structure

Although different proteins have different amino acid compositions (e.g. Gly_{20}, Ala_{15}, Tyr_7, Glu_5, Cys_2 ...), it is the *order* of the amino acid residues that confers unique biological properties to any individual protein. The order or sequence of amino acid residues is called the **primary structure**. The amino acid acids in the primary structure are numbered from the amino terminus ($^+NH_3-$) to the carboxyl terminus ($-COO^-$) as shown in Fig. 2.9.

The primary structure of a protein is specified by the order of bases in the genomic DNA. The cell has a complex metabolic machinery for converting the information contained in the base sequence of the DNA (the gene) into the amino acid sequences in a protein that is the gene product. Small peptides are somewhat unusual in that their syntheses seem usually to occur by one of two mechanisms. For example, neural peptides, such as those derived from propriomelanocortin, seem to be formed by synthesizing an extended polypeptide chain and then hydrolysing it into small pieces, which form the neural peptides (Fig. 2.12). Other peptides, such as glutathione are synthesized by directly joining amino acids together. Each amino acid being added in a separate enzyme-catalysed reaction (Fig. 2.13). Note the non-standard (γ) peptide bond in gluthathione. DNA is not directly involved in the formation of these types of peptides.

Small peptides probably have flexible structures, but polypeptides and proteins in the native state have other, higher orders of structure that are

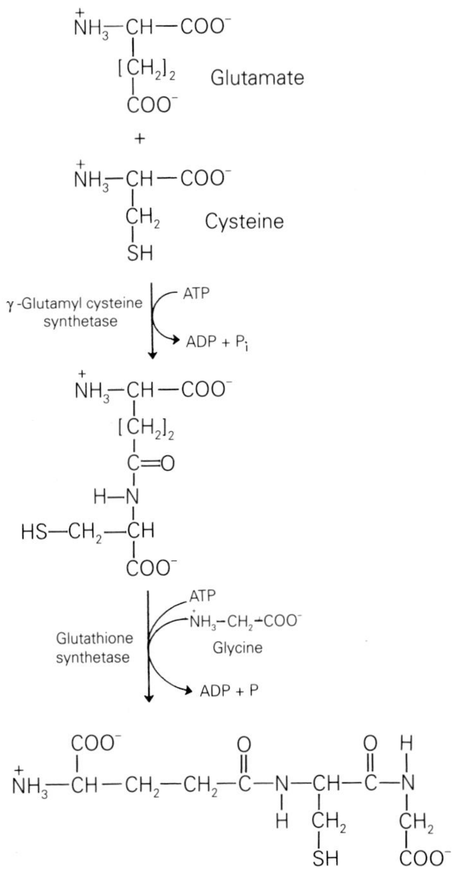

Fig. 2.11 Protonation of histidine.

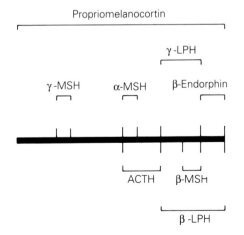

Fig. 2.12 Neural peptides produced by the specific hydrolyses of propriomelanocortin. MSH, melanocyte-stimulating hormone; ACTH, adrenocorticotrophic hormone; LPH, lipotropin.

Fig. 2.13 Synthesis of glutathione from its constituent amino acids.

Glutathione is the tripeptide γ-Glu–Cys–Gly. It is synthesized according to the scheme given in Figure 2.13 (note that the bond joining the glutamate and cysteine residues is not a normal peptide bond linking α carbons).

Glutathione can be reduced using NADPH in a reaction catalysed by the enzyme **glutathione reductase**. The reduced form (G–SH) is present in a 500-fold excess over the oxidized form (G–SS–G) in most cells. G–SH occurs at concentrations of about 5 mmol dm^{-3}, and serves as a **sulphydryl buffer** maintaining a reserve of reducing power in the cytoplasm.

$$
\begin{array}{c}
\text{γ-Glu–Cys–Gly} \\
| \\
\text{S} \\
| \\
\text{S} \\
| \\
\text{γ-Glu–Cys–Gly}
\end{array}
\quad + \text{ NADPH}_2
$$

$$\downarrow$$

$$
2 \left[\begin{array}{c} \text{γ-Glu–Cys–Gly} \\ | \\ \text{SH} \end{array} \right] + \text{NADP}^+
$$

Reduction of glutathione by NADPH.

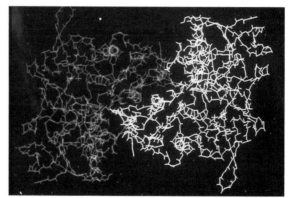

Molecular model of glutathione reductase. Note the two identical subunits. Courtesy of Drs A. Berry and R.N. Perham, Department of Biochemistry, University of Cambridge.

Glutathione is crucial to the erythrocyte functions of maintaining haemoglobin in a reduced state and iron in a ferrous state. It also has a metabolic role in detoxification. For example, it reduces organic peroxides to less dangerous alcohols, itself being oxidized in the process:

$$2\text{G–SH} + \text{ROOH} \rightarrow \text{G–S–S–G} + \text{H}_2\text{O} + \text{R–OH}$$

relatively fixed. Several types of higher-ordered structures are recognized (see later), but all are specified by the primary structure.

Peptide bonds

Two possible arrangements of peptide bonds, called the *cis* and *trans* forms, are possible (Fig. 2.14). However, steric hindrance between side groups and α-carbon atoms of adjacent amino acid residues mean that the *trans* form is favoured about 1000-fold over the *cis* arrangement. In the case of proline, steric hindrance is less than with *amino* acids because of the atypical side group, and proline residues are freer to adopt the *cis* configuration (Fig. 2.15).

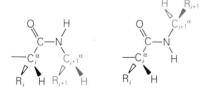

Fig. 2.14 Cis/trans forms of peptide bond.

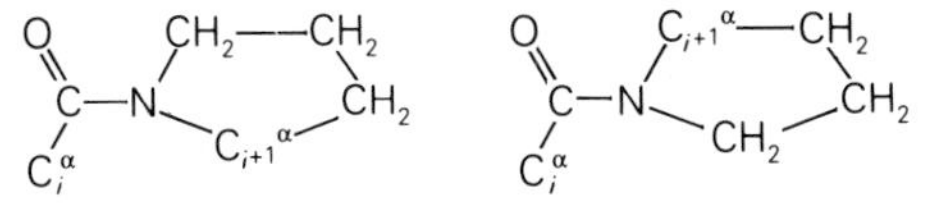

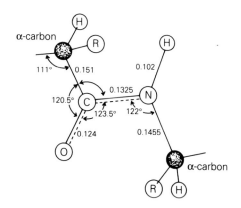

Fig. 2.15 Cis/trans forms of proline.

Peptide bonds are stabilized by resonance. The bond is best thought of as having an intermediate form between the extremes shown in Fig. 2.16, with the bond electrons **delocalized**, that is spread over the peptide linkage (Fig. 2.17). This has important consequences for the structure of the proteins, since the partial double-bond nature of the C=N bond in a peptide means that rotation is restricted. Thus, the group of atoms outlined in Fig. 2.18, often called a peptide unit, is an essentially rigid planar structure. Rotation about the **N–Cα** bond is denoted by ϕ, Cα–C' by ψ, and the C'–N (usually fixed, as explained above) by ω (Fig. 2.19).

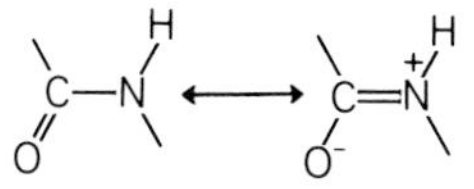

□ The delocalization of the electrons in the peptide bond means that they occupy more space, effectively lowering their energy. This feature contributes about 120 kJ mol^{-1} of resonance stabilization energy to the peptide bond.

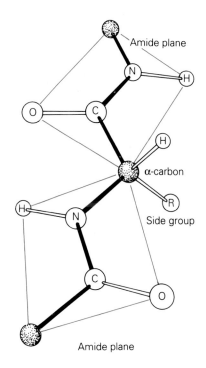

Fig. 2.16 Extreme resonance forms a peptide bond.

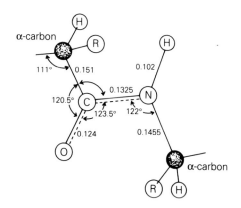

Fig. 2.17 'Delocalized' form of peptide bond, with bond angles shown and dimensions in nanometres.

If amino acids are in the *cis* configuration, then $\phi = \psi = \omega$ and their rotations are all given the arbitrary value of 0°. Rotation from this position (when viewed from behind the rotated bond) in a counterclockwise direction gives *negative* values, and if clockwise *positive* values. A maximum rotation of 180° is possible in either direction, with −180° corresponding to 180° (Fig. 2.19). A list of ϕ, ψ and the relatively invariant ω angles for all residues would completely define the conformation of the backbone, or main chain, of the polypeptide (Table 2.2). This does not, of course, say anything about the orientation of the side groups of the residues.

It might be thought that ϕ and ψ could adopt virtually any combination of angles, but this is not the case. The majority of combinations are excluded because they bring atoms so close together that steric hindrance occurs. Certain combinations of ϕ, ψ form particularly stable, regularly-shaped backbone structures called **secondary structures**.

Secondary structures

Three secondary structures have been particularly well described: α-**helix**, β-**pleated sheets** and β-**turns**; and are common in proteins. Stretches of polypeptide that do not have obviously recognizable secondary structural features are often called **random coils**. However, there is nothing random about their structure, although they may have relatively more flexibility of movement in solution, and there is an increasing tendency to refer to them merely as **coils**.

Fig. 2.18 Two peptide units (each outlined in red).

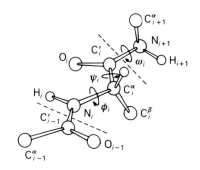

Fig. 2.19 Definition of ϕ, ψ and ω angles (all 180°).

N, Cα, C': the atoms of an amino acid residue that contribute to the backbone of the polypeptide, i.e. the amino nitrogen, the α-carbon and carboxyl carbon atoms respectively.

Table 2.2 *Conformational angles of bovine pancreatic trypsin inhibitor, which describe the folding of the polypeptide backbone. Note that although ϕ, ψ combinations vary widely, those of ω are relatively invariant*

Residue		ϕ	ψ	ω
Arg	1		137	178
Pro	2	−60	149	180
Asp	3	−58	−33	179
Phe	4	−63	−16	177
Cys	5	−70	−19	177
Leu	6	−89	−3	178
Glu	7	−78	150	177
Pro	8	−72	158	−177
Pro	9	−67	143	−176
Tyr	10	−121	115	−177
Thr	11	−77	−36	175
Gly	12	90	176	−176
Pro	13	−83	−10	−177
Cys	14	−92	159	−170
Lys	15	−119	32	169
Ala	16	−77	173	−175
Arg	17	−130	81	−175
Ile	18	−109	111	−178
Ile	19	−78	120	176
Arg	20	−122	179	172
Tyr	21	−114	147	168
Phe	22	−130	160	173
Tyr	23	−87	127	176
Asn	24	−100	97	−172
Ala	25	−59	−28	173
Lys	26	−70	−35	−174
Ala	27	−90	−20	−179
Gly	28	79	13	−178
Leu	29	−156	171	−173
Cys	30	−95	146	−174
Gln	31	−131	161	179
Thr	32	−89	156	175
Phe	33	−151	164	178
Val	34	−97	113	174
Tyr	35	−86	135	180
Gly	36	−79	−9	171
Gly	37	99	7	172
Cys	38	−146	157	172
Arg	39	62	36	−176
Ala	40	−56	156	164
Lys	41	−104	172	−168
Arg	42	−73	−26	179
Asn	43	−82	74	−168
Asn	44	−163	102	−169
Phe	45	−123	155	179
Lys	46	−86	−8	177
Ser	47	−151	158	174
Ala	48	−65	−30	180
Glu	49	−71	−39	175
Asp	50	−70	−38	179
Cys	51	−62	−45	179
Met	52	−72	−35	−179
Arg	53	−62	−38	180
Thr	54	−82	−42	−166
Cys	55	−111	−8	−167
Gly	56	−76	−6	164
Gly	57	84	168	179
Ala	58	−65		

Adapted from Creighton, T.E. (1984) *Proteins: Structure and Molecular Principles.* W.H. Freeman & Co., New York.

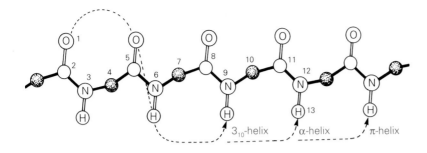

Fig. 2.21 Hydrogen bonding relationships between the 3_{10}, α- and π-helices.

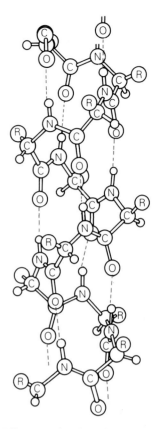

Fig. 2.20 Representation of an α-helix.

Secondary structures are stabilized by favourable hydrogen bonding between residues that are closely situated in the primary structure, and have been brought into close juxtaposition by folding or coiling of the primary structure.

α-HELICES consist of right-handed helices containing 3.6 amino acid residues per complete turn (Fig. 2.20). This arrangement allows hydrogen bonds of optimum length and direction to be formed between the >N–H of the amino acid residue *i* and the carboxyl oxygen of the residue *i*–4. This means the hydrogen bonds connect a stretch of 13 atoms together, hence the α-helix is described as a 3.6_{13} helix (Fig. 2.21). The side groups of residues project outwards from the helix, leaving it with a hollow core. The geometry and the lack of bonding involving the side groups mean that virtually any amino acid can participate in α-helix, except for proline where the 'side group' is a rigid ring. Nevertheless, some amino acids are more likely to occur in α-helices than others (see later).

If a helix were formed such that hydrogen bonding occurred between a residue (*i*) and one four residues before it is the primary structure (*i*–5) it would form a 4.4_{16} or π-helix, while 'tightening' up the α-helix by a single residue would give a 3_{10} helix (Fig. 2.21). π-Helices have not been observed in proteins: sometimes single turns of 3_{10} helices are found at the end of α-helices, while some forms of β-turns (see below) consist of four residues in a 3_{10} conformation.

α-Helices are found in both fibrous and globular proteins

β-PLEATED SHEETS consist of relatively extended lengths of polypeptide chains in which adjacent peptide groups are tilted in alternate directions: hence 'pleated'. The >N–H and >C=O groups point outwards at roughly right angles to the extended backbone and form effective hydrogen bonds between two chains, which run almost parallel to one another. Two polypeptide chains running in the *same* direction (i.e. both N → C) are described as parallel β-pleated sheets; if they run in *opposite* directions they are antiparallel β-pleated sheets (Fig. 2.22). Hydrogen bonding is more effective in antiparallel sheets and these are the more stable of the two. Like α-helices, β-sheets are found in both fibrous and globular proteins. Most sheets that have been observed are not exactly planar, but have a somewhat twisted conformation.

α-Helices and β-sheets are essentially linear structures. Globular proteins, as implied by their name, have molecules that are roughly spherical. Thus, changes in the direction of the polypeptide backbone or 'loop' structures are necessary to form compact globular shapes (Fig. 2.23).

☐ The structure of both the α-helix and β-pleated sheet were proposed by Pauling and Cory in 1951. In 1954 the many contributions of Pauling to chemistry and biochemistry were recognized with the award of the Nobel Prize in Chemistry. In 1962 the contribution of Pauling to the test ban treaty of 1962 was rewarded with the Nobel Peace Prize.

☐ The α-helix is the major secondary structural feature of α-keratins, the major proteins of skin, hair, horn and hoof, etc. β-Pleated sheets are found extensively in the silks, a group of fibrous proteins produced by many insects and spiders (see Chapter 3).

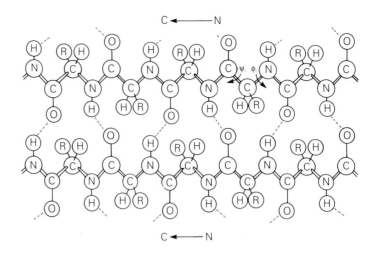

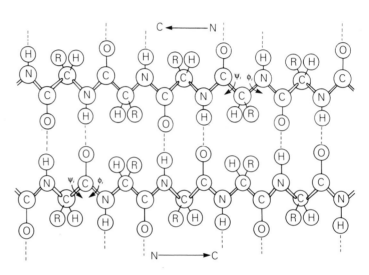

(a)

Fig. 2.23 Secondary structure of the enzyme dihydrofolate reductase. Note the loops (red), which connect the extended lengths of α-helices (spirals) and β-sheets (arrows) together. Original computer-generated representation courtesy of Dr J.M. Burridge, IBM, UK.

(b)

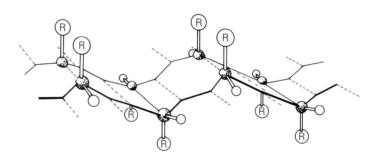

Fig. 2.22 (a) Parallel, and (b) two views of antiparallel, β-sheets.

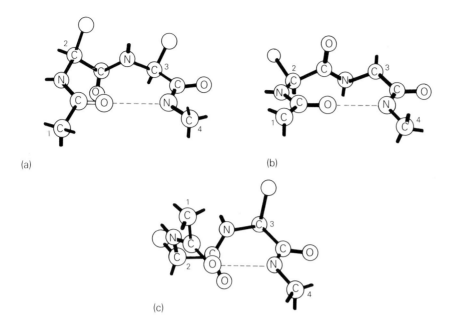

(a)

(b)

(c)

Fig. 2.24 β-Turns: (a) type I, (b) type II (R₃ is a glycine residue), (c) type III (four residues in a 3₁₀ helical conformation).

□ Glycine and proline are prominent amino acid residues in β-turns. Their shape and size makes them ideally suited to forming tight turns. The ring structure of proline imposes a distinctive twist to the polypeptide backbone, while the absence of a side group in glycine means it can occupy the restricted space available.

β-TURNS are the commonest type of loop structure found in globular proteins. They are also called **reverse turns,** β-**bends** and **hairpin bends/loops.** Three general types of β-turns are known (Fig. 2.24). All three contain four amino acid residues, with a hydrogen bond occurring between the carboxyl oxygen ($>C=O$) of the first residue and the amide nitrogen (N–H) of the fourth. Type I can contain any residues, except that proline cannot occur at position 3; type II *requires* glycine at position 3 with proline occurring almost exclusively at position 2. Type III is a portion of 3₁₀ helix; any amino acids are permissible. Other variants on these turns have been described, as have a variety of other types of loops.

Fig. 2.25 Secondary structure of the protein myoglobin. The associated haem group is coloured, the open circle represents Fe^{2+}. Redrawn from Robson, B. and Garnier, J. (1986). *Introduction to Proteins and Protein Engineering*. Elsevier, Amsterdam.

Proportions of secondary structures

The proportions of different types of secondary structures vary considerably between different globular proteins. For example, myoglobin and haemoglobin have approximately 75–80% α-helical structure (Fig. 2.25), while others, such as concanavalin A, rubredoxin and immunoglobulins, lack α-helices but contain a considerable quantity of β-sheets (Fig. 2.26). In general, helical and sheet regions tend to constitute something like 60–70% of the polypeptide, with about 30% of residues occurring in so-called 'random' coil regions (see Fig. 2.23). Typically the interior of globular protein molecules is a bundle of β-sheets with the surface covered with α-helices (Fig. 2.27). Coil regions also usually occur on the surface of the molecule and often constitute binding or antigenic recognition sites.

Usually, secondary structures consist on average of about 10 amino acid residues, although they can be longer. Thus, a polypeptide containing 150 residues would typically have a total of 15 sections of α-helix, relatively extended lengths and coil regions.

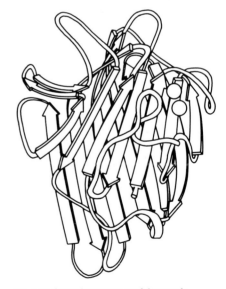

Fig. 2.26 Secondary structure of the protein concanavalin A. Redrawn from Richardson, J.S. (1981) *Advances in Protein Chemistry*, **34**, 167–339.

Reference Milner-White, E.J. and Poet, R. (1987) Loops, bulges, turns and hairpins in proteins. *Trends in Biochemical Sciences*, **12**, 189–92. Introduction to the structures that allow polypeptides to change direction.

Supersecondary structure and domains

Several consecutive sections of secondary structures often occur as a closely associated structure. These types of arrangements have been observed in a number of proteins, and are recognized as a further level of protein structure called **supersecondary** structure. Several types of supersecondary structure are common (Fig. 2.28). For example, $\beta\alpha\beta$ consists of two β-sheets connected by an intervening length of α-helix. β-**Meanders** consist of sequential lengths of antiparallel β-sheets connected by relatively tight β-turns. Lengths of β-sheets that are *not* sequential in the primary structure can associated to form so-called *Greek keys*.

DOMAINS are recognizable as roughly spherical or egg-shaped clusters of 50–150 residues formed by local compaction of the polypeptide chain (Fig. 2.29). A protein of more than 200 residues might consist of two or three domains.

It is difficult to say what the difference is between a supersecondary structure and a domain, and in many cases there is not a generally accepted answer. Indeed, a domain could be a single supersecondary structure, or a set of supersecondary structures combined together as a compact cluster of residues. Nevertheless, the idea of domains is useful since many *dis*similar proteins possess similar domains. For example, many proteins that bind NAD^+ and related molecules have a domain that binds NAD^+. This domain is called the **mononucleotide fold** (Fig. 2.30) and *may* be a protein structure that originated early in the evolution of organisms. Thus, domains appear to be fundamental structural units that are acted upon by natural selection during the evolution of proteins.

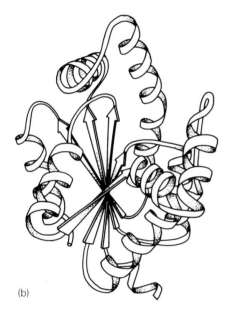

Fig. 2.27 Arrangements of secondary structures in the enzymes (a) triose-phosphate isomerase and (b) adenylate kinase. Redrawn from Richardson, J.S. (1981) *Advances in Protein Chemistry*, **34**, 167–339.

 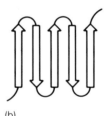

(a)　　　　　(b)

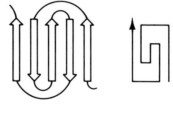 

(c)

Fig. 2.28 Examples of supersecondary structures: (a) $\beta\alpha\beta$ unit, (b) β-meander, (c) Greek key. Greek key pattern on a Greek amphora *c.* 460 BC, Manchester Museum specimen no. III i 30. Courtesy of Dr A.J.N.W. Prag.

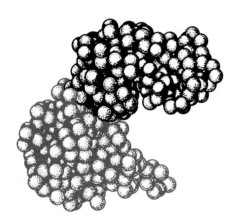

Fig. 2.29 The two well-separated domains of phosphoglycerate kinase. Redrawn from Richardson, J.S. (1981) *Advances in Protein Chemistry*, **34**, 167–339.

Greek key: *a supersecondary structure named after its resemblance to a pattern common on ancient Greek pottery.*

Tertiary structure

The **tertiary structure** of a protein is the overall three-dimensional shape of the folded polypeptide (Fig. 2.31). Each molecule of a particular protein has the *same* conformation and this differs from molecules of other proteins. Protein molecules fold into their specific shape with the result that hydrophobic side groups are buried in the interior of the molecule and not on the outside in contact with water.

The functioning of proteins depends upon their molecules folding into *specific* conformations, forming sites able to recognize the molecule with which the protein associates or reacts during metabolism. Thus, enzymes have sites, called *active sites*, which are *complementary* to their substrates and coenzymes (Figs 2.1 and 2.30). In fact, protein molecules are to some extent flexible and binding usually depends upon both protein and the small molecules altering their shapes dynamically to fit together. In other words, association is more like the fitting of a hand in a glove, both of which are flexible structures, than a key in a lock.

Quaternary structure

Some proteins are made up of subunits, each subunit consisting of a single polypeptide chain. The subunit structure is referred to as the **quaternary structure**. Proteins that lack quaternary structure, i.e. are composed of a single polypeptide, are sometimes described as **monomeric**. Proteins composed of two subunits are **dimeric**, three **trimeric**, four **tetrameric**, and so on. If all the subunits are identical, then **homomeric** proteins would be formed. Dissimilar subunits would form **heteromeric** proteins. Subunits are specified by giving them Greek letters, the number of each subunit being indicated by a subscript. For instance, adult haemoglobin has the quaternary structure $\alpha_2\beta_2$. Table 2.3 summarizes the quaternary structures of some proteins. The interactions that hold the subunits together are similar to those that govern the formation of tertiary structure, i.e. weak interactions, such as hydrophobic effects and van der Waals forces (see later).

Table 2.3 *Quaternary structures of selected proteins*

Name	Subunit composition
Lactose synthase	$\alpha\beta$
Haemoglobin	
Embryonic	$\xi_2\varepsilon_2$
Fetal	$\alpha_2\gamma_2$
Adult	$\alpha_2\beta_2$
cAMP-dependent protein kinase	$\alpha_2\beta_2$
Phosphorylase kinase	$\alpha_4\beta_4\gamma_4\delta_4$
Ribulose bisphosphate carboxylase	$\alpha_8\beta_8$
RNA nucleotidyltransferase	$\alpha_2\beta\beta'\sigma$

Exercise 3

From an examination of the structures of the side chains of amino acids (Table 2.1), which residues would you expect to find in the interior of a globular protein?

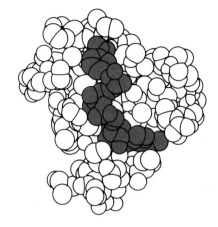

Fig. 2.30 The mononucleotide fold domain of the liver enzyme alcohol dehydrogenase with bound NAD$^+$ coenzyme (red). Redrawn from a photograph kindly supplied by Dr J.M. Burridge, IBM, UK.

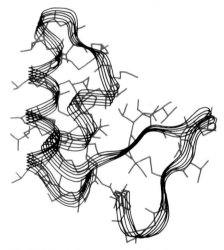

Fig. 2.31 The tertiary structure of the plant seed protein, crambin. The polypeptide backbone is represented as a ribbon from which the residue side chains project. Original computer-generated representation courtesy of Dr C. Freeman, Polygen, University of York.

Exercise 4

The protein yeast alcohol dehydrogenase has an M_r of 145 000, as determined by ultracentrifugation. However, under denaturing conditions a value of 36 000 was obtained. Give the most probable quaternary structure of this protein.

Reference Van Brunt, J. (1988) Beta barrels, helix bundles, hairpin turns and pleated sheets. *Bio/technology*, **6**, 655–61. A short, readable and nicely illustrated introduction to secondary and higher levels of protein structures.

Reference Richardson, J.S. (1981) The anatomy and taxonomy of protein structure. *Advances in Protein Chemistry*, **34**, 167–339. Splendid and beautifully illustrated account of the structure of proteins. Emphasis on supersecondary structure and domains.

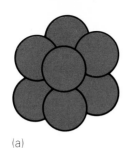

(a)

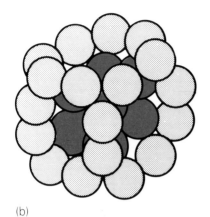

(b)

(c)

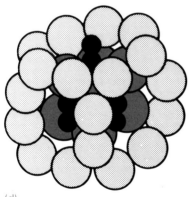

(d)

Fig. 2.32 The molecular anatomy of the pyruvate dehydrogenase complex (PDC) from *Escherichia coli*. (a) It consists of a cubic cluster of 24 transacetylase molecules (red) possibly arranged as eight trimers. (b) Twelve pyruvate dehydrogenase dimers (grey tone) arranged around the transacetylase core. (c) Six dimers (black) of dihydrolipoyl dehydrogenase shown on each face of the transacetylase core. (d) Complete pyruvate dehydrogenase aggregate. Redrawn from Bohinski, R.C. (1987) *Modern Concepts in Biochemistry*. Allyn & Bacon, Boston.

Multiprotein complexes and aggregates

Some proteins are involved in interactions with other molecules to produce extremely complex structures. Such interactions may involve fibrous proteins, such as the contractile machinery of skeletal muscle or the cytoskeleton. However, **multiprotein complexes** are also constructed from some globular proteins. For example, the production of the coats of the polio and Semliki Forest viruses involves the biosynthesis of a single large polypeptide. Subsequently, this is hydrolysed to produce individual identical globular proteins, which aggregate symmetrically to form the virus coat. Note that aggregation is a spontaneous process; the information directing the self-assembly of the coat is contained in the primary structure of the proteins.

SEVERAL ENZYMES OCCUR AS MULTIENZYME COMPLEXES. Two of the best studied are the **fatty acid synthetase complex** of yeast and the **pyruvate dehydrogenase complex** (PDC). PDC catalyses the conversion of pyruvate to acetyl CoA, with the loss of CO_2:

$$\text{pyruvate} + \text{CoASH} + \text{NAD}^+ \rightarrow \text{acetyl CoA} + CO_2 + \text{NADH}$$

The PDC contains three different types of enzymes, a **pyruvate dehydrogenase/decarboxylase**, a **transacetylase** and a **dihydrolipoyl dehydrogenase**, each of which may be separated and purified independently. Each enzyme catalyses a separate reaction and requires a different coenzyme for activity. The PDC of *Escherichia coli* consists of 12 dimeric molecules of pyruvate dehydrogenase, 24 molecules of transacetylase and six dimers of dihydrolipoyl dehydrogenase, giving 60 separate polypeptides. These are arranged in a *polyhedral* structure of 30 nm diameter with an M_r of about 4.6×10^6 (Fig. 2.32). The PDC of bovine heart is even larger, with an M_r of about 8.5×10^6.

Multiprotein complexes need not necessarily form symmetrical structures like the PDC. Ribosomes, the sites of protein synthesis, are asymmetric assemblies (Fig. 2.33). Presumably the functions of ribosomes in protein synthesis are so complex that a large array of different proteins is required: non-identical proteins can only coalesce in an asymmetric fashion.

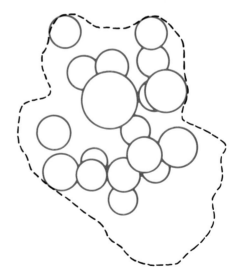

Fig. 2.33 Location of the 21 proteins (red) in the small subunit of a bacterial ribosome. The dotted line depicts the surface of the subunit. Redrawn from Zubay, G. (1988) *Biochemistry* (2nd edn), MacMillan Pub. Co., New York, p. 933.

Polyhedron: a regularly shaped solid bounded by plane faces. From the Greek polys, *many, and* hedra, *side.*

Reference Doolittle, R.F. (1985) Proteins. *Scientific American*, **253(4)**, 74–83. Excellent introduction to the structure of proteins. Particularly interesting account of how proteins evolve.

2.4 Forces that stabilize protein molecules

The forces that stabilize and direct a polypeptide to form a native conformation are mostly weak and non-covalent. The most important are hydrophobic interactions, hydrogen bonds, van der Waals and electrostatic forces.

HYDROPHOBIC INTERACTIONS arise from the hydrophobic nature of many amino acid side chains, such as those of valine and leucine. The enclosure of these groups into the interior of the molecule means they avoid thermodynamically unfavourable interactions with water molecules, since putting these hydrophobic portions into a body of water would involve re-arranging the structure of the water and this would require energy. Hydrophobic interactions are the most important contributors to the stability of the native protein. It is, however, important to appreciate that it is the removal of the non-polar side chains from water that gives the stability, *not* any interactions between the groups themselves.

HYDROGEN BONDS are stronger than hydrophobic interactions with **bond energies** of 10–20 kJ mol^{-1}. However, although proteins have many hydrogen bonds, it is unlikely they make any substantial contribution to the stability of the native conformation. If the polypeptide were unravelled, the amide and carboxyl groups would form effective hydrogen bonds with the surrounding water molecules. Thus, there is little energy difference between the folded and unfolded states as regards hydrogen bonding. Hydrogen bonds are, however, significant in ensuring the protein folds into a biologically appropriate conformation, since they are strongest when the atoms concerned are arranged *linearly*, e.g.

$$\overset{|}{\underset{|}{C}}=O^{\delta-} ----^{\delta'} H-\overset{|}{\underset{|}{N}}$$

Thus, the multitude of hydrogen bonds in a protein molecule impose geometrical constraints directing a polypeptide to fold appropriately.

VAN DER WAALS FORCES are weak attractions that occur between atoms and molecules when they are an optimum distance apart (said to be in van der Waals contact); this is about 0.4 nm for atoms in protein molecules. These interactions arise because the negatively charged electrons of an atom are attracted to the positively charged nuclei of other atoms, i.e. the atoms exist as dipoles. Although each van der Waals interaction contributes less than 10 kJ mol^{-1} of stabilizing energy, the extremely large number of interactions means van der Waals forces contribute significantly to the stability of protein molecules.

ELECTROSTATIC FORCES or **salt bridges** occur between ions. For example, attractive forces exist between the ε-NH$_3^+$ of lysine residues and the γ-COO$^-$ of glutamate (Fig. 2.34) if the groups are close together. The force between the two charges is inversely proportional to the **dielectric constant** of the medium that separates them. Detailed knowledge of the dielectric constant in the interior of the proteins is lacking. It is usually considered to be similar to that of a typical hydrocarbon with magnitude 2–4 compared with a value of about 80 for water. Although salt bridges are known in a number of proteins, e.g. chymotrypsinogen and haemoglobin (see later) in general, proteins have a limited number of salt bridges and therefore their overall contribution to stability is probably small.

☐ Membrane proteins are often exceptions to the usual arrangement of hydrophobic residues, being restricted to the interior of the molecule. Membranes have hydrophobic interiors composed of extended hydrocarbon chains of lipid molecules (see also *Cell Biology*, Chapter 4). Integral membrane proteins are characterized by continuous regions of hydrophobic amino acid residues in the primary structure. These regions typically form α-helices, which anchor the protein in the hydrophobic core of the membrane. For example, the membrane protein rhodopsin has seven α-helices, which criss-cross back and forth across the membrane discs of rod cells.

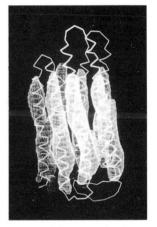

Computer-generated picture showing the membrane–spanning α-helices of the protein rhodopsin. Courtesy of Professor J.B.C. Findlay, Department of Biochemistry, University of Leeds.

Exercise 5

The globular proteins human carbonic anhydrase and equine cytochrome *c* have M_r of 27 000 and 12 000 respectively. Which protein has the higher proportion of polar (hydrophilic) amino acid residues?

Fig. 2.34 Salt bridge (ionic bond) between lysine and glutamate residues.

Bond energy: *the energy required to break a chemical bond. Hydrophobic interactions and hydrogen bonds are weak, with bond energies of about 4–8 and less than 20 kJ mol^{-1} respectively. Covalent bonds are much stronger. Disulphide bonds have energies of about 200 kJ mol^{-1}.*

Dielectric constant: *a measure of the relative effect a substance has on the force with which two oppositely charged groups attract one another. The larger the dielectric constant, the weaker the force.*

COVALENT BONDS, such as the disulphide bonds that occur between two cysteine residues, Fig. 2.9a, b, do exist in proteins. However, these bonds only form after the protein has assumed its three-dimensional shape. Thus, disulphide bonds appear to stabilize structures that have already formed, rather than directing folding. In fact, disulphide bridges are comparatively rare and seem to be generally restricted to extracellular and outer membrane proteins.

2.5 Protein folding

What determines how a polypeptide assumes a single native (functional) conformation? This shape is specified by the primary structure. Proteins with different primary structures fold to give differently shaped molecules. This shape is, of course, only produced under defined conditions, e.g. pH, ionic strength, temperature. That is, the primary structure only specifies a particular higher-ordered structure, with the required biological properties, if the molecule exists in appropriate physiological conditions. The 'correct' folded state is formed because it has a lower free energy than alternative states. However, the difference in energy is small.

In general, the compact, ordered globular structure of a protein is thermodynamically more stable than the open unfolded structure by only 15–60 kJ mol^{-1}. Several reasons contribute to this small value. It has already been mentioned that hydrogen bonds could occur between polar amino acid residues and water. Furthermore, secondary structures are ordered, a feature that increases their free energy. Artificial polypeptides containing random sequences of amino acids usually form loose, flexible chains, which do not exist in a defined conformation. It has therefore been suggested that the proteins of biological systems have been selected by evolutionary pressures because of their stability.

Box 2.4
Predicting higher ordered structure from the primary structure

Since the folding of a protein into its unique, native conformation is determined by its primary structure, in theory it should be possible to predict the secondary and tertiary structures of a protein molecule from its primary structure only.

Three main approaches have been used to predict secondary structures: (1) **probabilistic** or **statistical**, (2) **physicochemical** and (3) **statistical mechanical**. The simplest are the probabilistic methods. Observations of the relative frequencies of amino acid residues in known secondary structures (see table) are used as a basis for predicting unknown secondary structures. In the physicochemical methods, predictions are based on observations that non-polar residues within an α-helical region tend to form non-polar clusters or that the potential starting sites for polypeptide chain folding can be identified in a given sequence from the relative positions of particular types of amino acid residues. Statistical mechanics allows the calculations of observed properties of a system as an average or 'expected' property. Thus, the conformational properties of amino acids can be incorporated into the elements of a *relative statistical weight matrix*. Calculation of the functions of this matrix for a given sequence allows the conformational probability (i.e. secondary structure) to be predicted for that sequence.

The relative propensities for amino acids to be found in the indicated secondary structures. Values greater than unity are indicative of a greater than average frequency for a residue to be found in a particular secondary structure. Data based on 66 proteins.

Reference Karpus, M. and McCammon, J.A. (1986) The dynamics of proteins. *Scientific American*, **254(4)**, 30–9. A clear, informative account of the dynamics of protein molecules and how their motions can be explored by computer simulations. Illustrated by many beautiful pictures.

Secondary structure	Amino acid	α-helix	β-sheet	β-turn
Helical-preferring	Alanine	1.29	0.90	0.77
	Cystine	1.11	0.74	0.81
	Leucine	1.30	1.02	0.58
	Methionine	1.47	0.97	0.41
	Glutamate	1.44	0.75	0.99
	Glutamine	1.27	0.80	0.98
	Histidine	1.22	1.08	0.68
	Lysine	1.23	0.77	0.96
Sheet-preferring	Valine	0.91	1.49	0.47
	Isoleucine	0.97	1.45	0.51
	Phenylalanine	1.07	1.32	0.59
	Tyrosine	0.72	1.25	1.05
	Tryptophan	0.99	1.14	0.76
	Threonine	0.82	1.21	1.04
Reverse turn-preferring	Glycine	0.56	0.92	1.64
	Serine	0.82	0.95	1.32
	Aspartate	1.04	0.72	1.41
	Asparagine	0.90	0.76	1.28
	Proline	0.52	0.64	1.91
	Arginine	0.96	0.99	0.88

Adapted from Hider, R.C. and Hodges, S.J. (1984) *Biochem. Educ.*, **12**, 19–28.

Despite the large amount of effort and research into secondary structure prediction, all the methods currently in use are only about 60% accurate. Given that the chance of randomly assigning a residue to the correct structure is 25% (i.e. 1 in 4: coil, extended, α-helix and β-turn) there is clearly room for improvement.

Predictive methods for assigning tertiary from primary structures have also been devised. For example, the **thermodynamic** approaches attempt to calculate the absolute minimum free energy conformations arising from the interactions between all the atoms within the protein molecule as well as with the surrounding solvent molecules, counterions, etc. The complexity of these calculations is beyond the fastest, most-developed computing methods available today, and many simplifying assumptions have to be incorporated into the computer model. Thus, the technique has had limited success with molecules larger than small peptides. More successful has been the **building by homology** approach in which the structure of unknown polypeptide is determined by comparing it with an **homologous protein** (i.e. having sequences with many amino acid residues in common) of known tertiary structure. The most highly developed computer programs are able automatically to compensate for replacement, losses or insertions of residues between the two sequences during their separate evolutions. Clearly, however, the technique is only applicable if the three-dimensional structure of a suitable protein is known. Currently, the tertiary structure of only about 400 proteins is known, while a typical animal or plant cell probably contains about 30–50 000 different types of proteins!

How do proteins fold into their native conformation? If each rotatable bond in protein had only two possible conformations, then a typical protein with 500 rotatable bonds has 2^{500} possible conformations! This is a considerable underestimate, since there are more than two possible conformations per bond. In fact, protein molecules generally fold up into the 'right' conformation in about 0.1 to 10^3 seconds.

If the primary structure has all the information for protein structure, how is this rapid folding achieved? A random search for the most stable (lowest free energy), conformation would simply be too slow. Thus, the amino acid

☐ It has been estimated that a protein of 100 amino acid residues would take 10^{77} years to 'sample' all possible conformations. The age of the universe is estimated to be 'only' about 15×10^9 years.

Reference Goldenberg, D.P. and Creighton, T.E. (1985) Energetics of protein structure and folding. *Biopolymers*, **24**, 167–82. A short review describing, with the aid of a simple model system, how protein molecules fold into compact structures.

Exercise 7

What is the most likely secondary structure for the following section of a polypeptide?

... Ala–Val–Thr–Asp–Glu–Pro–Glu- Ile–Gly–Arg ...

---

sequence must contain *kinetic* information, which directs the folding process. The generally accepted idea is that folding occurs simultaneously in many places, giving stretches of secondary structure. Interactions then occur between these folded regions, forming supersecondary structures and domains. Finally, the domains associate to give the folded protein molecule.

2.6 Molecular activity of proteins: myoglobin and haemoglobin

It has already been noted that globular proteins are composed of somewhat flexible molecules, which function by forming reversible complexes with other molecules. The biological activity of proteins at the molecular and submolecular level is, however, understood for only a small number of proteins. The O_2-binding proteins, **myoglobin** and **haemoglobin** (Figs 2.25 and 2.35), are among the most extensively studied of proteins and, for this reason, their activities are among the best understood of all non-enzymic proteins.

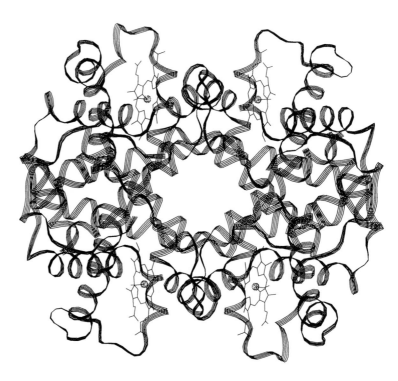

Fig. 2.35 Computer-drawn structure of haemoglobin. Courtesy of Professor A.C.T. North, Department of Biophysics, University of Leeds.

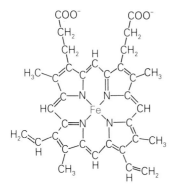

Fig. 2.36 Haem.

Oxygen transport

Organisms need O_2 to oxidize foods to supply energy. Oxygen must be picked up by carrier proteins at respiratory surfaces, such as the lung alveoli in mammals or the gill surfaces in fishes, and released to the active cells to support tissue respiration. In vertebrates, two well-characterized proteins are functional in the binding and release of O_2 to tissues: myoglobin and haemoglobin. Both proteins consist of a polypeptide portion and a non-protein iron(II) porphyrin ring system (Fig. 2.36). The polypeptide chains are wrapped around the haem and the O_2 is carried between the two (Fig. 2.37).

☐ The binding of haem to the polypeptide has a stabilizing effect. *Free* haem is unstable in contact with O_2 and the Fe(II)-porphyrin (haem) is oxidized to the Fe(III)-porphyrin (haemin).

Haemoglobin and myoglobin: *derived from the Greek words* haimo *and* myos *meaning blood and muscle respectively.*

Reference Kendrew, J.C. (1961) The three-dimensional structure of a protein molecule. *Scientific American,* **205(6),** 96–110. A lucid description of the tertiary structure of myoglobin: the first protein to be described at this level of structure.

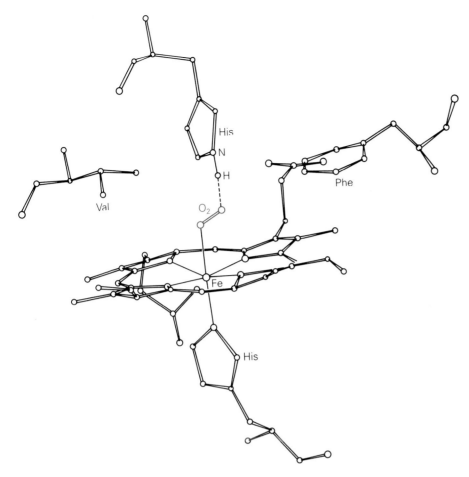

Fig. 2.37 Binding of an oxygen molecule in the crevice between the polypeptide chain and haem in myoglobin. Redrawn from Phillips, S.E.V. and Schoenborn, B.P. (1981) *Nature*, **292**, 81–2.

MYOGLOBIN consists of a single polypeptide chain and associated haem (Fig. 2.25). Adult human haemoglobin, however, consists of two identical α-subunits of 141 residues and two identical β-subunits of 146 residues (Fig. 2.35). Myoglobin is found intracellularly in muscles, while haemoglobin is found in erythrocytes. Both myoglobin (Mb) and haemoglobin (Hb) reversibly bind O_2.

$$Mb + O_2 \rightleftharpoons MbO_2$$

$$Hb + 4O_2 \rightleftharpoons Hb(O_2)_4$$

Myoglobin shows a relatively simple O_2-saturation curve (Fig. 2.38) picking up O_2 when the partial pressure (or concentration) of O_2 is high and releasing it as the partial pressure falls. However, the partial pressure must fall quite dramatically to release appreciable quantities of O_2 (Fig. 2.38). Indeed, myoglobin is still 50% saturated with O_2 when the partial pressure is as low as about 370 Pa (i.e. $P_{50} = 366$ Pa). Oxymyoglobin appears to function largely as a reservoir of O_2, releasing it only when extensive muscular activity depletes the cell of O_2.

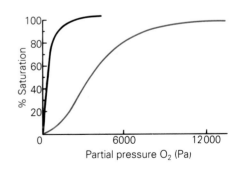

Fig. 2.38 Oxygen dissociation curves for myoglobin and haemoglobin (red).

HAEMOGLOBIN has a more complex O_2-dissociation curve than myoglobin in that it is **sigmoidal** (Fig. 2.38). The P_{50} for haemoglobin is about 3550 Pa, consequently oxyhaemoglobin can release oxygen to tissues, even when the concentration of O_2 is comparatively high. These advantageous features are a

---

P_{50}: *the partial pressure of oxygen at which the haemoglobin or myoglobin is 50% saturated with oxygen.*
Sigmoidal: *S-shaped.*

Box 2.5
Haemoglobinopathies

Survival of vertebrates is not possible without haemoglobin. However, many humans survive with partially defective haemoglobins. One such condition is sickle cell anaemia where, because of a mutation, an amino acid on the surface of the protein molecule is altered producing a haemoglobin that precipitates in the deoxy state and therefore does not transport O_2 effectively. This condition leads to deformation of the red cells ('sickling') which become trapped in the capillaries and haemolysis occurs, resulting in an anaemia.

Sickle cell disease is fairly common, especially amongst the North American black population, but it is rather unusual as a 'haemoglobinopathy' or haemoglobin disease. The amino acid change results in there being a 'sticky patch' on the β-polypeptide chain of deoxyhaemoglobin, leading to the aggregation and precipitation described above. The mutation arose by chance at some time in the past. Much more likely events to occur (and many hundreds of haemoglobin mutations are now known) are ones in which the haem pocket is modified so that haem does not bind or function properly, or ones in which the α or the β chains are not constructed properly.

As was mentioned in main text, the various parts of the haemoglobin, like all quaternary proteins, fit and stay together because they are complementary in shape, charge, hydrophobicity, etc. In particular, the Fe^{2+}-containing haem group is a highly hydrophobic molecule and requires to be placed in a hydrophobic pocket in the molecule, where it is held (Figs 2.25 and 2.35) and carries out its function of binding oxygen reversibly. Mutations that result in the amino acid residues lining the haem pocket being replaced by ones that are less hydrophobic or more bulky may result in a failure to bind haem or a failure to bind oxygen properly (i.e. not at all or irreversibly). Many such mutations are known and characterized. In the majority of cases only one type of subunit is affected. Thus, although the remaining unmutated subunits can potentially bind haem and oxygen normally, they often do not do so. Having only two oxygen-binding centres in the molecule, instead of four, does not allow for the usual subunit interactions which influence the binding and release of O_2. Instead of behaving in the required way generating a sigmoidal binding curve, the oxygen-binding curve may be much more like that of myoglobin (Fig. 2.38). Consequently, oxygen is not transported successfully.

Many patients with haemoglobinopathies are heterozygotic for the haemo-globinopathy in question; they have both a defective and a normal gene, so that effectively 50% of the haemoglobin they synthesize is normal. There may be a high rate of destruction of the abnormal haemoglobin which further lessens the problem. Also, several of the genes for the polypeptide chains of haemoglobin are present in multiple copies. Consequently, only one of the genes may have mutated, while the others, even in homozygotes, still produce normal polypeptides.

In some individuals the results of the mutation may be slight and not noticed until sensitive blood screening is carried out. In others, it may be sufficiently severe as to cause debilitating anaemia and other conditions. Many individuals probably do not survive because they are homozygous for the condition. However, this depends partly on how severe the defect is.

Haemoglobin variants are commonly detected by electrophoresis of a solution of the protein. When amino acids are changed as a result of a mutation, there may be a modification of the charge on the molecule, which may then display a higher or lower mobility than that of normal haemoglobin. Such screening may be done cheaply. Obviously, to determine *which* amino acid is altered requires a more extensive study, including peptide mapping and partial sequencing.

Haemoglobin variants are usually named from the town/hospital where the case was first detected (e.g. Hb 'Memphis'), although this gives the uninitiated little useful information. Hb Memphis is actually a rather unusual variant in which there are mutations in both the α and the β chains. It might more helpfully be described as:

$$\alpha_2^{\,23\text{Glu}\rightarrow\text{Gln}} \quad \beta_2^{\,6\text{Glu}\rightarrow\text{Val}}$$

Vella, F. (1980) Human haemoglobin and molecular disease. *Biochem. Educ.*, **8**, 41–53. An older review, but a very cursory read would give some impression of the range of different haemoglobinopathies known and the immense amount of research that has been done.

Stryer L. (1989) *Biochemistry*, W.H. Freeman and Company, New York, pp. 170–3. An easily accessible source of the sickle cell story and its biochemical details.

Weatherall, D.J. (1985) *The New Genetics and Clinical Practice*, Oxford University Press, Oxford, UK. More for the clinician, but gives interesting insights to some of the ethical issues.

Davies, K.E. (ed.) (1986) *Human Genetic Disease – A Practical Approach*, IRL Press, Oxford. A guide to some of the methods that are used to detect genetic diseases, including haemoglobinopathies.

Obviously, mutations do not necessarily have to be single amino acid substitutions, and do not have only to affect the haem pocket. Many mutations on the surface of the molecule are known, which have almost no effect on the properties of the molecule (sickle cell haemoglobin is the exception to this rule).

As well as single amino acid changes, there may be double changes, changes in both α and β chains, deletions resulting in a failure to make chains, mutations that change a stop codon so that a much larger than normal polypeptide is produced, and so on. It is probably true to say that almost all variations have been encountered. The present-day distribution of defective haemoglobins has arisen from the accumulation of harmless mutations, early death of individuals with harmful mutations, and survival of some individuals because although they have a harmful mutation, this confers a selective survival advantage such as increased resistance to malaria, as is the case with sickle cell disease.

As a result of a great deal of experimental work (protein sequencing and, later, DNA sequencing), an enormous amount is known about the haemoglobinopathies called **thalassaemias**. Almost all the possibilities that potentially could occur, do so. These include: deletion of one or more α chain genes per haploid genome; deletion of the β chain genes (unbalanced synthesis of chains may result in the production of homotetrameric molecules such as Hbα_4 in β-thalassaemia, which are unstable and precipitate or oxidize very rapidly); chain-termination mutation (e.g. Hb 'Seal Rock'); absent, reduced or inactive mRNA; gene fusion; and increased globin chain degradation.

Haemoglobin variants may now be detected in the fetus by molecular biology techniques and parents may be counselled about abortion. Although sickle cell disease may confer resistance to malaria, and consequently a selective survival advantage, there is usually little that can be done in any of the haemoglobinopathies in terms of medical treatment, other than to cope with crises and pain. Because there is an anaemia, blood transfusions may be used, but in the longer term, repeated blood transfusion is not helpful.

result of **co-operative** binding of O_2 and *allosteric* regulation of the binding.

If the haemoglobin molecule is dissociated into its subunits, each subunit binds O_2 rather like myoglobin does. However, in the intact tetramer the binding of an O_2 molecule to one subunit *enhances* the binding of O_2 molecules to the other subunits. Conversely, the loss of one molecule of O_2 from the O_2-saturated molecule ($Hb(O_2)_4$) makes the loss of additional O_2 molecules easier, i.e. the binding of other O_2 molecules is less strong. This **positive** co-operativity gives rise to the sigmoidal curve in Fig. 2.38.

Co-operativitiy offers several advantages. In tissues with a moderate O_2 demand, such as skin, the O_2 tension will be comparatively high and oxyhaemoglobin will release comparatively little O_2. In active tissues, such as contracting skeletal muscle, the partial pressure of O_2 will fall, allowing the oxygenated haemoglobin to release large quantitites (80% or more) of bound O_2. Thus, supply and demand are evenly matched.

BIOCHEMICAL BASIS OF CO-OPERATIVE BINDING lies in the ability of the haemoglobin molecule to recognize the binding of an O_2 molecule (and, indeed, other molecules) at one site, and communicate this to distal parts of the haemoglobin molecule.

In deoxyhaemoglobin, the Fe^{2+} is about 0.04 nm out of the plane of the porphyrin ring. The binding of an O_2 molecule causes the Fe^{2+} to move into the plane of the ring, altering the shape of the ring and pulling a bound histidine residue towards the porphyrin ring (Fig. 2.39a). The movement of the histidine residue is transmitted to the other subunits, which rotate relative to one another (Fig. 2.39b). This leads to conformation changes that enhance

Allosteric: *from the Greek* allos, other, *i.e. the binding of a ligand to a site on a protein molecule not associated with its normal biological activity changes the shape and therefore activity of the protein.*

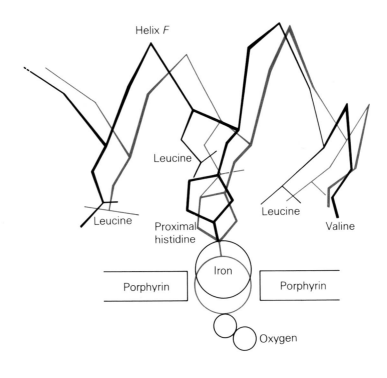

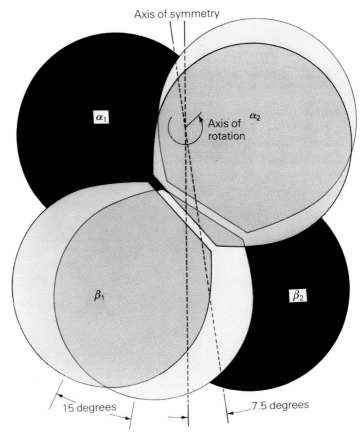

Fig. 2.39 Relative movements within the haemoglobin molecule on binding of an oxygen molecule. (a) Fe^{2+} and adjacent histidine residue, (b) the subunits. Redrawn from Perutz, M.F. (1978) *Scientific American*, **239(6)**, 68–86.

the binding of subsequent O_2 molecules. In essence, the subunits of deoxyhaemoglobin are held together by salt linkages giving a taut molecule (usually described as the **tense** (T) form. In contrast, in oxyhaemoglobin, the conformational changes described above, largely disrupt the salt links giving a more flexible (**relaxed**, R) molecule with a greater affinity for O_2.

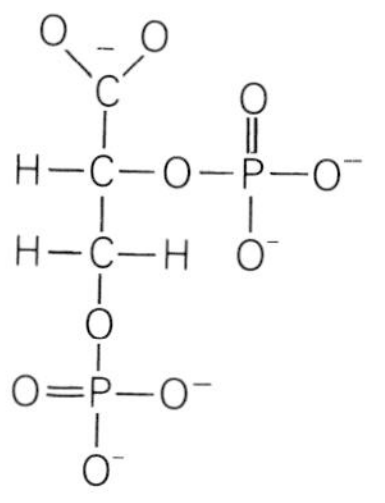

ALLOSTERIC REGULATION of haemoglobin is the result of the binding of one of several small molecules or ions to the haemoglobin molecule. This affects the affinity of haemoglobin for O_2. These allosteric effectors are **H^+**, **CO_2** and **2,3-bisphosphoglycerate** (BPG) (Fig. 2.40). The presence of any of these three components diminishes the affinity of haemoglobin for O_2, leading to its release.

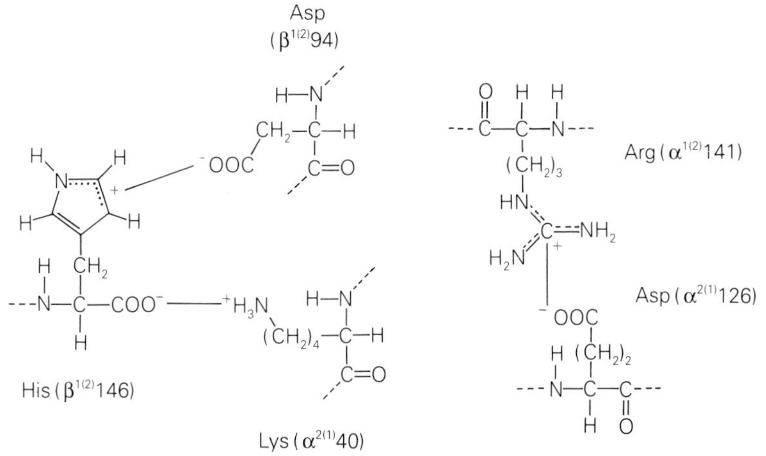

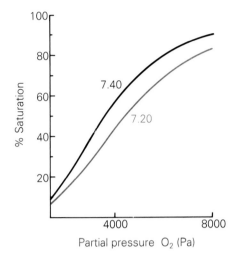

Fig. 2.40 2,3-Bisphosphoglycerate.

Asp
$(\beta^{1(2)}94)$

His $(\beta^{1(2)}146)$

Lys $(\alpha^{2(1)}40)$

Arg $(\alpha^{1(2)}141)$

Asp $(\alpha^{2(1)}126)$

Fig. 2.42 Salt links in haemoglobin favoured in the deoxygenated form. Residue numbers in the subunits are indicated. Redrawn from McGilvery, R.W. (1979) *Biochemistry: A Functional Approach*, W.B. Holt-Saunders, Philadelphia, p. 240.

Fig. 2.41 The decrease in oxygen-binding affinity of haemoglobin on acidification (pH 7.40 to 7.20, the Bohr effect). Redrawn from McGilvery, R.W. (1979) *Biochemistry: A Functional Approach*, W.B. Holt-Saunders, Philadelphia, p. 237.

The presence of H^+ moves the O_2-saturation curve of haemoglobin to the right (Fig. 2.41). This feature is called the **Bohr effect**. CO_2 has a similar effect. During rapid glycolysis and oxidation of fuel molecules, appreciable quantities of lactate and CO_2 are generated. Lactate production is accompanied by acidification of the serum. Thus, under conditions of O_2 stress, H^+ and CO_2 are produced and both promote the release of O_2 from oxyhaemoglobin. Release of O_2 occurs because protonation of specific amino acid residues enhances the formation of salt bridges (Fig. 2.42), shifting the R to the T conformation with the consequent release of O_2. CO_2 combines with terminal amino acid groups of the polypeptide chains to form **carbamates**:

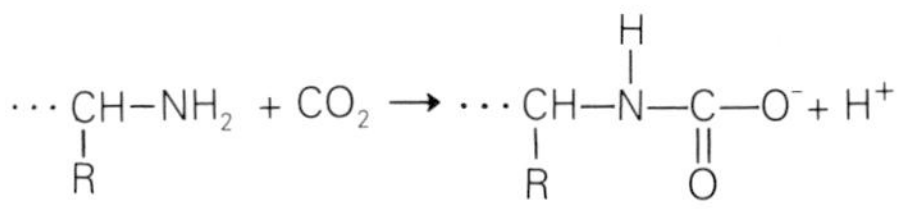

$$\cdots CH-NH_2 + CO_2 \rightarrow \cdots CH-N-C-O^- + H^+$$

The negatively charged carbamates form salt links, which stabilize the T conformation, again promoting the release of O_2.

BPG also modifies the O_2-binding by stabilizing the tense form of the haemoglobin molecule. The cleft between the β-subunits is lined with eight positive charges, arising from basic side groups or α-amino termini. BPG has five negative charges and its shape is complementary to that of the cleft. Thus, BPG fits into the cleft and forms stable electrostatic interactions linking the subunits (Fig. 2.43) and tending to fix the haemoglobin molecule in the T conformation with a corresponding diminished affinity for O_2.

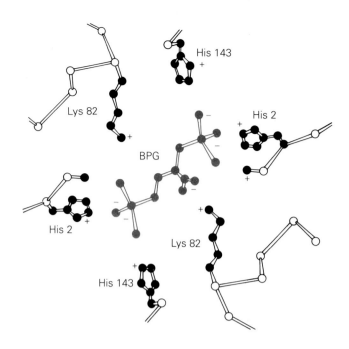

Fig. 2.43 Binding of 2,3-bisphosphoglycerate (BPG) in the cleft between the β-subunits. Redrawn from Arnone, A. (1972) *Nature*, **237**, 148.

BPG is retained in erythrocytes and significant changes in its concentration require several hours to be effected. Thus, unlike H^+ and CO_2, the concentration of BPG is used in longer term regulation of O_2-transport, for example, in response to environmental changes.

2.7 Protein denaturation

Given that the compact three-dimensional structure of a protein molecule is stabilized by weak interactions, and is only marginally more stable than an unfolded structure (see earlier) it is, perhaps, not surprising that many proteins are relatively easy to unfold, or, in the case of quaternary structures, to dissociate. This loss of higher-ordered structure is called **denaturation**. Loss of structure is accompanied by loss of biological activity. The effects of heavy-metal ions (Pb^{2+}, Hg^{2+}, etc.) as **protein denaturing agents** (although protein precipitants may be a more accurate description) has already been described.

Box 2.6
2,3-Bisphosphoglycerate (BPG) and O_2 supply

The regulation of O_2-release from haemoglobin by BPG has several interesting aspects. An impaired supply of O_2 to tissues leads to an elevated concentration of BPG to maximize the supply of O_2. Several situations contribute to such a condition. Anaemia (loss of haemoglobin), binding of CO to haemoglobin in cigarette smokers rendering it non-functional, or exposure to elevated altitudes, all reduce the O_2-carrying capacity of haemoglobin with a corresponding increase in the concentration of BPG.

Fetal haemoglobin (HbF) (Table 2.3) has a higher affinity for O_2 than adult haemoglobin, allowing the transfer of O_2 from the maternal to the fetal circulation across the placenta. In fact, the higher affinity is a consequence of BPG-binding less strongly to HbF, leading to a more relaxed molecule with an elevated affinity for O_2.

Reference Perutz, M.F. (1978) Haemoglobin structure and respiratory transport. *Scientific American*, **239(6),** 68–87. Clear, concise account of the structure of haemoglobin and the binding and transport of oxygen.

A variety of protein-denaturing procedures have found clinical applications. For example, surgical instruments are sterilized by autoclaving, that is heating the instruments in steam under pressure. This process kills microorganisms by denaturing their proteins. Alcohol (70%) is often used as a 'swab' to sterilize the skin before giving an injection or taking a blood sample. Again, alcohol is an effective protein denaturant.

Heavy-metal ions, e.g. Hg^{2+}, Pb^{2+}, Cu^{2+}, are poisonous because they can react with sulphydryl groups of proteins, cross-linking two groups and denaturing the protein (see Box 2.2). This feature is used to advantage when treating patients suffering from heavy-metal poisoning. Raw egg white and milk are administered. The proteins of the egg white and milk are denatured by the metal ions, forming an insoluble complex. The complex is then removed by pumping out the stomach contents or by inducing vomiting. This must be performed before digestive enzymes hydrolyse the proteins and release the heavy metals.

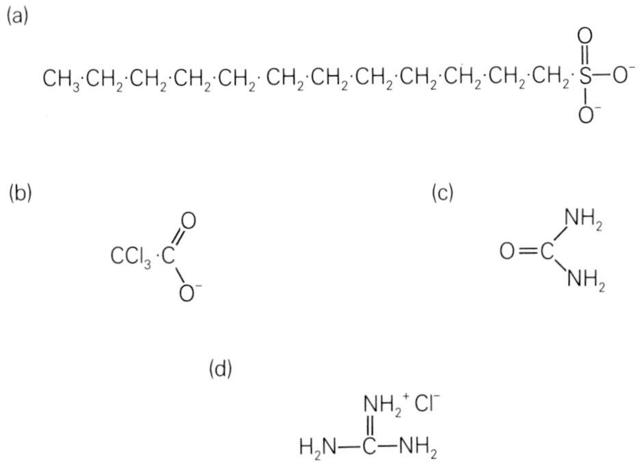

(a)

$CH_3 \cdot CH_2 \cdot CH_2 \cdot CH_2 \cdot CH_2 \cdot CH_2 \cdot CH_2 \cdot CH_2 \cdot CH_2 \cdot CH_2 \cdot CH_2 \cdot CH_2 \cdot \overset{\displaystyle O}{\underset{\displaystyle O^-}{\overset{\|}{\underset{\|}{S}}}}{-}O^-$

(b)

$CCl_3 \cdot C \overset{O}{\underset{O^-}{<}}$

(c)

$O{=}C \overset{NH_2}{\underset{NH_2}{<}}$

(d)

$H_2N{-}\overset{NH_2^+ \, Cl^-}{\overset{\|}{C}}{-}NH_2$

Fig. 2.44 Protein denaturing agents: (a) sodium dodecyl sulphate (SDS), (b) trichloracetate (TCA), (c) urea, and (d) guanidine hydrochloride (Gn-HCl).

Denaturation can be caused by a variety of other agents. Extremes of pH or temperature disrupt hydrogen bonding, leading to loss of secondary structure. The addition of organic solvents, such as ethanol or acetone, removes the sheath of water molecules, which surrounds the protein molecule, and also interferes with hydrogen bonding. Both these effects contribute to denaturation. Detergents, particularly ionic detergents (such as sodium dodecyl sulphate (SDS), *chaotropic agents* (for example, trichloracetate and $8 \, mol \, dm^{-3}$ urea or $6 \, mol \, dm^{-3}$ guanidine hydrochloride) (Fig. 2.44) all disrupt the hydrophobic effects largely responsible for stabilizing tertiary structure. All are therefore effective denaturing agents.

Detergents bind to and mask hydrophobic sites on proteins. In the case of SDS, the negative charges of the ionized sulphate groups mean that the protein forms a rigid rod structure with consequent loss of biological functions. This effect is often made use of in separating proteins by electrophoresis. Chaotropic agents are relatively large ions and are therefore normally described as having a low charge density. Such agents disrupt the normal highly ordered state of water, therefore facilitating the shift in position of hydrophobic amino acid residues from the interior of the protein molecule

---

Exercise 8

A protein consisting of 150 amino acid residues and containing five disulphide bonds was denatured in $8 \, mol \, dm^{-3}$ urea in the presence of excess 2-mercaptoethanol. If, following renaturation, the disulphide bonds reformed on a random basis what proportion of the original activity would be expected?

Chaotropic agents: those that increase disorder (chaos), in this case by destroying the highly ordered state that water molecules normally adopt.

□ Alfinson and co-workers used $8 \, \text{mol} \, \text{dm}^{-3}$ urea and β-mercaptoethanol to denature the enzyme **pancreatic ribonuclease**. β-Mercaptoethanol reduced the four disulphide bonds to sulphydryl groups allowing the molecule to unfold in the presence of urea. When the urea and β-mercaptoethanol were removed by dialysis, nearly 100% of the original enzyme activity was recovered. This was an important observation, since it confirmed that the primary structure contained the information necessary to specify the tertiary structure.

The renatured protein had folded to allow the correct four disulphide bonds to form. Since eight cysteines could form 105 different possible disulphide bonds ($8!/2^4 \times 4!$) if formation were random, recovery of less than 1% of original activity would be expected on this basis ($1 \times 100/105\%$). In fact, renaturation of the reduced, unfolded ribonuclease is a complex process. Initially, intermediates with different numbers of incorrect disulphide bonds are formed. These slowly rearrange to give the final, native molecule. *In vivo*, the enzyme **protein disulphide-isomerase** catalyses the conversion of protein molecules with inappropriate disulphide bonds to biologically functional types.

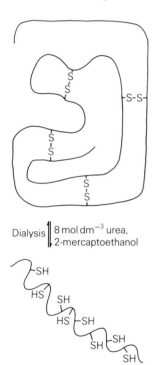

Dialysis $8 \, \text{mol} \, \text{dm}^{-3}$ urea, 2-mercaptoethanol

The denaturation and renaturation of pancreatic ribonuclease.

to the solvent. Urea and guanidine hydrochloride also appear to act by weakening intramolecular hydrophobic interactions, i.e. the non-polar side groups of buried residues 'dissolve' in concentrated solutions of these denaturants.

2.8 Overview

Proteins are formed from the condensation of large numbers of amino acids. Twenty standard α-amino acids are found in proteins, each of which differ only in their side groups. Thus, a variety of structural and chemical properties are contributed to the protein molecule. Further variety is provided by modification of the side groups and by the addition of organic non-protein groups in some cases.

Protein molecules show a hierarchy of structural levels, although all higher ordered levels of structure are determined by the primary structure, i.e. the sequence of amino acid residues in the polypeptide chain. The shape adopted by molecules of a particular protein has a lower free energy than alternative conformations, although the energy difference between these structures is small. The conformation adopted is stabilized by a variety of interactions, of which hydrophobic effects are the most important. The conformation of a protein underpins its biological activities. Since this particular structure is a co-operative structure determined by many individually weak interactions it is easily disrupted by a variety of denaturing agents, with a resulting loss of biological function.

The complete three-dimensional shape is known for only a few hundred proteins. Thus, the biological activities of only a handful of proteins is understood in terms of their molecular architecture. Among the best understood proteins at the submolecular level are the O_2-binding proteins, myoglobin and haemoglobin.

Reference Bradshaw, R.A. and Purton, M. (eds) (1990) *Proteins: Form and Function*. Elsevier Trends Journals, Cambridge, UK. A useful collection of 29 articles which have recently appeared in *Trends in Biochemical Sciences*. The book is divided into sections on 'Protein Primary Structure', 'Protein Conformation', 'Co- and Post-translational Modifications' and 'Molecular Recognition'.

1. Mean residues M_r (118.75) $\times$ 300 = 35–36 000.
2. Immediate environment surrounding that serine residue ensures it is particularly reactive. Therefore likely to be part of the active site.
3. Pro, Val, Ile, Leu, Met, Phe.

4. Four subunits possibly identical since all of same M_r.
5. Cytochrome c; smaller, therefore higher surface area to volume ratio.
6. (a) 1.803×10^{-11} N
   (b) 4.807×10^{-10} N

7. Coil. No continuous stretch of residues with 'preference' for any one type of secondary structure, therefore coil by default.
8. 0.11%.

QUESTIONS

FILL IN THE BLANKS

1. The forces that maintain the three-dimensional structure of a protein are mainly non-_____ . They include _____ interactions, _____ _____ _____ forces, _____ bonding and _____ _____ .

The structure of a protein molecule comprises a _____ of structural _____ . Secondary structures are _____ shaped conformations stabilized by _____ _____ . Folding of secondary structures generates _____ structures, which associate to form _____ . It is often difficult to distinguish between these last two levels of structure. The _____ structure of a _____ protein is the completely folded polypeptide chain and includes one or more _____ . Protein molecules are often composed of several polypeptides, each referred to as a _____ ; the overall assembly is called the _____ structure of the protein.

Choose from: covalent, domains (2 occurrences), globular, hierarchy, hydrogen, hydrogen bonds, hydrophobic, levels, quaternary, regularly, salt bridges, subunit, supersecondary, tertiary, van der Waals.

MULTIPLE-CHOICE QUESTION

2. Which of the following combinations of amino acid residues would be expected to be found on (i) the surface, (ii) the interior, or (iii) an indeterminate position of a protein molecule?
A. Lys, Arg, Asp, Asn
B. Ile, Val, Pro, Phe
C. Asp, Glu, His, Ser
D. Thr, Tyr, Gln, Cys

SHORT-ANSWER QUESTIONS

3. High concentrations of dithiothreitol cleave disulphide bonds. How many individual peptides would result from the treatment of the following peptide with dithiothreitol?

```
His—Cys—Ala—Gly—Phe
     |
     S
     |
     S
     |
    Cys—Tyr—Met—Cys—Glu—Gly—Cys
         |               |
         S———————————————S
```

4. How is it possible to identify membrane-spanning sites from an examination of the primary structure of a membrane protein?

5. Which of the following multimeric proteins would you expect to form symmetrical quaternary structures?
A. $\alpha_6\beta_6$
B. $\alpha_2\beta_2$
C. $\alpha_5\beta_3\gamma_2$
D. $\alpha_3\beta_3\gamma_5$

6. Which of the following statements are True.
A. The most stable conformation of a protein molecule is one with the lowest free energy.
B. Intramolecular hydrogen bonding occurs between hydrogen atoms on the surface of a protein molecule in solution.
C. The folding of a polypeptide chain is accompanied by a decrease in the entropy of the surroundings.
D. The formation of hydrogen bonds is the main type of interaction driving protein folding.
E. The minimum free energy conformation of a protein is sometimes reinforced by the formation of disulphide bonds.

7. Suggest a plausible supersecondary structure for the following portion of a polypeptide chain:

...Ala–Leu–Met–Glu–Glu–Lys–Leu–Ser–Gly–Ser–Asn–Val–Ile–Phe–Phe–Tyr–Val–Ile–Ile–Thr–Gly–Pro–Gly–Asp–Lys–His–Glu–Ala–Glu–Cys–Gln...

8. In paper electrophoresis at pH 8.6 normal human haemoglobin was run in comparison with four abnormal haemoglobins in which amino acid substitutions had occurred in the polypeptide chain of one type of subunit. The following result was obtained:

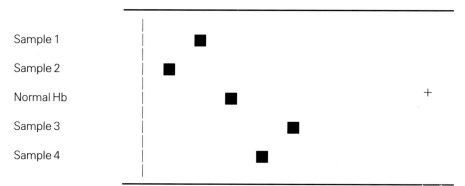

Amino acid analysis showed that in four haemoglobins the following amino acid substitutions had occurred:

A. Glutamate replaced lysine in the β-chains.
B. Lysine replaced glutamate in the β-chains.
C. Aspartate replaced glycine in the β-chains.
D. Lysine replaced asparagine in the β-chains.

(i) Say which of the haemoglobins a–d correspond with sample positions 1–4 on the electrophoresis strip.
(ii) Would it be possible to detect a haemoglobin variant in which a glycine in the β-chain was replaced by alanine?
(iii) Would you expect the charged residues referred to above to be found on the surface or in the interior of the protein molecule.

3

Fibrous proteins

Objectives

After reading this chapter you should be able to:

☐ describe the nature and functions of fibrous proteins;

☐ relate the structures of various fibrous proteins to their biological functions;

☐ explain the roles of the keratins and fibroins;

☐ relate some human diseases to a deficiency of fibrous proteins, particularly collagens.

3.1 Introduction

Proteins may be classified on the basis of their three-dimensional structure as either 'globular' or 'fibrous'. **Globular proteins** have attracted a lot of attention because most enzymes, many hormones and other proteins with known biological activity (such as antibodies) have this kind of structure. However, the **fibrous proteins** are equally important. Most have a structural or protective role and are key components in such biological structures as the extracellular matrix, cartilage, skin, tendons, hair, silk, insect exoskeletons and many more, and are also essential in forming the internal scaffold or cytoskeleton of cells. Some proteins, such as the muscle protein myosin, and certain membrane proteins, have both globular and fibrous domains. For many years it was believed that these structural proteins were inert, but from recent work it is apparent that they interact in many ways with each other, with cells and with their environment.

A characteristic feature of fibrous proteins is their *insolubility* in aqueous media, which makes it difficult to extract and study them because most of the available techniques were developed for studying globular proteins in solutions. Nevertheless, some of these problems are being overcome and details of their diverse structures and functions are now being elucidated.

The fibrous proteins show a wide range of structures. For example, some are rigid and rod-like, while others are flexible and sheet-like. Many are composed of small units that are extensively cross-linked to form fibrous structures. Some fibres are laid down in organic or inorganic matrices to provide structures that are both strong and resistant to the propagation of fractures. This structural diversity reflects the range of functions fibrous proteins perform. These functions range from the rigid tensile strength of insect exoskeletons (imparted by the protein **resilin**) to the elasticity of skin (imparted by **elastin**).

Reference Dickerson, R.E. and Geis, I. (1969) *The Structure and Action of Proteins*. Harper & Row, New York. Now somewhat dated, but still a good introduction to protein structure and function.

Many extracellular fibres are laid down as part of a matrix in which the combination of the interlinked fibres and their interaction with the other materials present form an extremely strong material, resistant to the propagation of fractures and able to bear a great deal of stress. This situation is not unique to biology. The same principle is used in many modern building materials, providing the great edifices prominent on so many modern skylines. Because their principal function is to provide support, the structures of fibrous proteins are much simpler than most globular proteins. They do not need to carry all the information of an enzyme which must interact with substrates, products and allosteric effectors as well as catalyse a chemical reaction.

United Nations Building, New York. Courtesy of S. Duckworth.

A list of some fibrous proteins and their functions is given in Table 3.1. It is not possible to describe all of these in detail and only a few representative examples are given.

Table 3.1 *Functions of some fibrous proteins*

Protein	Origin	Function
F-actin	Intracellular, all cells	Microfilament formation in cytoskeleton, contractile movement
Collagen	Extracellular matrix, bone, skin, blood vessels	Tensile strength
Desmin	Muscle cells	Scaffold structures inside cells
Elastin	Blood vessels, ligaments	Elasticity
Fibroin	Silk	Strength without stretch
Keratin	Skin, hair, etc., intra-cellular	Protective structures, tensile strength in epithelia
Lamin	Nuclear membrane	Membrane structures
Resilin	Insect wing hinges	Strength with flexibility
Sclerotin	Arthropod exoskeletons	Rigidity
Spectrin	Red blood cell membrane	Binds to F-actin allowing flexible membrane

3.2 Keratins and silks

Keratins and **silks** are found throughout the animal kingdom. Keratin's occur in tissues such as hair, horn, hoof and skin, while silks are produced by certain moths and spiders.

Keratins

The keratins are a group or family of related proteins characterized by possession of long sequences of α-helical secondary structure. An **α-helix** is a structure that is stabilized by hydrogen bonds parallel to the axis of the protein backbone. Each 'turn' of the helix contains 3.6 residues. In this form the helix is said to be **compact** or **relaxed**. It can be easily **extended** or **stretched** by applying tension to the ends. This breaks the hydrogen bonds and allows the backbone to be straightened. Under tension, the helix can reach twice its relaxed length. If the tension is increased beyond this point, the breaking strain of the helix is exceeded and the backbone snaps. If the tension is instead released, the helix relaxes and the hydrogen bonds re-form.

KERATINS OF EXTERNAL PROTECTION (skin, hair, nail, horn, feathers, etc.) are produced by epidermal cells. In these proteins, several α-helices are coiled together to form fibres. In hair, for example, supercoiled α-helices form a microfilament 'rope'. Eleven ropes join together to form a macrofilament, and a possible arrangement of macrofilaments in the cells of a single hair is shown in Fig. 3.1. As the cells develop, they become packed with keratin filaments and the cell membrane turns into a tough, cross-linked protein envelope. When the cells die, practically all of the organelles disappear and only the keratin remains (Fig. 3.2). Similarly, skin cells express a variety of keratins as they develop and progress from the basal layer of dividing cells of the epidermis to the squamous, dead, keratinized layers on the external surface (stratum corneum), in which the *squames* are simply toughened, flattened bags containing little except keratin.

The α-helices of hair keratins are relatively rich in the amino acid residue cysteine compared with many proteins. A typical globular protein has less

☐ Members of a family of proteins are closely related structurally. For example, they have similar amino acid sequences, similar three-dimensional structures, and often perform related functions. They may have developed by a series of gene duplication events.

☐ Human head hair grows about 15–20 cm per year. To achieve this, the cell must synthesize about 10^9 turns of keratin α-helix every second. The cells of the hair bulb are some of the most active in the body.

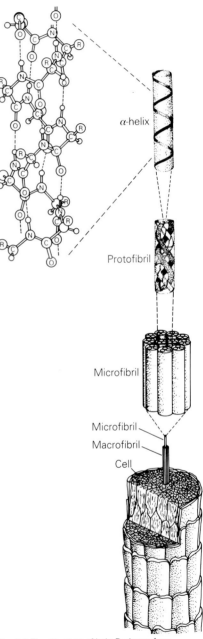

Fig. 3.1 The structure of hair. Redrawn from Bassindale, A.R. *et al.* (1977) *Systems Chemistry – Conformations of Proteins* (Unit 3.1). Open University Press, Milton Keynes, UK.

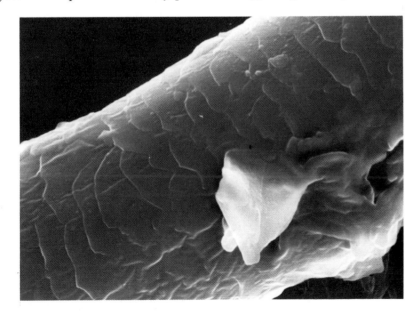

Fig. 3.2 Scanning electron micrograph of a human head hair. Magnification × 940. Courtesy of P.L. Carter, Department of Biological Sciences, Manchester Polytechnic, UK.

Keratins: *a family of proteins whose secondary structure is predominantly an α-helix, particularly found in external protections such as hair, and in epithelial cells.*
Squames: *the scale-like structures, actually dead epidermal cells, of which the stratum corneum of skin is composed.*

Reference Squire, J.M. and Vibert, P.J. (eds) (1987) *Fibrous Protein Structure.* Academic Press. Advanced up-to-date reviews of both specific proteins and techniques for studying them.

Exercise 1

Which amino acid residues are present in large amounts in nail and hair keratin?

than 1% of its residues as cysteine while, for example, wool keratin contains about 11% of its amino acid residues as cysteine (Table 3.2). These amino acid side-chains react to form disulphide bonds and so the individual α-helices can become cross-linked. Disulphide bonds are covalent and more difficult to break than hydrogen bonds. Hence cross-linking reduces the extensibility of the keratins. Disulphide bonds as well as hydrogen bonds are responsible for ensuring that the relaxed protein regains its original conformation. The more cross-linking, the greater the rigidity of the resulting structure. It is hard to believe that the principle component of human hair, tortoise shell, bird's feathers and fish scales is the same, but this is true. All are constructed of cell envelopes containing almost pure keratin, the difference being in the number of cysteine residues and the degree of cross-linking. In tortoiseshell, nearly one amino acid residue in every five is cysteine and most of these are involved in disulphide bond formation, producing a very hard and tough material. The amino acid sequence of all proteins is vital to their structure and function, and cysteine in keratins is just one example of this.

Another characteristic of keratins is their insolubility in water. This too can be explained at least partly in terms of the amino acid composition. In the

Table 3.2 *Amino acid composition of some fibrous and globular proteins. (Note that a few amino acid residues predominate in the fibrous proteins while this is less true of the globular proteins. Major amino acid residues in the three fibrous proteins are highlighted)*

Amino acids	Fibrous			Globular	
	Wool keratin from merino sheep	Silkworm fibroin	Collagen from rat tail tendon	Human haemoglobin	Bovine ribonuclease
Ala	5.0	29.4	10.7	12.5	9.7
Arg	7.2	0.5	5.0	2.1	3.2
Asx*	6.0	1.3	4.5	8.7	12.1
Cys	11.2	0	0	1.1	6.5
Glx*	12.1	1.0	7.1	5.6	9.7
Gly	8.1	44.6	33.0	7.0	2.4
His	0.7	0.2	0.4	6.6	3.2
Ile	2.8	0.7	0.9	0	2.4
Leu	6.9	0.5	2.4	12.5	1.6
Lys	2.3	0.3	3.4	7.7	8.1
Met	0.5	0	0.8	1.1	3.2
Phe	2.5	0.5	1.2	5.2	2.4
Pro	7.5	0.3	21.6	4.9	3.2
Ser	10.2	12.2	4.3	5.6	12.1
Thr	6.5	0.9	2.0	5.6	8.1
Trp	1.2	0.2	0	1.1	0
Tyr	4.2	5.2	0.4	2.1	4.8
Val	5.1	2.2	2.3	10.8	7.3

Amino acid contents are given as mol%.
* Asx is the sum of aspartate and asparagine contents, and Glx that of glutamate and glutamine contents.

Box 3.2
The 'permanent wave'

In many cultures it is popular for people to style their hair. Such is the perversity of human nature that it seems that almost everyone would like to have their hair with more or less curl! The nature of keratins allows some degree of compliance with such desires. Heat and chemical treatment of hair break the hydrogen bonds and reduce the disulphide bonds allowing the hair to be shaped. (The sulphurous reducing agents used to break the disulphide bonds are responsible for the obnoxious smell associated with perming hair!) Cross-linking by the formation of different patterns of disulphide bonds retains the new structure. This is the basis of the 'permanent wave'.

α-helix, the side-chains of the amino acid residues are exposed on the surface of the helix. In keratins, a high proportion of side-chains are hydrophobic groups, which do not readily interact with water. In contrast, in globular proteins, which tend to be water soluble, most hydrophobic groups occur hidden in the interior of the molecule. Denatured globular proteins are often insoluble because these hydrophobic groups become exposed to water in the process of denaturation.

However, the keratins also contain many polar amino acid residues which stabilize the α-helices. In the primary structure there is a repeat of seven residues, representing two turns of the helix, in which the first and fourth residues are hydrophobic and the fifth and seventh residues are polar or charged. The hydrophobic residues form a surface, which interacts with similar surfaces on other α-helices in the microfilament.

CYTOKERATINS have a major structural role within epithelial cells. The structures of these keratins are similar to those of the microfilaments of hair. Helical segments of the protein are separated by non-helical segments, allowing some flexibility and end-to-end aggregation of the molecules into keratin filaments, the major components of **tonofilaments**. These criss-cross the cytoplasm, anchored to structures in the cell membrane called **desmosomes**. Individual cells are linked via other filamentous proteins anchored on the external surfaces of desmosomes. This continuous structure of filaments helps provide the tensile strength of the epithelial sheet.

β-Sheet proteins or silks

Silks are β-sheet proteins or **fibroins**, which, in contrast to keratins, are extracellular. They are characterized by great strength and flexibility, but low extensibility, that is they cannot be stretched very much. Tension is resisted without the structure changing its conformation until breakage occurs, and the breaking strain can indeed be very high. Spider silk can withstand greater tension than many steels of equal thickness (Fig. 3.3).

These properties may be explained by the nature of the **primary and secondary structure** of the protein. In a typical silk fibroin, nearly half the amino acid residues are glycines, another third are alanines, and most of the rest are serines. Sulphur-containing amino acids are absent. Much of the primary structure consists of the repeated hexapeptide sequence –[Gly–Ala–Gly–Ala–Gly–Ser]–. In one sheet, polypeptide strands lie adjacent to each other and 0.47 nm apart so that alternate strands run from *N*-terminal to *C*-terminal and from *C*-terminal to *N*-terminal. The backbones are joined by hydrogen bonds between atoms of the peptide bonds. The spacing between residues and their sequence in the pleated sheet means that all the glycines are on one side and all the alanines and serine are on the other. The protein is built up of stacks of sheets, glycine to glycine and serine/alanine to serine/alanine residues 0.35 nm and 0.57 nm apart respectively (Fig. 3.4). The forces holding the sheets together are relatively weak van der Waals interactions, so that sheets can slide across each other, imparting flexibility and 'silkiness'. However, because the β-sheet is almost fully extended (a two-residue repeat of 0.70 nm), it cannot be stretched appreciably, and applied tension is bourne by the covalent backbone, making the fibres extremely strong. What little elasticity silk has is due to the less ordered sequences between pleated regions, where other bulkier amino acids residues are found. It is the relative proportions of sheet to non-sheet in the proteins that give different characteristics to different silks. Bulky amino acids vary from about 5 to 30% of the total in fibroins from different silkworms, and the resulting silks vary from almost no elasticity to an elasticity approaching that of wool.

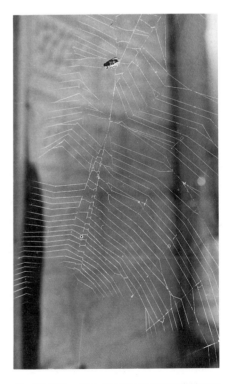

Fig. 3.3 Spiders Web. The effectiveness of this trap depends upon the great strength of the silk fibres. Courtesy of Winston J. Bailey.

Cytokeratins: *keratin-like proteins found as part of the intracellular scaffold, the cytoskeleton.*

Fibroin: *the main protein of silk, largely composed of β-pleated sheets, making the structure strong and inelastic.*

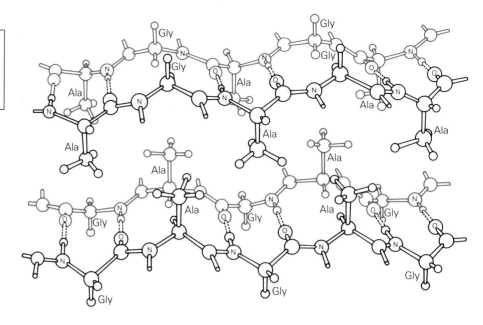

Exercise 2

Wool garments may either stretch or
shrink if washed and dried carelessly,
but garments made from silk do
neither. Why?

Fig. 3.4 The three-dimensional structure of silk. Redrawn from Dickerson, R.E. and Geis, I. (1969) *The Structure and Action of Proteins* Harper & Row, New York.

Fibroins are synthesized in highly specialized silk-producing cells, which contain large amounts of rough endoplasmic reticulum and are dedicated to the single function of producing fibroin. They carry out only such other functions as are necessary to keep them alive and produce the energy required for protein synthesis.

Not all β-sheet proteins are silks. **Resilin**, a protein resembling fibroin in composition and structure, is a major component of the wing hinge region of the exoskeleton of flying insects. It is covalently cross-linked via di-tyrosine residues, and is much more elastic than typical silks, being similar in this respect to elastin.

☐ Each silk gland cell has only one copy of the gene coding for fibroin but from this 10^8–10^9 molecules of fibroin are synthesized daily.

3.3 Collagen

Collagen is the major protein component of vertebrate organisms, and is found extensively in connective tissue, bone, cartilage and dermis.

Connective tissue

About one-quarter of the total protein in the mammalian body is collagen. It is the major protein of the ***extracellular matrix***, the material that surrounds and connects all cells of the body and which, together with cells such as fibroblasts (and the corresponding **chondroblasts** in cartilage, and **osteoblasts** in bone), forms ***basement membranes*** and connective tissue. Collagen interacts with cell-surface glycoproteins such as fibronectin allowing cell–cell contact, adhesion to surfaces, and other membrane-related interactions.

Different organs of the body contain differing amounts of connective tissue. For example, brain and spinal cord have little, whereas bone and cartilage are made up mostly of matrix with few cells. An interesting example is skin, where the epidermis is almost completely cellular with only a thin basement membrane of connective tissue, while the dermis is mostly non-cellular, being composed of connective tissue containing comparatively few cells. Collagen is secreted by fibroblasts into the extracellular matrix where it is processed

Fig. 3.5 Electron micrograph of a section of tissue showing collagen fibre.

Basement membrane: *layer of connective tissue to which cells are attached.*
Collagens: *a family of glycoproteins found in connective tissue, composed of cross-linked triple helices.*

Extracellular matrix: *the substance in which connective tissue cells, like all cells, are distributed, being largely a mixture of proteins and complex carbohydrates secreted by the cells.*

Box 3.3
Silk and silkworms

egg → larvae (caterpillar) → chrysalis (pupa) → imago (adult moth) → egg

(a)

(b)

(c)

Silk is a luxury material, still much prized despite competition from artificial fibres such as rayon, nylon and terylene. Its fibres are strong yet soft. It is easily dyed and is rarely attacked by mildew, making it a desirable product. Silk is made into sewing and embroidery threads, and is also woven and knitted into fabrics such as satin, chiffon and velvet.

The major silk-producing countries are Japan, China and the USSR, although several Mediterranean countries produce appreciable quantities. Silk is a product of the silkworm, which is not a worm but the larval stage of the moth, *Bombyx mori*, one of the few insects to have been domesticated. The 'silkworm' is the caterpillar stage, but it is the cocoon that is the source of commercial silk. Sericulture, the raising and care of silkworms, originated in China about 4000 years ago. During domestication, selected breeding has given rise to caterpillars that can move no more than about 30 cm at any one time and adults whose weak wings are incapable of supporting sustained flight. Thus *B. mori* is dependent upon human beings for food (solely mulberry leaves) and conditions for reproduction and, indeed, is no longer found in the wild.

Following hatching, the silkworms feed and grow attaining a final length of about 7.5 cm. After about five weeks they stop eating and begin spinning silk to form their pupal cocoons. At this stage, the caterpillars are put on cardboard cocoon frames which provide the necessary support. The silk is produced by two glands which open inside the mouth. The silken fibre, a protein called *fibroin*, is secreted into the mouth as two fine liquid filaments which are immediately joined together and coated with a sticky protein substance, *sericin*, produced from two further glands. (The amino acid *serine* was initially isolated in hydrolysates of sericin, hence its name.) The resulting single strand of silk is squeezed from the mouth through a fine exit tube called a spinneret.

The soluble fibroin dries and denatures on extrusion to form insoluble threads of silk. As it spins the threads, the silkworm rotates its head, winding the silk fibre around its body in figure-of-eight patterns. The cocoon is built in layers, the outer fibres being deposited first, the inner last. The first strands of silk produced are coarse and are used to secure the caterpillar to its support. Later fibres are clearer and softer. After three to five days spinning, the caterpillar is a soft egg-shaped case. The inside of the case is then smeared with a gummy substance which binds the thread together, finishing the cocoon. The cocoon consists of a single thread of silk, 1600 m or more long.

Some cocoons are allowed to complete the life-cycle to provide adult moths for further breeding. Those for silk production are heated in ovens or over dry steam or refrigerated to kill the pupae. The coarse outer fibres are removed and the cocoons placed in large pans of boiling water which softens the sericin. This treatment loosens the strands so that the long continuous filament of each cocoon can be wound off on to reels. A single thread would be too thin to use as thread and normally up to ten cocoons are reeled simultaneously, the threads being twisted together to form a stronger strand. These strands are later doubled and twisted making a thicker, stronger thread, which can be woven into cloth. Raw silk thread contains appreciable quantities of residual sericin, which is removed at the yarn or fabric stage by boiling it in soapy water. The final silk is soft, white and lustrous.

(d)

Representative stages in the production of silk. (a) Feeding silkworms with mulberry leaves. (b) Cocoon cut in two to show pupating silkworm. (c) Celled trays containing individual cocoons. (d) Cocoons being unwound while immersed in hot water.

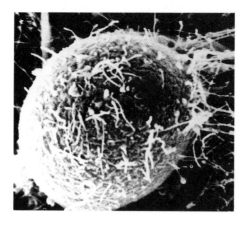
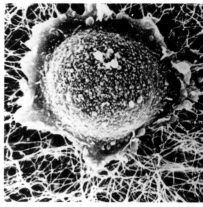

Fig. 3.6 Scanning electron micrograph of capilliary endothelial cells growing on a collagen substratum. (a) Spherical cell with microvilli (magnification × 3600); (b) semiflattened cell with microvilli and ruffled cytoplasm (magnification × 2300). Courtesy of Dr P. Kumar, Department of Biological Sciences, Manchester Polytechnic, UK. Reproduced from Kumar, S. *et al.* (1900) Responses of tissue cultured cells to angiogenesis factors – a review. In *Progress in Applied Microcirculation* (eds F. Hammerson and O. Hudlicka), pp. 54–75.

☐ Collagen will not form α-helices owing to the high proportion of proline and hydroxyproline which, because of the shape of their molecules, disrupt α-helical conformations. However, the collagen *triple helix* is very stable.

and formed into fibres that are rigid and impart great strength to the overall structure. The matrix also contains other fibres, which provide some flexibility and elasticity (Figs 3.5 and 3.6). To a large extent, it is the relative proportions of rigid and flexible elements that determine the different properties of tissues such as cartilage and skin.

STRUCTURE OF COLLAGEN. Like other fibrous proteins, collagen has an amino acid composition in which a few particular types of amino acid residues predominate. The primary structure contains several repeats of a short sequence of amino acids. Compared with typical soluble proteins and with many other fibrous proteins, the composition of collagen is unusual (Table 3.2). Typically about one-third of the residues are the smallest amino acid glycine, and there is also a very high content of proline residues, about one-quarter of the total. While glycine is a common amino acid residue in proteins, the content of proline typically tends to be quite low. Additionally, and uniquely, a large proportion of the proline in collagen is hydroxylated on either the 3 or 4 position (Fig. 3.7). Some of the lysine residues are also hydroxylated (5-hydroxylysine) and to these may be attached either the disaccharide glucosyl(β1–2)galactose or galactose so that collagen is a **glycoprotein**. However, the pattern of glycosylation is unique and is quite different from that in typical soluble, globular glycoproteins. Another unusual feature is that most collagens lack cysteine residues and so disulphide bond formation is not possible. However, this does not prevent collagen polypeptides from being cross-linked.

Box 3.4 Hydroxyproline

The amino acid hydroxyproline (Fig. 3.7) is found in very few proteins. However, it is present in a diversity of organisms, and always occurs as part of a structural protein. As well as being abundant in collagens, hydroxyproline provides up to one-third of the amino acid residues of the protein **extensin** from the cell walls of higher plants where, through the sugar arabinose, it may be involved in the cross-linking of cellulose chains. Both 3- and 4-hydroxyproline, together with 3,4-dihydroxyproline, methylhydroxylysines and other less common modified amino acid residues are found in the skeletal proteins of certain marine diatoms.

Box 3.5
Scurvy

Scurvy is the disease caused by deficiency of ascorbic acid, vitamin C. It is characterized by problems associated with connective tissues, including failure of normal bone formation in children, poor wound healing, and increased capillary fragility leading to haemorrhage (especially in the skin), together with effects on teeth and gums. Ascorbic acid is required for the formation of hydroxyproline, and collagen with insufficient hydroxyproline is unstable. The triple helical structure of the protein falls apart and the protein is rapidly degraded. This results in loss of connective tissue. It was the use of citrus fruit including limes, which contain vitamin C, to prevent scurvy in sailors which earned the British the nickname 'limeys'.

AMINO ACID HYDROXYLATION is a ***post-translational modification*** involving several enzymes. Proline is acted on by prolyl-3-hydroxylase and prolyl-4-hydroxylase, and lysine by the copper-containing enzyme lysyl hydroxylase. All three enzymes require Fe^{2+}, ascorbic acid and oxygen, and catalyse the general reaction:

$$\text{amino acid} + O_2 + \alpha\text{-oxoglutarate} \rightarrow \text{hydroxyamino acid} + CO_2 + \text{succinate}$$

Lysyl hydroxylase and prolyl-4-hydroxylase act on appropriate residue Y in the sequence Gly–X–Y, while prolyl-3-hydroxylase acts on X in the sequence X–Hyp–Gly (Hyp represents hydroxyproline).

Sequence analysis shows that the primary structure of collagen contains a three amino acid residue repeat, Gly–X–Y, where Y, and to a lesser extent, X, is frequently a proline (or hydroxyproline) residue. A typical sequence would be as follows:

–Gly–Pro–Hyp–Gly–Pro–Met–Gly–Pro–Hyp–Gly–Leu–Ala

Figure 3.8 shows the *N*-terminal sequence of a collagen polypeptide from skin.

Each fibre of mature collagen is a complex structure consisting of overlapping microfibrils, each containing several collagen polypeptides. The polypeptides are organized in helical rod-like structures. Mature collagen is built up as described below, and as summarized in Figure 3.9.

☐ Most animals and plants can synthesize vitamin C (ascorbic acid) from glucose via glucuronic acid. Humans, however, together with some apes and guinea pigs, lack the penultimate enzyme in the biosynthetic pathway, gulonolactone oxidase. For these species, vitamin C is an essential dietary component.

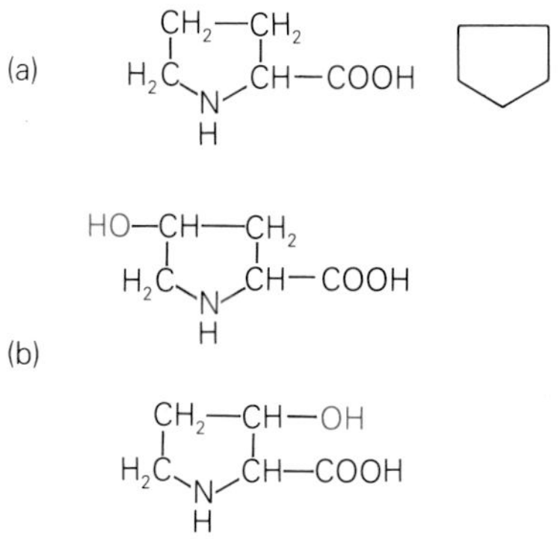

Fig. 3.7 Structures of (a) proline, (b) 3-hydroxyproline and 4-hydroxyproline.

pGlu	-Met	-Ser	-Tyr	-Gly	-Tyr	-Asp	-Glu	-Lys	-Ser	-Ala	-Gly	-Val	-Ser	-Val	-15
Pro	-Gly	-Pro	-Met	-Gly	-Pro	-Ser	-Gly	-Pro	-Arg	-Gly	-Leu	-Hyp	-Gly	-Pro	-30
Hyp	-Gly	-Ala	-Hyp	-Gly	-Pro	-Gln	-Gly	-Phe	-Gln	-Gly	-Pro	-Hyp	-Gly	-Glu	-45
Hyp	-Gly	-Glu	-Hyp	-Gly	-Ala	-Ser	-Gly	-Pro	-Met	-Gly	-Pro	-Arg	-Gly	-Pro	-60
Hyp	-Gly	-Pro	-Hyp	-Gly	-Lys	-Asn	-Gly	-Asp	-Asp	-Gly	-Glu	-Ala	-Gly	-Lys	-75
Pro	-Gly	-Arg	-Hyp	-Gly	-Gly	-Arg	-Gly	-Pro	-Hyp	-Gly	-Pro	-Gln	-Gly	-Ala	-90
Arg	-Gly	-Leu	-Hyp	-Gly	-Thr	-Ala	-Gly	-Leu	-Hyp	-Gly	-Met	-Hyl	-Gly	-His	-105
Arg	-Gly	-Phe	-Ser	-Gly	-Leu	-Asp	-Gly	-Ala	-Lys	-Gly	-Asn	-Thr	-Gly	-Pro	-120
Ala	-Gly	-Pro	-Lys	-Gly	-Gly	-Hyp	-Gly	-Ser	-Hyp	-Gly	-Glx	-Asx	-Gly	-Ala	-135
Hyp	-Gly	-Gln	-Met												

Fig. 3.8 Amino acid sequence of the first 139 residues from the amino-terminal end of the α1 chain of rat skin collagen. Note that the triplet –[Gly–X–Y]– is repeated many time. Hyp represents hydroxyproline.

THE TRIPLE HELIX is a structure seemingly unique to collagen although a similar structure is found in a protein of the immune defences, called complement component C1q. Collagen polypeptides are synthesized on the rough endoplasmic reticulum of fibroblasts as propolypeptides over 1000 amino acid residues long. These **pro-α-chains** are modified after synthesis by hydroxylation and glycosylation to form an **α-chain**. Each α-chain has an amino terminal sequence of about 180 amino acid residues and a carboxy terminal sequence of about 300 residues, flanking a middle section consisting

☐ Collagen is synthesized as preprocollagen, the 'pre' here referring to the short signal sequence (see also *Molecular Biology and Biotechnology*, Chapter 5) that enables the nascent polypeptides to cross into the endoplasmic reticulum.

Post-translational modifications: *changes made to the structure of a protein after the polypeptide has been synthesized, e.g. chemical changes to amino acid residues or cross-linking of residues.*

Reference Eyre, D.R. (1980) Collagen. *Science*, **207**, 1315–21. Review of the structure of collagen. Now a bit dated in some areas but well worth reading.

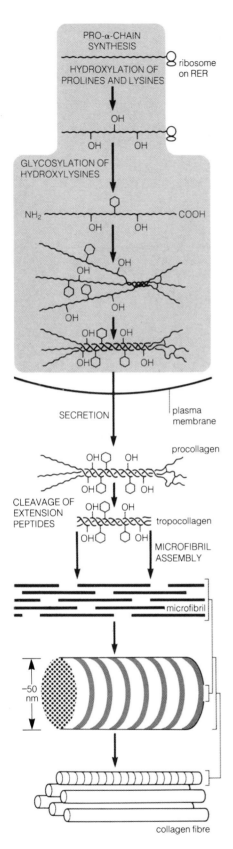

Fig. 3.9 Overview of collagen fibre formation.

Labels in figure (top to bottom): PRO-α-CHAIN SYNTHESIS; ribosome on RER; HYDROXYLATION OF PROLINES AND LYSINES; GLYCOSYLATION OF HYDROXYLYSINES; SECRETION; plasma membrane; procollagen; CLEAVAGE OF EXTENSION PEPTIDES; tropocollagen; MICROFIBRIL ASSEMBLY; microfibril; ~50 nm; collagen fibre

mainly of the three amino acid residue repeat sequence outlined above. The amino and carboxy terminal sequences are known as **extension peptides**, and these direct the spontaneous assembly of the middle sections of the α-chains into helices. Each helix contains three collagen α-chains and is therefore called a **triple helix**. In most, but not all, collagens, the three polypeptides are the same (Table 3.3). The whole molecule, triple helix and extension peptides, is known as **procollagen**.

Within the helical section of procollagen, each polypeptide chain has a left-handed twist. There is a rise of 0.29 nm per residue and three residues per turn (i.e. the *periodicity* is 0.87 nm), so the triple helix is much more extended and open than an α-helix. The three chains are coiled around each other in such a way that the triple helix has a right-handed twist. Only glycine has a side-chain sufficiently small to fit within the helix, while the bulkier side-chains project out from the helix without distorting it. The amide and carbonyl groups from the backbone are perpendicular to the helix, and allow hydrogen bond formation between the three chains of the helix. Proline and hydroxyproline cannot hydrogen bond in this way of course, but for most of the helix, two out of every three residues are involved in hydrogen bonds and this makes the triple helix a strong and quite rigid structure (Fig. 3.10).

HIGHER-ORDER STRUCTURE is shown by the collagen molecule. Procollagen is processed in the Golgi apparatus and secreted into the extracellular matrix by exocytosis. The fibroblast also secretes the enzymes, procollagen aminopeptidase and procollagen carboxypeptidase, which catalyse the hydrolysis of the extension peptides from procollagen to produce the triple helix called ***tropocollagen***. It is a rigid rod 300 nm long and 1.5 nm in diameter. If denatured, the triple helix cannot be re-formed, showing the importance of the extension peptides in directing the formation of the triple helix.

Table 3.3 Types of collagen

Type	Molecular structure	Distinctive features	Tissue distribution
I	$[\alpha1(I)]_2\alpha2(I)$	Low hydroxylysine, low carbohydrate, heteropolymer, broad fibres	Skin, tendon, bone, scar tissue, heart interstitial connective tissue (widespread)
II	$[\alpha1(II)]_3$	50% lysine hydroxylated, high carbohydrate, homopolymer, thin fibres	Cartilage, vitreous body, intervertebral disc, notochord
III	$[\alpha1(III)]_3$	Most Pro hydroxylated, low hydroxylysine, contains cysteine, high glycine, low carbohydrate, homopolymer	Skin, especially newborn, blood vessel walls, interstitial connective tissue, scar tissue
IV	$[\alpha1(IV)]_3$ or $[\alpha2(IV)]_3$	Very high hydroxylysine, rich in 3-hydroxyproline, contains cysteine, low in alanine, low in arginine, very high carbohydrate, other sugars than Glc and Gal, homopolymer, two isoproteins, probably retains procollagen extension peptides	Basement membrane, lens capsule, glomeruli
V	$[\alpha(V)]_2\alpha2(V)$ or $[\alpha1(V)]_3$ or $\alpha1(V)\alpha2(V)\alpha3(V)$	High hydroxylysine, low alanine, high glycine, high carbohydrate	Cell surfaces, exoskeleton, basement membranes (widespread in small amounts)

Tropocollagen: *the triple helical structure of collagen remaining after the terminal propeptides have been removed.*

Gelatin: *the soluble product formed when collagen is boiled in water; the triple helices are disrupted by the heat and do not reform in the absence of the extension peptides.*

Tropocollagen molecules spontaneously associate, all aligning in the same direction, to form **microfibrils**. This structure maximizes the favourable interactions that can occur between residues of the chains. In some types of collagen (Table 3.3), the extension peptides are not cleaved, and such molecules do not self-associate to form fibrils. As well as directing triple helix formation, it is likely that the extension peptides also prevent the formation of collagen aggregates within the fibroblast, which would damage the cells.

Microfibrils pack closely together to form **mature collagen** fibres 50 nm thick. Adjacent tropocollagen molecules are displaced by 64 nm which about one-quarter of the length of a single molecule (Fig. 3.11). This 64 nm repeat is the most obvious pattern seen when collagen is viewed at high magnification in the electron microscope (Fig. 3.12) but the precise three-dimensional arrangement of fibres is uncertain.

The structure of the mature collagen fibre is stabilized by the formation of covalent cross-links both within and between microfibrils. Lysine side-chains may be deaminated to give an aldehyde group forming the derivative allysine. Two allysines or allysine and lysine can interact to form strong cross-links (Fig. 3.13).

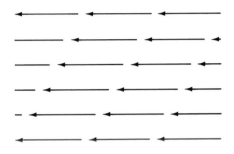

Fig. 3.11 In a collagen fibril, individual tropocollagen molecules (300 nm long) are arranged in an overlapping array with a stagger of 64 nm from the end of one monomer to that of the monomer lying next to it. Thus every sixth row is a repeat of the first.

Fig. 3.12 Electron micrograph of intact calf-skin collagen fibrils, deposited from suspension and shadowed with chromium. The magnification is × 30 000.

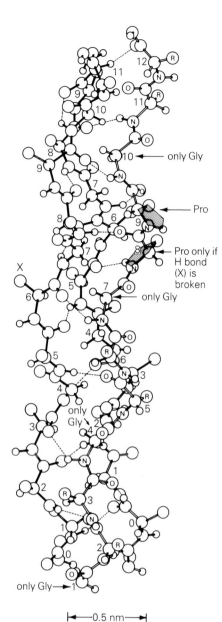

Fig. 3.10 The structure of the collagen triple helix.

Box 3.6
Collagen in food

It is the presence of collagen in the connective tissue and blood vessels of muscle which makes meat tough and in general the older the animal, the more cross-links in the collagen, and the tougher the meat. To make the flesh edible it must be cooked; the process breaks some of the bonds in collagen. Boiling collagen in water hydrolyses some of the bonds and disrupts the triple helix. This results in the soluble mixture of polypeptides known as gelatin, which forms the basis of many processed foods.

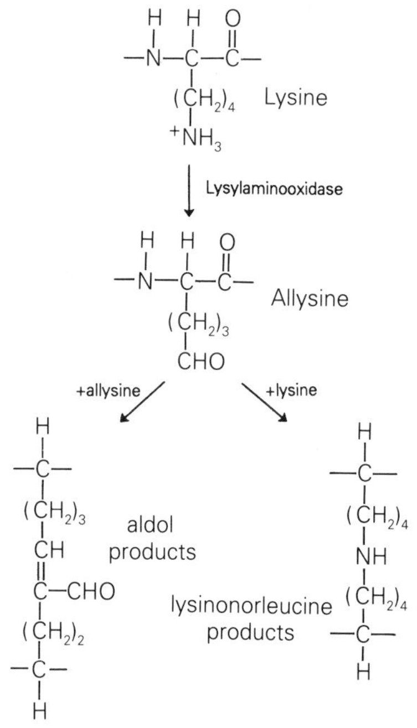

Fig. 3.13 Structures of lysine and allysine and formation of cross-links in collagen.

□ Collagenase can be used to separate individual cells from an isolated organ, without damaging the cells, which can then be studied *in vitro*. The enzyme is also used in harvesting cells from cell culture systems (see also *Molecular Biology and Biotechnology*, Chapter 10).

Exercise 4

From the data given in Table 3.3 and the other information you have concerning collagen, predict the likely effects of an inability to make any one kind of collagen chain.

Types of collagen and tissue distribution

Each α-chain is encoded by a single gene containing many coding portions called introns. More than seven distinct chains have been identified from related, but different, genes. This would, in theory, give a total of 343 possible combinations of triple helices but so far only about 10–15 types of collagen have been identified, and several of these occur in only small amounts. Many have 'homohelices' with three identical chains, others have two identical and one different, while some contain three non-identical chains. The different collagens are numbered by Roman numerals: type I, type II and so on. The α-chains are named according to the type of collagen in which they are found, e.g. 1(III) from type III, 2(I) from type I. Some types of collagen have more than one possible subunit composition. The chains differ somewhat in amino acid composition and sequence, in degree of glycosylation, in the cleavage or retention of the extension peptides, and in solubility. The characteristics of the major forms are summarized in Table 3.3.

Although many cells can synthesize more than one type of collagen, most types are typical of a particular or few tissues and in most tissues, one or two types of collagen predominate, e.g. type II in cartilage, type III in cartilage, scar and soft tissue, type IV in basement membranes. The cells synthesizing and secreting collagen are usually fibroblasts or other cells derived from the same progenitors such as chondroblasts in cartilage and osteoblasts in bone. There is some evidence that the nature of the extracellular matrix, in other words the environment of the cell, can affect which types of collagen are synthesized. In embryonic and fetal development and during, for example, the conversion of cartilage to bone, changes in the predominant form of collagen occur. Also, ageing is accompanied by increasing numbers of cross-links, which is part of the reason why bones become brittle and skin becomes more fragile and less flexible in older people.

COLLAGEN TURNOVER is slow most of the time, as is the case with most other structural proteins. However, turnover becomes greater when tissue structures are changing, such as during embryonic development (when cartilage is being replaced by bone) or during wound healing. Several **collagenases** are known. These enzymes disrupt the triple helix by cleavage of each α-chain into two peptides, which are susceptible to attack by other extracellular or lysosomal proteases. In response to injury, the body produces several materials known as 'acute-phase proteins', among which are inhibitors of collagenases. These promote wound healing or the formation of scar tissue by allowing collagen synthesis to increase while preventing its degradation. In certain chronic inflammatory conditions such as systemic lupus erythymatosis and rheumatoid arthritis, antibodies (Chapter 3 : 10) to collagen may be present, which play a part in the pathological destruction of the inflamed tissue. Under normal conditions, collagen fragments from damaged tissue are important initiators of inflammatory responses, but in these diseases the process gets out of hand.

FUNCTIONS OF COLLAGEN IN SOME SPECIALIZED TISSUES. Collagen is an important component of many structures of the eye. It provides strength in the cornea, where the uniform distribution of fine fibrils scatters light in such a way as to retain transparency of the tissue. In the sclera ('white of the eye'), by contrast, the fibres are thicker and arranged at random, and scattering of light renders the tissue opaque. Collagen is also a major component of the protective membrane around the lens. In the vitreous humour, together with structural polysaccharides and osmotic pressure, collagen is responsible for maintaining the shape of the eyeball.

Collagenases: *enzymes that degrade collagen. Notable examples occur during the formation of gas gangrene (causative organism,* Clostridium histolyticum*) and in the metamorphosis of tadpole tails.*

Reference Mayne, R. and Burgeson, R.E. (1987) *Structure and Function of Collagen Types.* Academic Press, New York. Reviews all types of collagen I–XI. Very advanced but contains interesting clinical details.

In mature bone, nearly all the organic material is type I collagen, embedded in an inorganic matrix. This produces a very inelastic tissue capable of taking a great deal of stress without breaking, but unable to bend or cope with sharp lateral attacks or twists. Bone collagen contributes to these properties by being more dense and more highly cross-linked than in other tissues, and (particularly in adult bone) has a very regular array of collagen fibres. As bone develops, the collagen fibres form the nucleus for deposition of the inorganic material, which fills in the framework they provide. Related processes form cartilage and teeth. The general name for this type of process, where mineral is deposited in an organic matrix to create a rigid structure, is **calcification**. Another, familiar example of this process is in crab shell, where the organic matrix is the polysaccharide, chitin.

3.4 Elastin

Collagen is by no means the only protein present in connective tissues, although it is the most abundant. Another protein with important properties is *elastin*. As its name suggests, elastin is especially abundant in those tissues where **elasticity** is required, such as the walls of arteries, the skin, in ligaments and in the lungs. Its main function is to impart to a tissue the ability to be stretched and to relax to the original state without damage. This type of structural role is different from the resistance to deformation characteristic of collagen. In many tissues there is a network of collagen fibres, giving strength

☐ Elastin is exceedingly insoluble and can be prepared from connective tissue by heating or mild alkaline hydrolysis, processes that remove collagen and other proteins and proteoglycans.

Box 3.8
Collagen diseases

The following are some clinical conditions associated with defects in collagen.

Disease	Biochemical defect	Feature
Osteogenesis imperfecta congenita	Decreased synthesis of type I collagen	Bone fragility and deformity
Ehlers–Danlos IV	Decreased synthesis of type III collagen	Rupture of arteries, intestine and uterus; thin, easily bruised skin
Cutis laxa and Ehlers–Danlos V	Deficiency of lysylaminooxidase (also affects elastin)	Increased stretch and decreased elasticity of skin; joint hyper-mobility; skeletal deformity; poor wound healing
Ehlers–Danlos VI	Deficiency of lysylhydroxylase	Joint hypermobility; increased stretch and fragility of skin; damage to eye; poor wound healing
Marfan's syndrome	Altered α-chains may prevent cross-linking of type I collagen	Skeletal, ocular and cardiovascular disorders

Prockop, D.J., Kivirikko, K.I., Tuderman, L. and Guzman, N. (1979) The biosynthesis of collagen and its disorders. *New England Journal of Medicine*, **301**, 77–85.

Tsipouras, P. and Ramirez, F. (1987) Genetic disorders of collagen. *Journal of Medical Genetics*, **24**, 2–8. Extensive up-to-date review.

Elastin: connective tissue protein cross-linked in such a way as to provide elasticity.

Reference Scott, J.E. (1987) Molecules for strength and shape: our fibre-reinforced composite bodies. *Trends in Biochemical Sciences*, **12**, 318–21. Covers the interactions of collagen with proteoglycans in connective tissue.

and rigidity, with smaller amounts of elastin which give elasticity. This is reflected in the structural differences between the two proteins.

Elastin, like collagen, is synthesized by fibroblasts. In tissues such as the yellow elastic tissue of ligaments and the elastic connective tissue layer of the major arteries, elastin is a more important contributor to the extracellular matrix than collagen. Elsewhere, the balance may either lie towards collagen synthesis (e.g. in bone) or both proteins being produced in significant amounts.

THE AMINOACID COMPOSITION OF ELASTIN is typical of fibrous proteins, with the predominance of a few types of amino acid residues. As is the case with collagen, in elastin nearly a third of the residues are glycines. There is a high content of proline and alanine, together with an overall weighting towards hydrophobic residues. A small proportion of the prolines are hydroxylated at the 4 position, many fewer than in collagen, and as there is no hydroxylysine glycosylation cannot occur on lysine residues. Mature elastin appears to be closely associated with lipids, including sphingolipids.

When elastin is viewed by electron microscopy, no ordered ultrastructure can be observed. There is a lack of the regular repeating structure shown by the other fibrous proteins so far described. The structure varies, too, from tissue to tissue, appearing as fine interconnected threads in the skin, as massive fibres in ligaments and as overlapping fenestrated plates in blood vessel walls (Fig. 3.14). Physical and spectral evidence suggests that in contrast to the well-defined secondary structures of keratins and collagens, elastin exists as a random coil containing open helices. The position of individual amino acids with respect to each other in elastin is not restricted by secondary structure.

Despite the apparent lack of order in its three-dimensional structure, elastin has residue sequences that are repeated often, if not regularly, throughout the protein. These common sequences are rich in lysine and alanine residues. Examples are:

Lys–Ala–Ala

Lys–Ala–Ala–Lys

Lys–Ala–Ala–Ala–Lys

Such sequences are substrates for the enzyme lysylaminooxidase, which catalyses the deamination of lysine to form allysine. As in collagen, allysine is extensively involved in cross-linking elastin. Both the aldol and lysinonorleucine products are formed, but the major derivative involves three allysines and one lysine in a structure that cross-links two or four chains. These heterocyclic derivations are called **desmosine** and **isodesmosine** (Fig. 3.15). The regions between cross-links are rich in glycine, proline and valine residues, and form extensible helices.

SYNTHESIS AND SECRETION of soluble proelastin chains occurs by fibro-blasts, with cross-linking occurring within the extracellular matrix. It is not known whether the cleavage that converts proelastin to **tropoelastin**, the molecule of elastin fibres, occurs before or after secretion. In the absence of the oxidase enzyme or in the case of copper deficiency, single proelastin polypeptides lacking cross-links occur. These have an M_r of 72 000 (about three-quarters the size of a collagen α-chain) and lack modified lysine residues. In immature connective tissues, elastin is initially laid down within a meshwork of microfibrils of an unidentified glycoprotein.

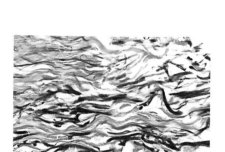

Fig. 3.14 Elastic fibres in the wall of a large artery (magnification × 250). The elastic recoil of these fibres, following the pulse, smoothes the flow of blood. Courtesy of Biophoto Associates.

☐ Since it is resistant to drying and the action of proteolytic enzymes, elastin has survived in good condition in Egyptian mummies and other preserved bodies.

☐ Elastase is an enzyme which is secreted as a pancreatic zymogen and which cleaves bonds on the carboxy terminal side of alanine, glycine, and serine residues. Its function is to help in the digestion of connective tissue and other dietary proteins.

Desmosine and isodesmosine: *derivatives of lysine, which occur in elastin.*
Tropoelastin: *monomer of elastin.*
Elastase: *an enzyme that degrades elastin; specific for neutral aliphatic amino acids.*

Reference Eyre, D.R., Paz, M.A. and Gallop, P.M. (1984) Crosslinking in collagen and elastin. *Annual Review of Biochemistry*, **53**, 718–48. Advanced, extensive review with good references. Short section on abnormal cross-links in disease.

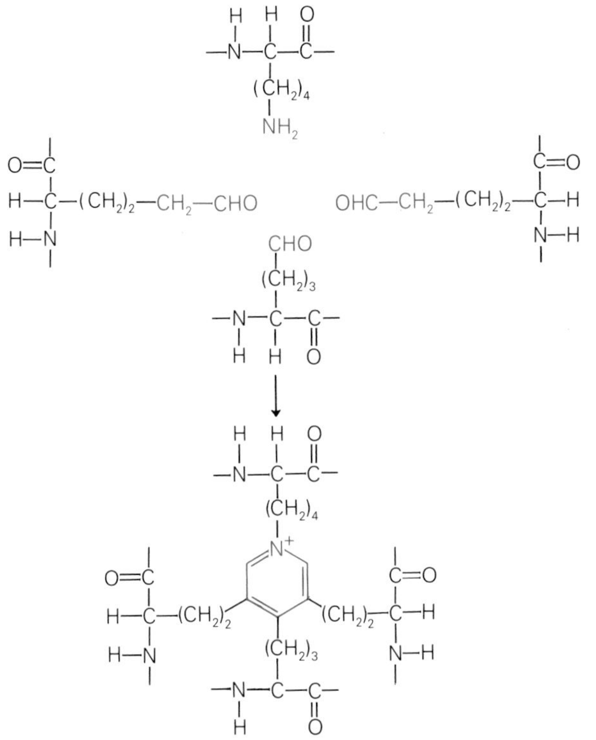

Fig. 3.15 Formation of demosine from three molecules of allysine and one of lysine.

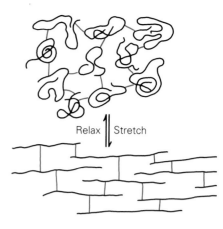

Fig. 3.16 The cross-linked random coils of elastin monomers form a three-dimensional matrix that can stretch and relax to its original structure.

THE ELASTIC PROPERTIES of connective tissues are based on the cross-linked random coils of elastin. When fully hydrated the spherical monomers can stretch under tension to form ellipsoid structures, almost straight chains, with about 70% extensibility (Fig. 3.16). The water is necessary to act as an 'energy-sink' for the conformational change. The cross-links ensure that the structure relaxes when tension is released. The elastin network is interwoven with long collagen fibres, which limit the degree of stretching and prevent damage to the tissues. In this way, both the rhythmic stretching and relaxing of blood vessels, and the more unpredictable stresses on ligaments can be accommodated. The uniqueness of elastin lies in its ability to stretch in all dimensions.

3.5 Overview

'Fibrous proteins' is a term used to cover a range of proteins with widely differing structures. These structures do, however, have some features in common. They have distinctive amino acid compositions, in which a very few amino acid residues predominate, and usually there is a short sequence of residues that is repeated many times. They often contain unusual, modified amino acid residues and tend to be highly cross-linked, giving them strength.

Extracellular fibrous proteins are often synthesized in a precursor form, and their active structure is polymeric. Whether intra- or extracellular, they usually serve either a structural or protective function. Because of their wide distribution and important roles, deficiencies of fibrous proteins usually result in serious clinical conditions.

Exercise 6

Elastin is a protein that is extremely stretchable in all three dimensions, rapidly relaxing to its unstretched form when the stress is released. What kind of structure do you think would have these properties? Can you think of any biological situations, other than eukaryotic connective tissues, where materials with elastic properties are found?

Exercise 7

Lysosomes contain an enzyme, elastase, that degrades proteins and peptides rich in neutral amino acids, such as elastin. How do you think this enzyme might be involved in mediating rheumatic disorders and other inflammatory conditions?

Reference Steinert, P.M. (1990) The two-chain coiled-coil molecule of native epidermal keratin intermediate filaments is a type I–type II heterodimer. *Journal of Biological Chemistry*, **265**, 8766–74. Research paper describing an investigation and the resolution of higher order structure of intermediate keratin filaments.

Reference Gray, W.R., Sandberg, L.B. and Foster, J.A. (1973) Molecular model for elastin structure and function. *Nature*, **246**, 461–6. Original description of the model of elastin structure.

1. Keratins contain quite large amounts of cysteine in comparison with most proteins; cysteine is involved in cross-linkages. There are also substantial amounts of 'helix-promoting' amino acid residues such as serine. They also have many hydrophobic residues - alanine, isoleucine, phenylalanine and valine are particularly important. Also in this group are leucine, methionine, proline and tryptophan. However, proline is a 'helix breaker' and is not present in keratin in significant amounts.

2. The weight of water can cause the garment to stretch, breaking the hydrogen bonds in the keratin, which do not re-form upon drying. The α-helices also stretch, and some fibres may snap. Heat from hot water and/or electric drying can also break hydrogen bonds and may allow new disulphide bonds to form. Extended α-helices, which are present in wool, may be converted into compact α-helices and locked in position by reformation of disulphide bonds. Silk, on the other hand, is fully extended, and overcoming the hydrogen bonds, which are interstrand rather than intrastrand, does not allow further stretching or contraction of the structure.

3. The rat is unable to incorporate hydroxyproline directly into collagen, but can incorporate proline. The inference is that proline is hydroxylated to form hydroxyproline only after it has been incorporated. The machinery for polypeptide synthesis (see *Molecular Biology and Biotechnology*, Chapter 4) is well understood. There is no code to specify the incorporation of hydroxyproline into proteins.

4. See section on genetic deficiencies in collagen metabolism, and Table 4.4.

5. Meat contains collagen and poor-quality (tough) cuts are often used for processed food in which the collagen may be extensively cross-linked. To render it palatable, enzymes are added to break down some of the cross-links and peptide bonds, and so convert the collagen into more digestible peptides.

6. Such a structure must be compact to allow stretching with the tendency to regain its compact form. Such structures often contain helices (see the text for structures of elastin and keratin). A protein with properties of both elastin and fibroin (see text), resilin, is found in arthropod exoskeletons. Certain plants contain the elastic material, latex, from which rubber is made. Other plants are the source of chicle, used in the manufacture of chewing-gum.

7. When cells are damaged, lysosomal enzymes are released. These can be either tissue cells or immune response cells such as phagocytes. The enzymes act in their environment and degrade the extracellular matrix. They also attack the surface molecules of nearby cells, thus extending the amount of damaged tissue and leading eventually to a chronic inflammatory condition. In rheumatoid arthritis, the tissues most badly damaged are joints and blood vessel walls.

QUESTIONS

FILL IN THE BLANKS

1. Keratins are a _____ of _____ proteins found in _____ , _____ and _____ . Their main functions are the _____ of the external surfaces of the animal but they also have an important intracellular role in the _____ . Their amino acid composition shows a relatively high content of _____ residues and amino acid residues with _____ side-chains as well as polar side-chains. Their primary structure shows a _____ amino acid repeat in which the first and fourth residues are _____ , the fifth and seventh _____ . The secondary structure is _____ and _____ _____ _____ helices interact to form a _____ . Individual _____ are cross-linked both within and between _____ by _____ _____ .

Choose from: cysteine, cytoskeleton, disulphide bonds, family, fibrous, hair, α-helical, α-helices, horn, hydrophobic (2 occurrences), microfilament, microfilaments, nail, polar, protection, seven, three or four.

2. The predominant secondary structure in silk is the _____ _____ . This structure is resistant to _____ because the tension is borne by _____ _____ in a backbone that is nearly fully _____ . On one side of the sheet _____ is the exposed amino acid side-chain, and on the other _____ and _____ residues. Adjacent sheets lie with _____ amino acid residues together.

Choose from: alanine, covalent bonds, extended, glycine, β-pleated sheet, serine, similar, stretching.

3.   The amino acid composition of collagen is unusual in that it contains the residues

_____ and _____ . Assembly of collagen polypeptides to form a _____ helix

requires the presence of _____ peptides, later cleaved to form _____ . Molecules

form a _____ structure called a _____ and several of these form a mature _____

_____ . The strength of fibres is increased by _____ involving the amino acids

_____ and _____ . Collagen is also glycosylated by addition of _____ and _____

sugar residues to _____ in the polypeptide chain.

Choose from: allysine, collagen fibre, cross-links, extension, galactose, glucose, hydroxylysine (2 occurences), hydroxyproline, lysine, microfibril, staggered, triple, tropocollagen.

4.   The other major protein of connective tissue is _____ . Like collagen, it contains

a high proportion of _____ and _____ residues, but has less _____ . It forms a

three-dimensional network of monomers cross-linked through _____ and _____

residues, forming the unusual amino acid structures of _____ and _____ .

Choose from: allysine, desmosine, elastin, glycine, hydroxyproline, isodesmosine, lysine, proline.

MULTIPLE-CHOICE QUESTIONS

In all the following questions, select those options (any number may be correct) that correctly complete the initial statement, or correctly answer the question posed.

5.   In keratins:
A.   there are extensive regions of β-sheet structure
B. α-helices are mainly cross-linked through cysteine residues
C.   the repeat sequence is Lys–Ala–Ala–Lys
D.   the structure can be permanently altered by heat
E.   microfilaments contain five interlinked α-helices

6.   Fibroin:
A.   has high tensile strength
B.   is supple
C.   contains triple helices
D.   contains desmosine
E.   lacks cysteine

7.   Collagen:
A.   is secreted by red blood cells
B.   polymerizes intracellularly
C.   has different structures in different tissues
D.   has an extensible structure
E.   contains isodesmosine

8.   The collagen triple helix:
A.   forms within fibroblasts
B.   is present in procollagen
C.   requires extension peptides to be present for its formation
D.   contains a high proportion of hydroxyproline and hydroxylysine
E.   is glycosylated

9.   Unusual features of collagen, when compared with a typical globular protein, include:
A.   a triple helical structure

B. the presence of the amino acid residue hydroxyproline
C. the presence of a repeating unit –[Gly–X–Y]– for most of its length
D. an extremely low glycine content
E. the presence of covalent cross-links between amino acid residues other than cysteine

10.   Type I collagen:
A. is found mainly in skin
B. is the predominant form in bone
C. contains only α1(I) polypeptides
D. is deficient in cutis laxa
E. has few cross-links

11.   Proline:
A. may be hydroxylated in the 3 position
B. may be hydroxylated in the 4 position
C. may be dihydroxylated
D. is largely present in the hydroxylated form in elastin
E. is involved in cross-linking collagen

12.   Elastin monomers are cross-linked through which of the following residues?
A. lysine
B. hydroxylysine
C. allysine
D. hydroxyproline
E. cysteine

13.   Fibrous proteins:
A. have similar amino acid compositions to globular proteins
B. have similar amino acid compositions to each other
C. usually have a few types of amino acid residue present in large amounts
D. usually contain covalent cross-links
E. usually contain modified amino acid residues

SHORT-ANSWER QUESTIONS

14.   What are the characteristics of keratins?

15.   Make a list of those features of mature collagen by which it differs from a typical globular protein.

16.   Draw a diagram to show the process of collagen formation indicating which processes are intra- and which are extracellular.

17.   Describe three diseases associated with collagen.

18.   Why do most fibrous proteins have a low nutritional value?

ESSAY QUESTIONS

19.   What are the typical properties of fibrous proteins? How do these properties enable them to carry out their functions?

20.   Compare the structure and properties of hair and silk.

21.   Compare the structures and properties of the two major connective tissue proteins, collagen and elastin.

4

Enzyme kinetics

Objectives

After reading this chapter you should be able to:

☐ describe the principles of catalysis;

☐ outline the theory of enzyme kinetics;

☐ explain how the kinetic parameters of enzymes are experimentally obtained;

☐ explain the significance of kinetic constants in the metabolic role of enzymes;

☐ summarize how enzyme activity can be controlled.

4.1 Introduction

Almost all of the chemical reactions that occur in biological systems are catalysed by proteins called **enzymes**, and life is only possible because of the controlled action of these catalysts. An enzyme may consist of a single polypeptide chain, may be composed of multiple subunits, or may form part of a multienzyme complex. In addition to the polypeptide chain(s), many enzymes also require components (cofactors) that are crucial to the catalytic process. For example, some enzymes require the participation in the reaction mechanism of small organic molecules called **coenzymes**, e.g. NAD^+ or ATP, which act as donors or acceptors of functional groups such as protons or phosphate. Others may contain bound non-protein components or prosthetic groups such as complexed metal ions.

In general, enzyme-catalysed reactions occur at moderate pressures, temperatures and pH values. Similar chemically-catalysed reactions often require much harsher conditions. Furthermore, enzymes usually show a remarkable specificity, for the reactants and for the reaction that is catalysed, including the ability to distinguish between optical isomers. Therefore, side-reactions are kept to a minimum. This is important to the organism as there is no opportunity to purify the product of one reaction before it is utilized by the next enzyme in a metabolic pathway, as might be done in a chemical laboratory. Enzyme activity is also tightly regulated in an organism as part of its homoeostatic mechanisms.

Nearly 2500 different enzyme-catalysed reactions have been catalogued by the Enzyme Commission of the International Union of Biochemistry and there are often several different enzymes that promote a given reaction. Surprisingly, however, a few straightforward rules are sufficient to describe the basic reaction kinetics of most enzymes. This chapter presents these rules and explains how they may be applied to the study of enzymes.

Enzymes: *macromolecular biological catalysts. The majority of enzymes are proteins, but recently certain RNA molecules have been discovered to possess catalytic activity. From the Greek* en zyme, *ferment.*
Coenzyme: *an organic molecule, usually a vitamin derivative, that participates in, and is*

essential for, an enzyme-catalysed reaction.
Cofactor: *a non-protein compound that is essential for the catalytic activity of an enzyme. Cofactors that are tightly bound to enzymes are called prosthetic groups. Cofactors may be small organic molecules (that is coenzymes) or metal ions, e.g. Mg^{2+}, Ca^{2+}.*

Box 4.1
A history of enzymes

1833 Payen and Persoz purified amylase from wheat.
1877 Kuhne coined the term enzyme meaning 'in yeast'.
1893 Ostwald showed that enzymes are catalysts.
1894 Fischer proposed the lock and key theory for substrate specificity.
1905 Harden and Young isolated a coenzyme that was subsequently identified as NAD^+.
1909 Sorensen demonstrated the effects of pH on enzyme action.
1912 Batelli and Stern discovered dehydrogenases.
1913 Michaelis and Menten developed the theory of enzyme kinetics.
1925 Briggs and Haldane proposed important modifications to the theory of enzyme kinetics.
1926 Sumner crystallized urease.
1931 Lineweaver and Burk published their method for the determination of the kinetic constants of enzymes.
1943 Chance proved the existence of enzyme–substrate complexes by spectrophotometry.
1960 Hirs, Moore and Stein determined the primary amino acid sequence of ribonuclease.
1962 Phillips produced a low-resolution three-dimensional structure for lysozyme.
1970 Chang and Cohen cloned a gene encoding the enzyme β-lactamase from *Staphylococcus aureus* into a strain of *Escherichia coli*.
1982 A mutant form of the enzyme tyrosyl tRNA synthetase was produced by site-directed mutagenesis.

Box 4.2
Biological versus chemical catalysts

The efficiency of enzyme-catalysed reactions is well demonstrated for the process of nitrogen fixation. The chemical industry uses finely divided iron as a catalyst to fix atmospheric nitrogen at a temperature of 550°C and a pressure of about 25×10^6 Pa (250 atm). These high temperatures and pressures are required to ensure a sufficient number of high-energy collisions between the gas molecules for the reaction to occur. In contrast, microorganisms are capable of using enzymes to fix nitrogen at ambient soil temperature (approximately 5–30°C) and a pressure of about 10^5 Pa (one atm). See *Biosynthesis*, Chapter 5.

Reference Webb, E.C. (ed.) (1984), *Enzyme Nomenclature*, Academic Press, London, UK. The definitive classification of enzyme-catalysed reactions.

4.2 The principles of catalysis

Reaction equilibrium

Catalysts cannot alter the equilibrium constant of a reaction because of the restrictions implicit in the laws of thermodynamics, although they do **enhance the rate** at which equilibrium is reached. Therefore, the final equilibrium mixture resulting from a reaction will be the same regardless of whether or not it is enzyme catalysed. This arises because the catalyst works by accelerating both forward and the reverse reactions by precisely the same factor, and so the equilibrium constant will remain the same. For the reaction:

$$A + B \underset{k_b}{\overset{k_f}{\rightleftharpoons}} C + D$$

the equilibrium constant, K_{eq}, is defined by the equation:

$$K_{eq} = \frac{[C][D]}{[A][B]} = \frac{k_f}{k_b}$$

where [A], [B], [C] and [D] represent the concentrations of A, B, C and D respectively, and k_f and k_b are the rate constants for the forward and back reactions respectively.

Transition states and activation energy

The interconversion of two compounds, X and Y, during a chemical reaction involves a **transition state** which has a **higher** free energy than either type of

---
Exercise 1

At equilibrium a reaction mixture comprises one part of substrate to ten parts of product. What is the equilibrium constant for this reaction?

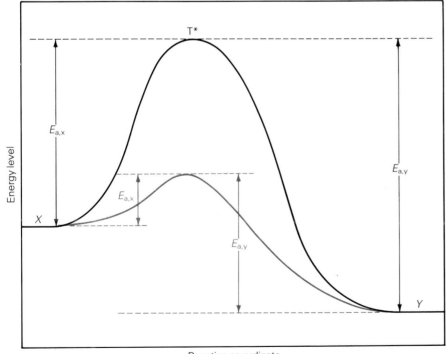

Reaction co-ordinate

Fig. 4.1 The energy profile of an uncatalysed and a catalysed reaction. The uncatalysed reaction for the conversion of X to Y via an activated intermediate T* has an activation energy of $E_{a,x}$ for the forward direction and $E_{a,y}$ for the reverse direction (shown in black). Enzyme catalysis of the reaction reduces the activation energy of both the forward and reverse reactions by the same factor (shown in red). The absolute energies of X and Y remain unchanged and therefore the equilibrium position of the reaction is the same in the presence and absence of enzyme.

participating molecule (Fig. 4.1). The difference in free energy between X and the transition state (T*) is the **activation energy** for the forward reaction:

$$X \rightarrow T^* \rightarrow Y$$

The greater the activation energy, the slower the rate of the reaction. As there is no activation energy barrier in progressing from T* to Y, it follows that the rate governing this part of the reaction will be maximal. Therefore, in any reaction the rate of activation of X, i.e. X → T*, will be the rate-limiting step and will dictate the overall rate of conversion of X to Y.

There will also be a reverse reaction:

$$Y \rightarrow T^* \rightarrow X$$

with an activation energy equal to the difference between the free energies of T* and Y. Similarly, the step from Y to T* will be rate limiting and will be dictated by the activation energy.

If the free energies of X and Y are the same (note that this would make the activation energies the same for the forward and reverse reactions), then the rate constants of the forward and reverse reaction will be equal. Therefore, at equilibrium the concentration of X will equal that of Y and the rates of the forward and reverse reaction will be the same.

Enzymes, like all catalysts, achieve their effect by reducing the free energy of the transition state. Therefore, the activation energy is reduced (Table 4.1) and the rate of attainment of equilibrium (regardless of the starting point) will be greater in the presence of catalyst. To summarize, enzymes are regulatable catalysts that accelerate reactions with a high specificity for the reaction catalysed.

Table 4.1 *The effect of catalysis on the activation energy for the decomposition of hydrogen peroxide*

Catalyst	Activation energy $(J\,mol^{-1})$
None	75 600
Colloidal platinum (non-enzyme catalyst)	49 140
Catalase (enzyme)	8 400

Kinetics of reactions and rate constants

Chemical reactions may be classified according to their kinetic behaviour. Each *order* of chemical reaction is described by a simple equation that relates the **rate** of reaction to the **concentration** of the reactants. The simplest example to analyse is the first-order reaction. In this case the rate of progress for the reaction:

$$A \rightarrow X$$

is directly proportional to the concentration of A, [A]. Thus, the rate of disappearance of A with time, $-d[A]/dt$, may be defined by:

$$-d[A]/dt = k[A]$$

□ $-d[A]/dt$ is simply a convenient way of expressing the rate of decrease in the concentration of A with time.

where k is the rate constant for the reaction. First-order rate constants have dimensions of $time^{-1}$. The rate of disappearance of A will decrease exponentially as the reactant, A, is consumed (Fig. 4.2).

A second-order reaction is one in which the rate is proportional to the concentration of two reactants. In the simplest case of a bimolecular reaction in which:

$$A + A \rightarrow X$$

the rate of disappearance of A, $-d[A]/dt$ will be described by:

$$-d[A]/dt = k[A][A] = k[A]^2$$

A second-order rate constant will have units of $concentration^{-1}\ time^{-1}$. An alternative example of a second-order reaction is:

$$A + B \rightarrow X \qquad (4.1)$$

Reference Cornish-Bowden, A. (1981) *Basic Mathematics for Biochemists*, Chapman and Hall, London, UK. An inexpensive and readable paperback for those who need to brush up on their mathematics.

in which:

$$-d[A]/dt = k[A][B]$$

Note that a bimolecular reaction need not necessarily appear to be second order under all conditions. Furthermore, the orders of the forward and reverse reactions need not be the same. For example, in Reaction 4.1 the forward reaction could be second order whereas the reverse reaction might be first order. However, if in Reaction 4.1 the concentration of B greatly exceeds that of A, then it is likely that the rate would be independent of the depletion of B and dependent only on the concentration of A. Under these conditions, the rate may be described as:

$$-d[A]/dt = k[A]$$

i.e. a first-order reaction. This ability of bimolecular reactions to behave kinetically as first order under certain conditions can be of importance in considering enzyme kinetics.

The rate of a zero-order reaction is independent of reactant concentration and the rate constant will have units of concentration time^{-1}. Enzyme-catalysed reactions may be zero order under certain circumstances. This happens when the rate is independent of the concentration of the reactant(s) (see later).

It is common for reactions to be 'mixed order'. This phrase means that the reaction order changes as the reaction progresses.

4.3 Kinetics of enzyme-catalysed reactions

The kinetic behaviour of enzyme-catalysed reactions may be analysed in a similar manner to that of any catalysed chemical transformation. A major difference, however, is that the enzyme will remain unchanged by the reaction and therefore will appear on both sides of a chemical equation:

$$E + S \rightarrow E + P$$

where E is the enzyme, S is the **substrate** and P is the product. The enzyme, just like any catalyst, can be saturated with substrate, and this means that at low substrate concentrations the rate of reaction is proportional to the substrate concentration, [S]. Indeed, the reaction is first order with respect to [S] but at higher concentrations of substrate the reaction will become zero order (Fig. 4.3). These observations are best explained by proposing the formation of an enzyme–substrate complex, ES, as described by the reaction sequence:

$$E + S \underset{k_2}{\overset{k_1}{\rightleftharpoons}} ES \underset{k_4}{\overset{k_3}{\rightleftharpoons}} E + P \qquad (4.2)$$

The Michaelis–Menten equation

A kinetic equation describing the enzyme-catalysed conversion of substrate to product (called the Michaelis–Menten equation after its two proposers) may be derived simply, providing certain assumptions are made. These are that:

1. Only initial rates of reaction will be studied. Therefore, as [P] is small, the reverse Reaction 4.2:

$$E + P \overset{k_4}{\rightarrow} ES$$

will be insignificant and may be ignored.

Substrate: *the compound that an enzyme converts into product.*

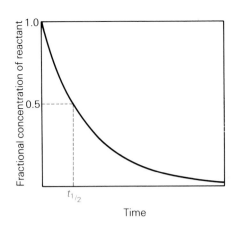

Fig. 4.2 The time course of a first-order reaction. The time taken for half of the reactant to disappear is called the half-life, $t_{1/2}$, of the reaction and is independent of initial concentration. The half-life is determined from the first-order rate constant, k, using the equation $t_{1/2} = \ln 2/k$.

See Chapter 2

See Chapter 2

Exercise 2

Many everyday products such as detergents, toothpastes, soft-centred chocolates and beer contain added enzymes. Which enzymes are likely to be present in these products and what are their functions?

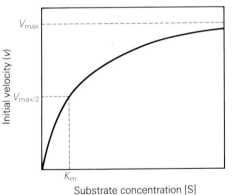

Fig. 4.3 The initial velocity of an enzyme-catalysed reaction as a function of substrate concentration. Note that the maximum velocity, V_{max}, is only attained at an infinite concentration of substrate. The Michaelis constant, K_m, is the substrate concentration at which the initial rate, v, is half V_{max}.

2. After an initial transient state (usually lasting for milliseconds) the concentration of ES will remain constant. This is called the **steady state.**
3. The concentration of substrate is much greater than the concentration of enzyme (this is convenient to arrange in a laboratory experiment and helps to simplify the derivation of the equations but rarely occurs in living cells).

The first step in the derivation is to analyse the rates of formation and breakdown of ES. Thus, the rate of formation of ES is:

$$d[ES]/dt = k_1[E][S]$$

Note that because of assumption 1 the term $(k_4[E][P])$ is not included in this equation. The rate of breakdown of ES is:

$$-d[ES]/dt = k_2[ES] + k_3[ES] = (k_2 + k_3)[ES]$$

As the overall concentration of ES remains constant (assumption 2):

$$d[ES]/dt = -d[ES]/dt$$

therefore

$$k_1[E][S] = (k_2 + k_3)[ES] \qquad (4.3)$$

Rearranging Equation (4.3) gives:

$$\frac{k_2 + k_3}{k_1} = \frac{[E][S]}{[ES]} \qquad (4.4)$$

The left-hand term of Equation (4.4) comprised of the three rate constants is called the **Michaelis constant,** K_m. The right-hand term may be rewritten by describing [E] in terms of the total concentration of enzyme $[E_t]$ and the concentration of enzyme in the complex, ES. This is done because it is difficult experimentally to measure the proportion of enzyme associated with substrate. Thus:

$$[E] = [E_t] - [ES] \qquad (4.5)$$

Substituting the right-hand term of Equation 4.5 into Equation 4.4 gives:

$$K_m = \left(\frac{[E_t] - [ES]}{[ES]} \right) [S]$$

therefore,

$$K_m = \left(\frac{[E_t]}{[ES]} - 1 \right) [S] \qquad (4.6)$$

Box 4.3
Dimensions of kinetic constants

The Michaelis constant, K_m, has units of substrate concentration and is usually quoted in terms of μmol dm^{-3} or mmol dm^{-3}.

The maximum velocity, V_{max}, has the same units as the initial velocity measurements. Usually, this is expressed in terms of μmol of product produced per minute, often defined as units of activity. The concentration of enzyme will affect the absolute value of V_{max}. Therefore, for this constant to be useful for comparative purposes it should be expressed in terms of μmole of product produced per minute per milligram of enzyme, that is, units per milligram of enzyme.

One international unit of enzyme activity is defined as that amount of enzyme causing the transformation of one μmole of substrate per minute.

One katal (kat) is defined as the transformation of one mole of substrate per second.

When dealing with equations and graphs it is useful to be able to check whether algebraic manipulations have been performed correctly. One way of doing this is to check for dimensional consistency. The main rule of this method is that the units on the left-hand side of an equation must be the same as those on the right-hand side. Thus, in the Michaelis–Menten equation:

$$v = \frac{V_{max}\,[S]}{K_m + [S]}$$

If the appropriate units are substituted into the equations, i.e. $[S]$ and K_m have units of (concentration) whereas v and V_{max} have units of (concentration) (time^{-1}), then:

$$(\text{concentration})\,(\text{time}^{-1}) = \frac{(\text{concentration})\,(\text{time}^{-1})\,(\text{concentration})}{(\text{concentration}) + (\text{concentration})}$$

Cancelling through on the right-hand side of the equation gives:

$$(\text{concentration})\,(\text{time})^{-1} = (\text{concentration})\,(\text{time})^{-1}$$

i.e. the units on both sides are the same. If this had not been so then it would have been clear that an error had been made in the derivation of the original Michaelis–Menten equation.

The rate of formation of products or the velocity, v, of an enzyme-catalysed reaction will be:

$$v = k_3[ES]$$

Therefore, substituting $[ES]$ in Equation 4.6 gives:

$$K_m = \left(\frac{k_3\,[E_t][S]}{v} \right) - [S]$$

or

$$v = \frac{k_3\,[E_t][S]}{K_m + [S]} \qquad (4.7)$$

As the maximum initial velocity, V_{max}, of a reaction will occur when all of the enzymes are associated with ES complex, it follows that:

$$V_{max} = k_3[Et] \qquad (4.8)$$

Therefore, Equation 4.7 may be written as:

$$v = \frac{V_{max}[S]}{K_m + [S]} \qquad (4.9)$$

which is the **Michaelis–Menten equation**. It is this equation that defines the relationship between initial velocity, v, and substrate concentration, $[S]$, as illustrated in Fig. 4.3. Providing that the values of the constants K_m and V_{max} are known, then it is possible to predict the initial velocity of the enzyme reaction for any concentration of substrate. The initial velocity of the reaction will be directly proportional to total enzyme concentration (Fig. 4.4).

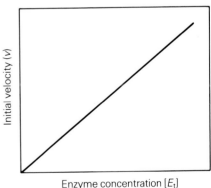

Fig. 4.4 The relationship between initial velocity and enzyme concentration. The velocity will increase linearly with enzyme concentration, providing that the substrate concentration is not rate-limiting.

Reference Cornish-Bowden, A. (1979) *Fundamentals of Enzyme Kinetics,* Butterworths, London, UK. This book covers all of the practical aspects of enzyme kinetics that undergraduates or postgraduates are likely to need.

The importance of kinetic parameters

It is important to understand what these kinetic parameters represent and why it is useful to know them for an enzyme. The Michaelis constant, K_m, is often assumed to be equivalent to the **dissociation constant** for the enzyme–substrate complex. However, this would only be so if the rate constant for the formation of products, k_3, was considerably smaller than k_2 (Eqn 4.2). In those cases where K_m is equivalent to the dissociation constant of the ES complex, that is k_2/k_1, then K_m is a direct measure of the **affinity** of a substrate for an enzyme. The lower the value of K_m, the greater the affinity of the substrate for the enzyme. Frequently, however, $k_3 \geqslant k_2$ in which case K_m is not directly equivalent to a dissociation complex for ES. Nevertheless, in either case, K_m is equal to the substrate concentration at which half the active sites on the enzyme are occupied.

The maximum velocity, V_{max}, is not by itself a very useful comparative parameter because of its dependence on enzyme concentration (Eqn 4.8). A more useful concept is the **turnover number** or k_{cat}, which is equivalent to the rate constant, k_3, for the degradation of the ES complex to product, when the enzyme is fully saturated with substrate (Eqn 4.2). Therefore, the turnover number, k_{cat}, is equal to $(V_{max}/[E_t])$ and, as a first-order rate constant, will have units of reciprocal time. The turnover number is a measure of the maximum potential catalytic activity of an enzyme. The reciprocal of the turnover number $(1/k_{cat})$ will be the time taken for a single round of catalysis to occur when the enzyme is saturated with substrate.

Another important parameter is the ratio of the turnover number to the Michaelis constant. The value of k_{cat}/K_m is the second-order rate constant (with units of concentration^{-1} time^{-1}) for the reaction of free enzyme and substrate to form products:

$$E + S \rightarrow E + P$$

Therefore:

$$v = \frac{-d[S]}{dt} = \frac{k_{cat}}{k_m} [E][S] \tag{4.10}$$

It is useful to think of k_{cat}/K_m as a **specificity constant** for an enzyme. For example, suppose that an enzyme can catalyse a reaction with either of two substrates A or B. Then it follows from Equation 4.10 that:

$$v_a = \frac{-d[A]}{dt} = \frac{k_{cat,a}}{K_{m,a}} [E][A]$$

and

$$v_b = \frac{-d[B]}{dt} = \frac{k_{cat,b}}{K_{m,b}} [E][B]$$

If the value of $k_{cat,a}/K_{m,a}$ is greater than $k_{cat,b}/K_{m,b}$, then it follows that substrate A will be utilized at a greater rate. In other words the enzyme has a greater 'specificity' for substrate A than for B. Enzyme specificity has a molecular as well as a kinetic basis: the concepts of molecular specificity of enzymes are dealt with in some detail in Chapter 5.

Ultimately, the maximum value of k_{cat}/K_m that is theoretically possible is one in which the rate of reaction is diffusion controlled, that is, when k_{cat}/K_m has the value of approximately 10^9 dm^3 mol^{-1} s^{-1}. Several enzymes have values of k_{cat}/K_m that are of this order (Table 4.2). This illustrates the remarkable power of enzymes not only to catalyse reactions with great specificity but also, in certain cases, to catalyse reactions at near to the maximum rate that is theoretically possible.

Table 4.2 *Kinetic constants of some enzymes*

Enzyme	Substrate	K_m(mol dm^{-3})	k_{cat}(s^{-1})	k_{cat}/K_m(dm^3 mol^{-1} s^{-1})
Carbonic anhydrase	CO_2	8×10^{-3}	600 000	7.5×10^7
Chymotrypsin	Acetyl L-trypto-phanamide*	5×10^{-3}	100	2.0×10^4
Lysozyme	Hexa-*N*-acetyl-glucosamine	6×10^{-6}	0.5	8.3×10^4
Penicillinase	Benzylpenicillin	5×10^{-5}	2000	4.0×10^7

* This substrate is a chemically synthesized compound, i.e. an artificial substrate.

In summary, the two cardinal properties of enzymes, high catalytic activity and high specificity for substrates, can be described kinetically by the constants k_{cat} and k_{cat}/K_m. Furthermore, K_m may be determined even when the concentration of enzyme is unknown, and so it is possible to determine the K_m of an enzyme in a homogenate or a crude preparation.

Exercise 3

Show that the units of the specificity constant, k_{cat}/K_m are dm^3 mol^{-1} s^{-1}

Estimation of kinetic constants

Clearly it is useful to know the values of K_m and V_{max} so that predictions may be made about the velocity of the reaction at a particular substrate concentration. The ability to predict reaction rates is important for several reasons. For example, if an enzyme is to be used as part of a laboratory assay method then a knowledge of the kinetic constants, and consequently the rate of reaction, would enable estimates to be made of assay parameters, such as the minimum incubation times necessary to ensure complete conversion of a given initial concentration of substrate. Also, a knowledge of the reaction kinetics enables predictions to be made about the possible flux of metabolites through a pathway within a cell.

Experimentally, the kinetic constants are estimated from initial reaction velocity data obtained over a restricted range of substrate concentrations. One way to obtain the value of V_{max} would be to use the data to draw a graph of v against [S] (Fig. 4.3). The value of V_{max} could be estimated directly by extrapolation of the curve. The value of K_m is the substrate concentration at which the velocity is $V_{max}/2$. The relationship between K_m and $V_{max}/2$ can be proved as follows. If $v = V_{max}/2$ then:

$$\frac{V_{max}}{2} = \frac{V_{max}[S]}{K_m + [S]}$$

Dividing both sides by V_{max} gives:

$$\frac{1}{2} = \frac{[S]}{K_m + [S]}$$

and rearranging:

$$\frac{K_m + [S]}{2} = [S]$$

therefore,

$$K_m = [S]$$

However, the major disadvantage of plotting v against [S] is that it is very difficult to judge accurately the value of V_{max} and this is almost invariably underestimated when determined directly from a graph. The advent of computer methods has now largely alleviated this problem (see later).

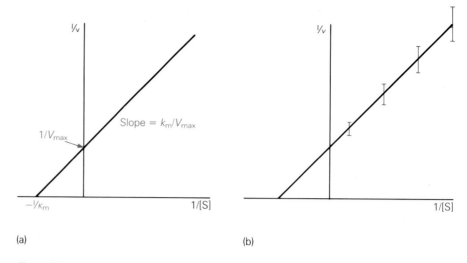

(a)   (b)

Fig. 4.5 (a) The Lineweaver–Burk plot. The reciprocal of the initial velocity is plotted against the reciprocal of the substrate concentration and kinetic constants are determined from the intercepts as indicated on the diagram. (b) The effect of experimental error on the Lineweaver–Burk plot. The bars represent a ± 10% error in the initial velocity values.

Over the years numerous graphical and other methods have been developed in an attempt to improve the accuracy with which the kinetic constants may be determined. Although various individuals have championed different methods, only three options will be considered here to illustrate some of the inherent problems in these analyses.

Perhaps the most widely used graphical method for the determination of K_m and V_{max} has been the **Lineweaver–Burk** plot, named after the two scientists who devised the technique. This is a **double-reciprocal plot**, derived simply by taking reciprocals of both sides of the Michaelis–Menten equation (Eqn 4.9) to give:

$$\frac{1}{v} = \frac{1}{[S]}\frac{K_m}{V_{max}} + \frac{1}{V_{max}}$$

Therefore, when $1/v$ is plotted against $1/[S]$ a straight line is obtained with a slope of K_m/V_{max} and an intercept of $1/V_{max}$ on the $1/v$ axis (Fig. 4.5). The advantage of this method is that because it is a straight line there is no guesswork involved in estimating V_{max} or K_m and deviations from normal Michaelis–Menten kinetics are usually fairly obvious. However, the Lineweaver–Burk plot does suffer from a very serious disadvantage that is often ignored or not appreciated. By taking reciprocals, most significance is placed on the rates obtained at *low* substrate concentrations and it is these that are most likely to be subject to the greatest experimental error. If the Lineweaver–Burk graph is plotted with constant percentage errors in the value of v the uneven distribution of error in the rate at each substrate concentration becomes obvious (Fig. 4.5). Therefore, although the double-reciprocal plot has been used widely, and is also useful for demonstrating the effects of deviations from normal kinetics, it cannot be recommended as a method for estimating K_m and V_{max}. Although there are other linear graphical methods that are less prone to bias, they all suffer from this problem to some extent.

An alternative approach is to substitute values of K_m and V_{max} into the Michaelis–Menten equation and to compare the calculated initial velocities with those obtained experimentally. The statistical technique that is normally used to determine the goodness of fit of the data is the sum of least-squares method. In this technique the values for the square of the difference between

---

Exercise 4

In addition to the Lineweaver–Burk plot there are two other linear transformations of the Michaelis–Menten equation. In these methods graphs are constructed of either [S]/v against [S] or v against v/[S]. Transform the Michaelis–Menten expression to obtain the equations that describe these graphs and establish how the kinetic parameters K_m and V_{max} would be estimated from the graphs.

each calculated and observed velocity, i.e. $(v_{calc} - v_{obs})^2$, at each concentration of substrate are added together (Fig. 4.6). The smaller the value of this sum of squares, the closer the estimates of K_m and V_{max} are to the real values. Therefore, the calculations are repeated many times with different estimates of K_m and V_{max} until a **minimum value** is obtained for the sum of squares. This is probably the best method available from the statistical point of view for estimating the values of the kinetic constants. Although this method is only practical if it is computer-based and the appropriate programs are available, the advent of cheap microcomputers and suitable software means that it should be accessible in any laboratory.

A technique for obtaining the kinetic constants that has the advantage of simplicity as well as being statistically acceptable is the **direct linear plot**. This method is a little unusual in that it treats V_{max} and K_m as variables and v and [S] as experimentally determined constants according to the rearranged Michaelis–Menten equation:

$$V_{max}/v - K_m/[S] = 1$$

A graph is constructed with the x-axis labelled K_m and the y-axis labelled V_{max}. For each experimental result a straight line is drawn from a value of $-[S]$ on the x-axis to the value, v, on the y-axis. This line is extended into the first quadrant of the graph and represents a series of values of V_{max}/K_m that would produce a rate, v_1, at a substrate concentration of $[S_1]$. A similar line is drawn for the second set of data $[S_2]$, v_2 and so on. The lines should all intersect at a single point with co-ordinates of K_m and V_{max} which represent the only values of K_m and V_{max} that satisfy all of the sets of data (Fig. 4.7).

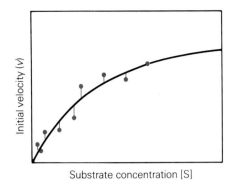

Fig. 4.6 Direct estimation of kinetic constants from a plot of v against [S]. Values of K_m and V_{max} are incorporated into the Michaelis–Menten equation and the corresponding curves calculated. The best-fit curve is one in which the sum of the residuals (shown in red) is a minimum.

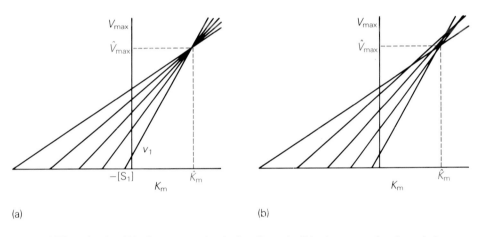

(a) (b)

Fig. 4.7 (a) The estimation of kinetic constants using the direct linear plot. If the data are error free then a single intersection point is obtained. (b) The estimation of kinetic constants using the direct linear plot. The effect of experimental error is to produce multiple intersections (ten intersections from five experimental results). The median intersection gives the best-estimate values of K_m and V_{max}.

In practice, the determination of initial rates will be subject to experimental error. This means that there is likely to be more than one intersection point. In fact, the potential number of intersection points is determined by the equation:

$$\text{number of intersections} = \tfrac{1}{2}n\,(n - 1)$$

where n is the number of experimental determinations. When there are multiple intersections the best estimates of the true values of K_m and V_{max} are determined by the co-ordinates of the median intersection point (Fig. 4.7). It is crucial to the accuracy of the method that it is the **median** and not the mean value that is used, although the reasons for this need not be of concern here.

Overall, the direct linear plot is probably the best method to use for the determination of the kinetic parameters, unless access is available to computer-fitting methods (see earlier). The method has the added advantage that once the first few initial rates have been measured, and the data plotted, preliminary values of K_m and V_{max} may be obtained. It is then possible to predict the substrate concentrations that will give the best spread of initial velocity values.

Where enzymes catalyse reactions with more than one substrate it will be necessary to determine more than one set of kinetic constants. In these instances, initial rates are measured at several different concentrations of the first substrate, S_1, while the concentration of the second substrate, S_2, is held constant. From these data it is possible to calculate the K_m for S_1. The experiment is then repeated with measurement of initial rates for different concentrations of S_2 at constant concentrations of S_1. Although the two substrates will probably have different values for K_m, the value for V_{max} must be the same for both.

Physiological significance of enzyme kinetics

By this stage it should be clear that in principle there is usually no difficulty in estimating the kinetic constants for an enzyme, given sufficient experimental data and (preferably) a computer. However, what may be less obvious is the physiological significance of the determined values of K_m and V_{max}. There is a danger in using results obtained *in vitro* to predict what may be happening in the organism, but nonetheless it is possible to use kinetic data to provide useful information about enzyme systems *in vivo*.

An important point can be made about the physiological implications of enzyme kinetic constants. If the substrate concentration is known, then it is possible to predict from the K_m value whether or not an enzyme is saturated. This, in turn, would indicate whether the enzyme was likely to be the rate-limiting step in a pathway. Also, a knowledge of the values of V_{max} for the various enzymes in a pathway would allow predictions to be made of the maximum obtainable flux of metabolites through a particular route. For example, many enzymes are substrate limited *in vivo*, that is the substrate concentration is low compared with that of the enzyme. However, if one of the enzymes in a pathway is fully saturated with substrate then the concentration of this enzyme will be the controlling influence on the flux through that pathway (assuming that there are no other forms of metabolic control).

An excellent illustration of the physiological importance of enzyme kinetics is the phosphorylation of glucose to glucose 6-phosphate. In all the glucose-utilizing cells of the body this reaction is catalysed by the enzyme **hexokinase** (K_m for glucose approximately 50 μmol dm^{-3}). However, the liver cells (**hepatocytes**) have a second enzyme, **glucokinase**, that catalyses the same reaction, but its K_m for glucose is about 10 mmol dm^{-3}. In addition, glucose 6-phosphate, the product of the reaction, inhibits hexokinase but not glucokinase.

Under fasting conditions the concentration of blood glucose is 4 mmol dm^{-3} and so hexokinase is essentially saturated ([S] is 80 times K_m). Therefore, hexokinase will work at maximum velocity until the concentration of glucose 6-phosphate builds up sufficiently in the cell to inhibit the enzyme. Immediately after a carbohydrate-rich meal the blood glucose concentration may increase to 10–12 mmol dm^{-3}. As the K_m for the liver glucokinase is 10 mmol dm^{-3}, the reaction velocity is able to increase as the glucose concentration increases. However, hexokinase with a low K_m of 50 μmol dm^{-3} is unable to increase the rate of glucose utilization. Therefore,

Hepatocyte: *a parenchymal liver cell. The word is derived from the Greek* hepar, *liver and* kytos, *cell.*

Reference Dixon, M. and Webb, E.C. (1979) *Enzymes*, 3rd edn, Longmans, London, UK. A comprehensive treatise on enzymes, covering aspects of kinetics, purification and mechanisms.

Box 4.5
RNA as a catalyst

Until recently it had been assumed that biological catalysts were invariably proteins. Recently it has been demonstrated that RNA is able to act catalytically in the splicing of introns. Also, one species of RNA, designated L19 RNA, has been shown to possess all of the typical properties of an enzyme – except that it is not a protein.

L19 RNA is 395 nucleotides long and is derived from an intron in the rRNA precursor of *Tetrahymena thermophila*. This catalytic RNA or **ribozyme** catalyses both the degradation and synthesis of small oligonucleotides. For instance, the single-stranded oligonucleotide pentacytidylate (C_5), i.e. five cytidine residues (C) joined together by $5'–3'$ phosphodiester bonds, is hydrolysed principally to the corresponding C_4 and C_1 compounds.

$$5'\text{-CCCCC-}3' \rightarrow 5'\text{-CCCC-}3' + 5'\text{-C-}3'$$

$$5'\text{-CCCCC-}3' + 5'\text{-C-}3' \rightarrow 5'\text{-CCCCCC-}3'$$

Sometimes the single residue of cytidine remains covalently attached to the enzyme. In this case a second C_5 oligonucleotide may bind to the enzyme and an addition reaction occurs in which C_1 is attached to C_5 to produce a new molecule.

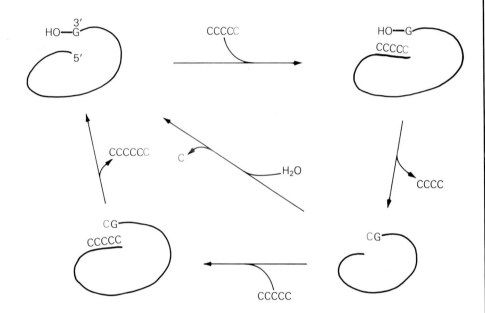

(It is assumed in the figure that the substrate is bound to the L19 RNA by base pairing between nucleotides. The 3' guanosine residue of L19 RNA forms a covalent linkage with the terminal nucleotide of the substrate. Redrawn from Zang, A.J. and Cech, T.R. (1985). The intervening sequence RNA of *Tetrahymena* is an enzyme. *Science*, **231**, 470–5.

L19 RNA shows Michaelis–Menten kinetics: thus, the hydrolysis of C_5 has a K_m of 42 μmol dm^{-3} and a k_{cat} of 0.033 s^{-1}. It is competitively inhibited by the deoxy analogue of the substrate and shows no activity with oligonucleotides of adenosine or guanosine. Thus, L19 RNA shows all the kinetic properties of an enzyme: Michaelis–Menten kinetics, substrate specificity and inhibition by substrate analogues.

the excess glucose will quickly accumulate in the liver, initially as glucose 6-phosphate, and eventually be converted into glycogen.

The beauty of this system is that the excess glucose is quickly removed from the blood to the liver, without the need for hormonal control, simply because of the kinetics of the enzymes involved in the system. Also, under conditions in which the blood glucose concentration is low (hypoglycaemia), organs such as the brain, which depend on glucose as a metabolic fuel, are assured an adequate supply of glucose 6-phosphate through the action of hexokinase.

The effects of temperature on enzyme reactions

All chemical reactions are influenced by temperature. As an approximation the rate of most reactions doubles for each 10°C rise in temperature. The reason for this is that as the temperature increases, so a greater proportion of the reactant molecules have an energy that exceeds the activation energy (E_a) for the reaction. This relationship was quantified by Arrhenius who derived an equation to relate the rate constant (k) of a reaction to the absolute temperature (T). Thus:

$$k = A\, e^{-E_a/RT}$$

where R is the gas constant and A is a constant whose value is dependent on the nature of the reaction.

The rate of an enzyme-catalysed reaction is influenced by temperature according to the Arrhenius equation. However, enzymes consist of relatively unstable molecules and an increase in temperature will also increase the rate of denaturation, with a consequent loss of activity. Therefore, the measured rates of enzyme action at various temperatures will be a combination of both of these factors. Consequently, attempts to measure a temperature optimum of an enzyme should not be made as the results will be dependent on the time over which the reaction is measured (Fig. 4.8). Furthermore, the rate of heat inactivation of an enzyme is often reduced in the presence of substrate or altered by other components in the system, and consequently the analysis becomes even more complicated.

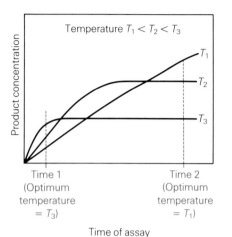

Fig. 4.8 The effect of temperature on enzyme-catalysed reactions. The apparent 'optimum temperature' will depend on the time of assay and therefore is not a suitable parameter to determine.

Box 4.6
Industrial uses of enzymes at high temperatures

Gacesa, P. and Hubble, J. (1987) *Enzyme Technology*, Open University Press, Milton Keynes, UK. An up-to-date account of the application of enzymes for medical and technological uses.

The ability to use enzymes at high temperatures is sometimes of great importance in industrial processes. The liquefaction or partial hydrolysis of starch with α-amylases is the first step in the production of sugar syrups for use in foodstuffs and soft drinks.

The α-amylase from pig pancreas cannot be used effectively above 50°C. This limits its usefulness in the industrial context because of the need to use high temperatures to maintain sterility and to help disrupt the starch granules (a process that occurs between 60 and 90°C). However, α-amylases from other sources are much more stable and one from the bacterium *Bacillus licheniformis* may be used at 110°C.

4.4 Inhibition of enzyme activity

The inhibition of enzymes is of practical importance for several reasons. First, the flux of metabolites through metabolic pathways can be controlled by selective enzyme inhibition. Second, the mechanism of action of many enzymes has been elucidated by the use of **inhibitors**. Third, enzyme inhibitors are used as drugs (Table 4.3), insecticides and chemical weapons.

Table 4.3 *Therapeutic agents that affect specific enzymes*

Type of action	Drug	Enzyme	Medical use	Comments
Substrate	Ethanol	Alcohol dehydrogenase	Methanol poisoning	Excess ethanol prevents the enzyme-catalysed oxidation of methanol
Substrate	Neostigmine	Acetylcholine esterase	Myasthenia gravis	Binds avidly to the enzyme but is hydrolysed very slowly
Inhibitor	Vinyl amino-hexanoate	GABA amino-transferase	Epilepsy	Bind irreversibly to the enzyme
Inhibitor	Edrophonium chloride	Acetylcholine esterase	Myasthenia gravis	Reversible inhibition causes transient (5–10 min) alleviation of symptoms; used to diagnose the disorder
Inhibitor	Demecarium bromide	Acetylcholine esterase	Glaucoma	Irreversible inhibition; a single dose may have effects for 14 days
Inhibitor	Pargyline HCl	Monoamine oxidase	Antidepressant	Suicide substrate
Inhibitor	Methotrexate	Dihydrofolate reductase	Cancer therapy	Tight-binding competitive inhibitor

The activity of enzymes can be inhibited by a variety of molecules, usually but not always small ones. Although there are many examples of inhibitors, the mechanisms of inhibition are conveniently divided into two major classes, **reversible** and **irreversible**. This widely used distinction, is to some extent artificial. There is really a continuous spectrum of inhibitor affinities for enzyme molecules.

Irreversible inhibition of enzymes

The definition of an irreversible inhibitor is one that is bound so tightly to an enzyme that the rate of dissociation of the complex is negligible. Often the inhibitor binds covalently to an amino acid side-chain. For example, the nerve gas, **diisopropylphosphofluoridate (DIPF)** reacts readily with the hydroxyl group of the serine residue in the active site of the enzyme acetylcholine esterase (Fig. 4.9) inhibiting it and thus compromising normal synapse function. However, this type of inhibitor will also react with any other enzyme that contains an active-site serine, and so it is relatively non-specific. Trypsin, chymotrypsin and a number of other proteinases have an active-site serine residue.

A more specific type of irreversible inhibitor is the 'suicide substrate'. Such a compound acts as a *substrate analogue* and will bind only to the *active site* of the appropriate enzyme. Once the suicide substrate is bound at the active site it will be activated by the normal catalytic process to form a reactive species that irreversibly modifies an active-site amino acid. An example of a suicide substrate is the anticancer drug **fluorouracil** which inhibits the enzyme **thymidylate synthetase** (Box 4.7).

Irreversible inhibitors have also been used successfully to determine which amino acid residues are present at the active site of an enzyme. This aspect is dealt with in more detail in Chapter 5. It is possible to use irreversible

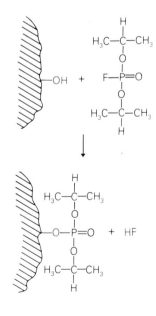

Fig. 4.9 The inactivation of acetylcholine esterase by diisopropylphosphofluoridate. The reagent will covalently modify the active-site serine residue and irreversibly inhibit the enzyme.

Substrate analogue: a compound with a similar chemical structure to the true substrate.
Active site of an enzyme: the part of the enzyme molecule to which the substrate binds and where catalytic processes occur.

Box 4.7
Anticancer drugs as suicide inhibitors of enzymes

Several of the drugs that have been developed for the treatment of cancer cells inhibit specific enzymes. For example, the drug fluorouracil (fluorodeoxyuridine) inhibits the enzyme thymidylate synthetase, a key enzyme for dTMP synthesis (see *Biosynthesis*, Chapter 10). Fluorouracil is converted in the body to fluorodeoxyuridylate. This compound acts as a normal substrate during the first part of the catalysis by thymidylate synthetase. However, once the enzyme–substrate complex is formed the fluorodeoxyuridylate becomes irreversibly bound to the active-site sulphydryl group (highlighted in red below) of the enzyme thus causing inactivation and loss of catalytic activity.

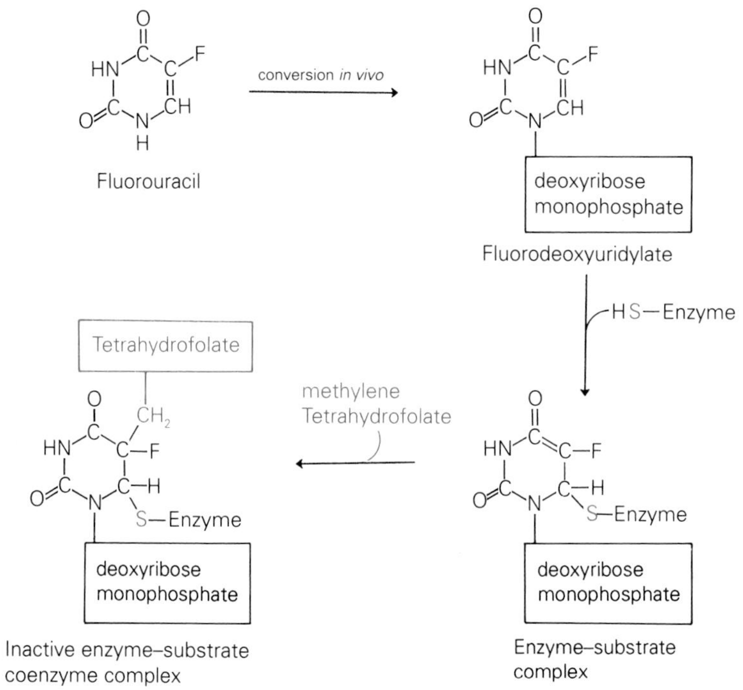

Fluorouracil

conversion *in vivo*

deoxyribose monophosphate

Fluorodeoxyuridylate

HS—Enzyme

Tetrahydrofolate

methylene Tetrahydrofolate

deoxyribose monophosphate

Inactive enzyme–substrate coenzyme complex

deoxyribose monophosphate

Enzyme–substrate complex

Box 4.8
Chemical weapons as enzyme inhibitors

One of the earlier chemical weapons developed in the USA during World War I was lewisite, a volatile arsenic-containing compound developed by W. Lee Lewis:

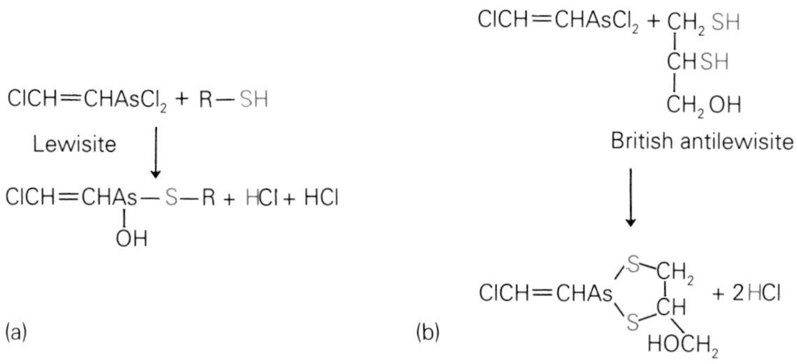

$ClCH{=}CHAsCl_2 + R{-}SH$

Lewisite

$ClCH{=}CHAs{-}S{-}R + HCl + HCl$
$|$
OH

(a)

$ClCH{=}CHAsCl_2 + CH_2\,SH$
$|$
$CHSH$
$|$
$CH_2\,OH$

British antilewisite

$ClCH{=}CHAs\begin{smallmatrix}S{-}CH_2\\S{-}CH\\\ \ HOCH_2\end{smallmatrix} + 2\,HCl$

(b)

This compound reacts readily with sulphydryl groups at the active site of enzymes such as glyceraldehyde 3-phosphate dehydrogenase. However, the compound was of limited use because of its susceptibility to hydrolysis (it cannot be used on a damp day) and because of the development of an effective antidote called British Anti-Lewisite or BAL.

inhibitors to titrate the quantity of active sites present in an enzyme preparation since in principle 1 mole of inhibitor reacts with '1 mole' of active sites.

Reversible inhibition of enzymes

Reversible inhibitors dissociate freely from the enzyme molecules. There are several kinetically distinct types of reversible inhibition of which the two extreme types are **competitive** and **non-competitive**. Here again the distinction is somewhat artificial. In practice, many inhibitors behave in an intermediate fashion.

COMPETITIVE INHIBITORS are structurally similar to substrate molecules and will bind to the active site of the enzyme molecule. Therefore, at any one time either a competitive inhibitor or a substrate molecule may be bound at the active site of the enzyme. As this binding is reversible it follows that if the substrate concentration is increased then most of the enzyme molecules will be present in the form of an ES complex, whereas if the inhibitor concentration is increased then the EI (enzyme–inhibitor) complex will predominate:

$$EI \rightleftharpoons E \rightleftharpoons ES \rightarrow E + P$$
$$[I] \gg [S] \quad [S] \gg [I]$$

A consequence of this is that such inhibition can be overcome by increasing the substrate concentration.

It is a simple matter to determine the equilibrium constant (K_i) for the reaction:

$$E + I \underset{k_2}{\overset{k_1}{\rightleftharpoons}} EI$$

it is $K_i = k_2/k_1$ which gives a measure of the affinity of binding of the inhibitor to the enzyme. As the inhibitor binds to the same site on the enzyme as the substrate it will only be K_m that is affected and not V_{max}. Therefore, if initial rates are measured at a variety of substrate concentrations in the presence and absence of inhibitor, K_i may be determined from the change in value of K_m. The apparent K_m will increase by the factor $(1 + [I]/K_i)$ and may be determined from direct linear plots or by other means (Fig. 4.10; Table 4.4).

Fig. 4.10 (a) A direct linear plot showing the effect of competitive inhibition. Initial rates are obtained in the absence (black) and in the presence (red) of inhibitor. (b) A direct linear plot showing the effect of non-competitive inhibition. Initial rates are obtained in the absence (black) and in the presence (red) of inhibitor.

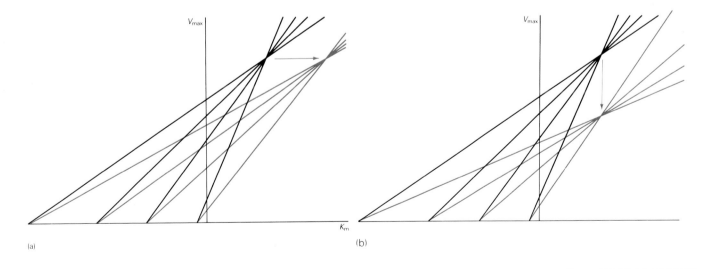

(a)   (b)

Box 4.9
Competitive inhibitors of succinate dehydrogenase

The enzyme succinate dehydrogenase catalyses the oxidation of succinate to fumarate.

$$
\begin{array}{c}
\text{COO}^- \\
| \\
\text{CH}_2 \\
| \\
\text{CH}_2 \\
| \\
\text{COO}^-
\end{array}
\; + \; \text{hydrogen acceptor}
\;\rightleftharpoons\;
\begin{array}{c}
\text{COO}^- \\
| \\
\text{CH} \\
|| \\
\text{CH} \\
| \\
\text{COO}^-
\end{array}
\; + \; \text{reduced acceptor}
$$

Succinate   Fumarate

Both of these compounds are dicarboxylic acids. Many of the competitive inhibitors of succinate dehydrogenase are also dicarboxylic acids with structures analogous to those of the substrate and products.

$$
\begin{array}{c}
\text{COO}^- \\
| \\
\text{CH}_2 \\
| \\
\text{COO}^-
\end{array}
\qquad
\begin{array}{c}
\text{COO}^- \\
| \\
\text{COO}^-
\end{array}
\qquad
\begin{array}{c}
\text{COO}^- \\
| \\
\text{CH}_2 \\
| \\
\text{C}=\text{O} \\
| \\
\text{COO}^-
\end{array}
$$

Malonate   Oxalate   Oxaloacetate

Table 4.4 *The effect of some types of inhibition on the value of the kinetic constants*

Type of inhibition	Michaelis constant	Maximum velocity
None	K_m	V_{max}
Competitive	$K_m(1 + [I]/K_i)$	V_{max}
Non-competitive	K_m	$V_{max}/(1 + [I]/K_i)$

Enzymes are often competitively inhibited by the **products** of the reactions that they catalyse. This might be predicted, as the structure of the product will be closely related to that of the substrate. Product inhibition will occur in the later stages of the reaction when product concentration is relatively high and substrate concentration is comparatively low.

NON-COMPETITIVE INHIBITORS also bind reversibly to the enzyme but at a site **separate** from that of the substrate. Therefore, the inhibitor will be able to bind regardless of whether or not substrate is bound to the enzyme. Consequently, increasing the substrate concentration will not diminish the extent of inhibition. A non-competitive inhibitor is able to bind to either the free enzyme or the enzyme–substrate complex to produce the species EI and EIS, respectively. Both EI and EIS are inactive and will not form product. The kinetic effect of non-competitive inhibition is that V_{max} will be altered but K_m will remain unchanged (Table 4.4). Therefore, the value of K_i may be determined readily from direct linear plots using data in the absence and presence of a known concentration of inhibitor (Fig. 4.10). Non-competitive inhibition is observed rarely and it is much more common to see mixed forms of inhibition.

It is not uncommon for various types of **mixed inhibition** to occur. In many cases these can be rather difficult to analyse and sometimes it may be impossible to ascertain the precise mechanisms of inhibition.

Exercise 5

Draw a Lineweaver–Burk plot showing the relationship between initial velocity and substrate concentration for an enzyme in the absence and presence of (a) a competitive inhibitor and (b) a non-competitive inhibitor.

4.5 Control of enzyme activity

In a physiological context it is important that the activities of enzymes are controlled. Many enzymes are regulated by molecules unrelated in structure to the substrate or the product. Typically, these **allosteric effectors** (they bind to sites independent of the active site) are end-products of metabolic pathways or molecules that reflect the 'energy status' of the cell, that is to say the relative concentrations of ATP, ADP and AMP. The attenuation of the catalytic activity of an enzyme by the final product of a metabolic pathway is called **feedback inhibition**. In general, it is only those enzymes at the start of metabolic pathways, or those located at strategic branch-points, that will be regulated by allosteric effectors. The other enzymes of a pathway will be regulated simply because the supply of available substrate is limited. In practice most enzymes in a cell will be substrate-limited for much of the time.

Another feature of allosteric enzymes is that they often operate in pairs. In a metabolic pathway such as that connecting glucose to the tricarboxylic acid (TCA) cycle the balance between glycolysis (degradation of glucose) and gluconeogenesis (formation of glucose) is controlled, at least in part, by the two enzymes **phosphofructokinase**:

$$\text{fructose 6-phosphate} + \text{ATP} \rightarrow \text{fructose 1,6-bisphosphate} + \text{ADP} + \text{Pi}$$

and **fructose-1,6-bisphosphatase**

$$\text{fructose 1,6-bisphosphate} \rightarrow \text{fructose 6-phosphate} + \text{Pi}$$

These enzymes catalyse different and essentially irreversible reactions. When the allosteric effector AMP binds to phosphofructokinase the enzyme is activated, whereas the converse is true for fructose-1,6-bisphosphatase. The fluxes through glycolysis and gluconeogenesis depend on the concentration of AMP and are therefore related to the energy needs of the cell.

The activity of enzymes may also be modified by covalent modification, particularly phosphorylation. Many enzymes, such as those concerned with glycogen metabolism become phosphorylated and are subsequently dephosphorylated under the indirect influence of hormones.

Models of allosteric enzymes

Allosteric enzymes are usually complex multisubunit proteins (Fig. 4.11). They have independent binding sites for the substrate(s) and for effector molecules. The binding of an activator or inhibitor to the allosteric site inevitably causes a conformational change in the protein structure and this change in conformation alters the activity of the enzyme. The flexibility of enzyme molecules is a vital aspect of allosteric control.

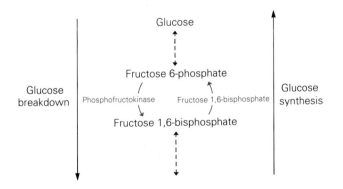

□ Enzymes can be controlled by allosteric effectors or by covalent modification, but basically they can only be turned on or turned off. In a metabolic pathway that is required to operate in either direction depending upon physiological circumstances, control of a single enzyme will not work. If the enzyme is 'on' then both pathways operate, if 'off' neither. For example, a well-fed animal should not be breaking glucose down but turning it into glycogen. Conversely, a starving animal should be breaking glucose down to supply metabolic energy, not storing it.

The way to achieve control is to have two enzymes rather than one, catalysing a particular step in the pathway, one for biosynthesis and one for degradation. Each needs to be controlled, and usually this is done by the same effectors. In the case of glucose breakdown (glycolysis) and glucose synthesis (gluconeogenesis), the two enzymes are phosphofructokinase and fructose 1,6-bisphosphatase. Turning one of them off allows the other to run, and *vice versa*, thus controlling the pathway. AMP stimulates glucose synthesis and inhibits glucose breakdown, but many other compounds also affect the activities of these two enzymes.

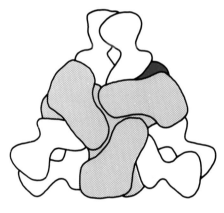

Fig. 4.11 The structure of aspartate trans-carbamoylase. This is a multisubunit enzyme that shows allosteric behaviour. The catalytic subunits are shown in black and the control subunits are shown in red. (Redrawn from Krause, K.L. *et al.* (1985) *Proceedings of the National Academy of Sciences,* **82**, 1643.

□ In solution, in cells, enzymes exist in a flexible dynamic state just on this side of stability. They are not far away from being denatured. This instability is a significant feature of their function so that induced fit and allosteric changes can occur. An illustration of the metastable, flexible state is trypsin:

$$\text{trypsin (native)} \rightleftharpoons \text{trypsin (denatured)}$$

At 37°C, K_{eq} for this equilibrium is about 0.1, 10:1 in favour of the native state. At 47°C it is about 2.5, favouring the denatured state slightly.

Covalent modification: *the adding of an additional group to the enzyme molecule by forming a covalent chemical bond. Alternatively, a covalent bond may be broken to remove part of the molecule.*

Reference Monod, J., Wyman, J. and Changeaux, J.-P. (1965) *Journal of Molecular Biology* **12**, 88–118. A classical paper proposing the concerted model of allosteric inhibition.

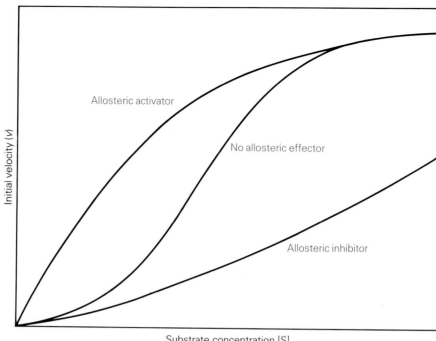

Fig. 4.12 The relationship between initial velocity and substrate concentration for an allosteric enzyme. Note that the presence of activator or inhibitor alters the shape of the curve.

Almost all allosteric enzymes show atypical kinetic behaviour and bind the substrate in a co-operative manner. These enzymes have multiple active sites and the binding of the first substrate molecule either encourages (**positive co-operativity**), or discourages (**negative co-operativity**), the binding of subsequent molecules of substrate. This phenomenon is analogous to the binding of oxygen to haemoglobin. The relationship between the initial velocity, v, and substrate concentration, [S], is **sigmoidal** rather than hyperbolic (Fig. 4.12). This is called a **homotropic effect.** Superimposed upon this is the ability of allosteric effectors to modify the degree of co-operativity and this is called the **heterotropic effect.** This sigmoidal relationship between substrate concentration and velocity is necessary for the efficient control of an enzyme. Very small changes in substrate concentration will result in enzyme activity being effectively switched on or off. In contrast, if the enzyme was displaying normal Michaelis–Menten kinetics then a comparatively large change in substrate concentration would be needed to cause a significant change in initial velocity.

There is no single kinetic model of co-operativity that is satisfactory for all cases. The Hill equation, originally developed to analyse the binding of O_2 to haemoglobin, has been used to describe the interactions of substrates with allosteric enzymes. However, its use has proved of limited value.

☐ The co-operative binding of a substrate to an allosteric enzyme may be described by the Hill equation:

$$\log (v/V_{max} - v) = x \log S - \log k$$

where k is a constant for a given system and x an interaction factor. The interaction factor, x, has a value of 1.0 in non-co-operative binding. In co-operative binding x would have a higher value of n, which was originally taken to be equal to the number of substrate binding sites per molecule of enzyme. In practice, the experimentally measured value of x is usually less than n, and does *not* therefore give an estimate of the actual number of binding sites.

4.6 Overview

Enzymes greatly enhance the rate of chemical reactions and also show remarkable selectivity for the reactants. The kinetics of many enzyme-catalysed reactions are described by the Michaelis–Menten equation. Kinetic constants are derived experimentally using a high concentration of substrate relative to enzyme. However, the values of K_m and V_{max} describe the relationship between initial velocity and substrate concentration of an

Reference Cohen, P. (1983) *Control of Enzymic Activity*, 2nd edn, Chapman and Hall, London, UK. A relatively cheap, short introduction to the topic.

enzyme-catalysed reaction at any concentration of substrate. In the cell, many enzymes are substrate-limited.

Inhibitors have been used both to gain information about the mechanisms of enzymes and as compounds of therapeutic or industrial use. Reversible enzyme inhibition is an important mechanism for the control of metabolic pathways.

Answers to Exercises

1. $K_{eq} = 10$.
2. Detergents may contain proteinases, e.g. subtilisin, lipases and α-amylases. These enzymes are present to remove 'biological stains'.

Toothpastes may contain amyloglucosidase and glucose oxidase. The former enzyme degrades starch to glucose which is used by the second enzyme to generate H_2O_2, a potent antibacterial agent.

Beer often contains proteinases to remove the haze (i.e. a fine protein precipitate) that sometimes occurs during cold storage.

Soft-centred chocolates contain invertase to convert crystalline sucrose, which is in the centre of the sweet, into a glucose/fructose syrup.

3. The units are rate (s^{-1})/concentration $(mol\ dm^{-3})$, i.e. $dm^3\ mol^{-1}\ s^{-1}$.
4. Michaelis–Menten equation:

$$v = \frac{V_{max}[S]}{K_m + [S]}$$

Take reciprocals of both sides:

$$\frac{1}{v} = \frac{K_m}{V_{max}}\frac{1}{[S]} + \frac{1}{V_{max}}$$

which gives the Lineweaver–Burk plot.

Multiply both sides of the equation by $[S]$:

$$\frac{[S]}{v} = \frac{K_m}{V_{max}} + \frac{[S]}{V_{max}} \qquad (1)$$

Multiply by V_{max} and rearrange:

$$v = -K_m\frac{v}{[S]} + V_{max} \qquad (2)$$

Both Equations 1 and 2 are suitable linear transformations that will allow the graphical determination of the kinetic constants.

With Equation 1 a plot of $[S]/v$ against $[S]$ is constructed. The slope of the graph is $1/V_{max}$ and the ordinate intercept is K_m/V_{max}.

With Equation 2 a plot of v against $v/[S]$ is constructed. The slope of the graph is $-K_m$, the abscissa intercept is V_{max}/K_m and the ordinate intercept is V_{max}.

5. (a) Competitive inhibition
   (b) Non-competitive inhibition

QUESTIONS

FILL IN THE BLANKS

1. Enzyme catalysis does not alter the _____ constant of a reaction but does reduce the _____ _____ barrier.

The Michaelis constant has dimensions of _____ _____ and represents the concentration of _____ at which the velocity is _____ _____ . The constant k_{cat} is derived from the maximum velocity by accounting for the total concentration of

_____ .

The value of k_{cat}/K_m represents the _____ order rate constant for the reaction _____ and is a measure of the _____ of an enzyme.

Choose from: activation energy, enzyme, equilibrium, half V_{max}, second, specificity, substrate, substrate concentration, $E+S \rightarrow E + P$

MULTIPLE-CHOICE QUESTIONS

2. State which of the following are correct.
A. Enzymes reduce the free energy change of a reaction.
B. Enzymes are always proteins.

C. Enzymes only catalyse reversible reactions.
D. Enzymes can be inhibited by substrate analogues.
E. Enzymes have no effect upon the equilibrium constant of a reaction.

3. State which of the following statements are correct.
A. Competitive inhibition results in a decrease in the K_m but not in the V_{max}.
B. Competitive inhibition does not affect the K_m.
C. Non-competitive inhibition results in an decreased V_{max} and increased K_m.
D. The activity of allosteric enzymes plotted against substrate concentration is a rectangular hyperbolum.
E. K_m values always have the dimensions of concentration.

SHORT-ANSWER QUESTIONS

4. A dipeptidase enzyme can hydrolyse two dipeptides, A and B, with k_{cat} and K_m values of $25\,s^{-1}$ and $1.5\,mmol\,dm^{-3}$ and $90\,s^{-1}$ and $5.5\,mmol\,dm^{-3}$ respectively. Which substrate is used most effectively?

5. Estimate the values of K_m and V_{max} (using a direct linear plot) for an enzyme-catalysed reaction using the data given below.

Substrate concentration (mmol dm^{-3})	Initial velocity (μmol min^{-1})
0.5	0.85
2.0	2.65
3.5	3.85
4.5	4.60
7.0	5.30

6. The initial velocities of an enzyme-catalysed reaction were measured in the presence and absence of an inhibitor. The data for these measurements are given below.

[S] (mmol dm^{-3})	Initial velocity (μmol min^{-1})	
	$-$Inhibitor	$+$Inhibitor (10 mmol dm^{-3})
5	13.0	9.5
10	18.9	14.9
15	22.3	18.4
20	24.6	20.9
30	27.3	24.1

A. Determine (using the Lineweaver–Burk plot) the K_m and V_{max} for the uninhibited reaction.
B. Is the inhibition competitive or non-competitive?
C. What is the K_i for the binding of the inhibitor to the enzyme? (Hint: consult Table 4.4.)

Enzyme mechanisms

Objectives

After reading this chapter you should be able to:

□ explain how enzymes catalyse reactions;

□ describe the basis of enzyme specificity;

□ outline how reaction mechanisms are elucidated;

□ explain why enzymes have pH optima;

□ describe the concept of enzyme families;

□ outline the concept of zymogen activation.

5.1 Introduction

The description of enzyme catalysis in kinetic terms was a relatively early development in the appreciation of biological catalysis. The major papers describing the essentials of enzyme kinetics were published within the first 25 years of this century. In contrast, the detailed chemical mechanisms of catalysis are still only understood for a fairly small proportion of enzymes. Although considerable insight into an enzyme mechanism can be gained by the use of substrate analogues and inhibitors, a definitive understanding requires detailed knowledge of the primary structure and conformation of the amino acid residues. Currently, the only way of obtaining detailed information about the three-dimensional structure of an enzyme is to use

---

Box 5.1
Synchrotron radiation

Synchrotron radiation is the electromagnetic radiation generated as a result of the acceleration in an electric field of charged particles, usually electrons, to velocities approaching the speed of light. Part of this radiation consists of a highly focused, intense beam of polarized X-rays which may be used for the analysis of crystal structures.

A major problem with X-ray crystallography is that irradiation of the protein over prolonged periods (hours) is required in order to collect the diffraction data and this causes damage to the crystal structure. The use of higher intensity radiation over a shorter period would be less destructive. The availability of high-intensity X-rays from the synchrotron means that data concerning protein structure may be obtained over very short periods, thus minimizing radiation damage. Furthermore, the broad-wavelength spectrum of synchrotron-derived X-rays allows much more information about the crystal structure of the protein to be obtained than is possible with conventional sources. Unfortunately, synchrotron X-ray facilities are available at only a few places in the world.

Box 5.2
*Two-dimensional nuclear
magnetic resonance*

When placed in a constant magnetic field certain atomic nuclei absorb energy from a radio-frequency field at characteristic frequencies. The frequency of absorption is dependent on both the type of atom involved and the influence of other neighbouring atoms. For example, a proton covalently bonded to a hydrocarbon chain will exhibit a different frequency of absorption to one bonded to a nitrogen atom. Therefore, using the technique of 'proton nuclear magnetic resonance' it is possible to determine to which atoms various protons are attached.

Two-dimensional nuclear magnetic resonance is a more sophisticated version of this technique in which it is possible to determine, for particular protons, both the atoms to which they are covalently attached (through-bond connectivities) and the atoms that are close-by but not directly connected by covalent bonds (through-space connectivities). Consequently, it is possible to gain information about the structure of proteins and the possible conformations of the macromolecule in solution.

See Section 8.13

X-ray crystallography. The difficulty inherent in obtaining suitable crystals of pure enzyme has impeded progress in this area. Consequently, it is advances in protein crystallography and the analysis of three-dimensional structure using newer X-ray sources, e.g. synchrotron radiation, that will do most to facilitate this work. Other techniques, such as two-dimensional proton nuclear magnetic resonance, are proving useful for solution studies. Molecular genetic techniques, e.g. alteration of a selected codon in the gene encoding the enzyme by site-directed mutagenesis, are allowing specific changes to be made to the primary structure of enzymes.

This chapter will largely concentrate upon the types of reactions catalysed by enzymes. The binding of substrates to enzymes, and the mechanisms of catalysis will be illustrated by describing the activities of selected representative enzymes.

5.2 *Enzyme structure and classification*

There is an enormous diversity of enzymes both in terms of structure and the reactions that are catalysed. Enzymes range in M_r from approximately 13 700 for ribonuclease, which has one active site, to several millions and more for multienzyme complexes that catalyse several discrete reactions. An example is yeast fatty acid synthetase, M_r 2.3×10^6 which is made up of 12 subunits and catalyses five partial reactions. Nevertheless, all of these enzymes utilize the properties of relatively few amino acid side-chains to effect catalysis (Table 5.1) although this small chemical repertoire is enhanced by the involvement of coenzymes or prosthetic groups.

It is useful at this stage to consider how enzymes are classified as this will help to put into perspective the types of reactions that are catalysed. Enzymes were classified by the Enzyme Commission of the International Union of Biochemistry into six major classes (Table 5.2). Each of the enzymes within a class catalyses a related reaction and therefore it is not unusual to find common mechanisms of catalysis although the various substrates may be very different from each other. For instance, all enzymes designated as Class 1 are involved in oxidation/reduction reactions. However, within that class there are subdivisions; for instance, one subgroup of enzymes utilize NAD^+ as the oxidizing agent whereas another subgroup uses molecular oxygen. Nevertheless, within these groups and subgroups a common theme of mechanism tends to emerge and some of these will be discussed later in this Chapter.

☐ Using the Enzyme Commission system of classification, enzymes are given a four-letter 'E.C.' number. Each number describes the reaction catalysed by a specific enzyme. For example, trypsin has the number E.C. 3.4.21.4. This denotes the following:

   3: hydrolase group enzymes
   4: acts on peptide bonds
   21: serine proteinase
   4: specific number designating trypsin activity

Reference Fersht, A. (1985) *Enzyme Structure and Mechanism*, 2nd edn, W.H. Freeman & Co., New York, USA. A thorough and comprehensive treatment of the study of enzyme structure and mechanisms.

Table 5.1 *Amino acid side-chains involved in catalysis*

Amino acid	Reactive group		Enzyme
Serine	Hydroxyl	–OH	Serine proteinases (e.g. trypsin) Acetylcholinesterase
Cysteine	Thiol	–SH	Thiol proteinases (e.g. papain) Glyceraldehyde-3-phosphate dehydrogenase
Aspartate	Carboxylate	–COO⁻ or –COOH	Lysozyme ATPase Acid proteinases (e.g. pepsin)
Glutamate	Carboxylate	–COOH	Lysozyme
Lysine	Amino	–NH₂ or –NH₃⁺	Aldolase Pyridoxal-dependent enzymes (e.g. transaminases)
Histidine	Imidazole	–C–NH ∥ ＼CH –C–NH	Phosphoglycerate mutase Malate dehydrogenase
Tyrosine	Hydroxyl	–OH	Topoisomerase

Table 5.2 *Enzyme Commission of the International Union of Biochemistry classification of enzymes*

Class	Type of reaction catalysed
1 **Oxidoreductases**	Synthesis of compounds via oxidative or reductive cleavage of a high-energy bond e.g. alcohol dehydrogenase alcohol + $NAD^+ \rightarrow$ aldehyde or ketone + NADH + H⁺
2 **Transferases**	Transfer of functional group from one molecule to another e.g. aspartate aminotransferase L-aspartate + 2-oxoglutarate $\rightarrow$ 2-oxaloacetate + L-glutamate
3 **Hydrolases**	Cleavage of bonds by hydrolysis e.g. acetylcholinesterase acetylcholine + $H_2O \rightarrow$ choline + acetate
4 **Lyases**	Cleavage of bonds by elimination e.g. pyruvate decarboxylase 2 oxoacid $\rightarrow$ aldehyde + CO_2
5 **Isomerases**	Change in the shape or spacial arrangement of molecules e.g. phosphoglycerate mutase 2-phosphoglycerate $\rightarrow$ 3-phosphoglycerate
6 **Ligases**	Joining of molecules using energy derived from the hydrolysis of high-energy bonds e.g. acetyl CoA ligase ATP + acetate + CoA $\rightarrow$ AMP + pyrophosphate + acetyl CoA

5.3 The active site

The active site of an enzyme is the region in which the substrate(s) and coenzymes (if any) are bound and where catalysis occurs. Active sites have several important features that are common to all enzymes. For example, they are situated away from the surface of enzyme molecules, in other words substrates typically are bound within clefts or pockets in the enzyme (Fig. 5.1). A major reason for this is that 'unwanted' potential reactants, especially water, can be prevented from binding and participating in side reactions. The second reason is that multiple interactions between parts of the enzyme and the substrate are essential if the catalysis is to be specific. Clearly, a substrate

---

Box 5.3
Coenzymes

Many enzymes only function if they contain a non-protein component or **cofactor** which may be a metal ion, or an organic molecule called a **coenzyme**.

Coenzymes function as carriers of small chemical groups, such as acetyl, carboxyl or methyl, or they may carry electrons, or protons *and* electrons. The combination of protein and coenzyme is called a ***holoenzyme***: if the coenzyme is removed the protein part is referred to as the **apoenzyme**. Coenzymes enable holoenzymes to carry out functions that could not be performed by the apoenzyme. For example, none of the amino acid side-chains in proteins is capable of carrying electrons, but the addition of a coenzyme such as FAD confers this function. Coenzymes that are very tightly bound to the apoenzyme are called ***prosthetic groups***.

Coenzymes function as acceptors or donors of functional groups or atoms removed from or added to a substrate by an enzyme. They are not consumed in the course of a reaction but are used again and again. This is because the group they carry does not become a permanent part of the coenzyme but is subsequently transferred to an acceptor molecule. Many coenzymes are derivatives of water-soluble ***vitamins***. These are not stored to any extent and are required constantly in the diet. The accompanying table lists the coenzyme derivatives of some water-soluble vitamins, together with their functions and common abbreviations. Because coenzymes play a central role in metabolism, a shortage of them typically leads to a range of apparently unconnected symptoms. For example, a lack of thiamin leads to **beri beri**, niacin, **pellegra** and folic acid **megablastic anaemia**. Although plants and microorganisms are the principal dietary sources of coenzymes for animals, it should be remembered that the coenzymes play an identical role in these organisms.

The water-soluble vitamins, their corresponding coenzymes and their functions

Vitamin	Coenzyme derivative	Abbreviations	Function
Thiamin (B_1)	Thiamin pyrophosphate	TPP	Decarboxylation and acyl transfer
Riboflavin (B_2)	Flavin mononucleotide	FMN	Hydrogen and electron carriers in oxidation/ reduction
	Flavin adenine dinucleotide	FAD	
Nicotinic acid	Nicotinamide adenine dinucleotide	NAD^+	Hydrogen and electron carriers in oxidation/ reduction
	Nicotanamide adenine dinucleotide phosphate	$NADP^+$	
Pyridoxine, pyridoxal and pyridoxamine (B_6)	Pyridoxal phosphate		Transamination and decarboxylation
Pantothenic acid	Coenzyme A	CoASH	Acyl transfer
Biotin	Covalently attached to carboxylases		Carboxylation
Folic acid	Tetrahydrofolate	TH_4	One-carbon transfer
Cobalamin (B_{12})	Cobamide coenzymes		Rearrangements, methyl transfer

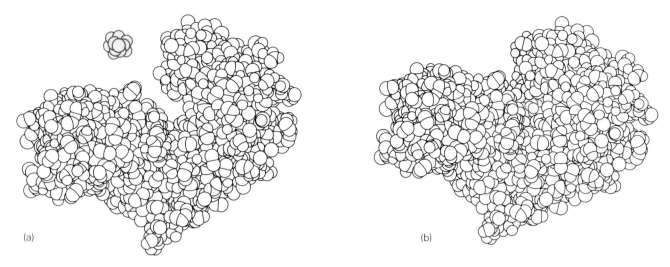

(a)

(b)

Fig. 5.1 A space-filling model of hexokinase illustrating the conformational change in the overall protein structure that occurs on binding of the substrate, glucose (shown in red). Redrawn from Bennett, W.S. and Steitz, T.A. (1980) *Journal of Molecular Biology*, **140**, 211–30.

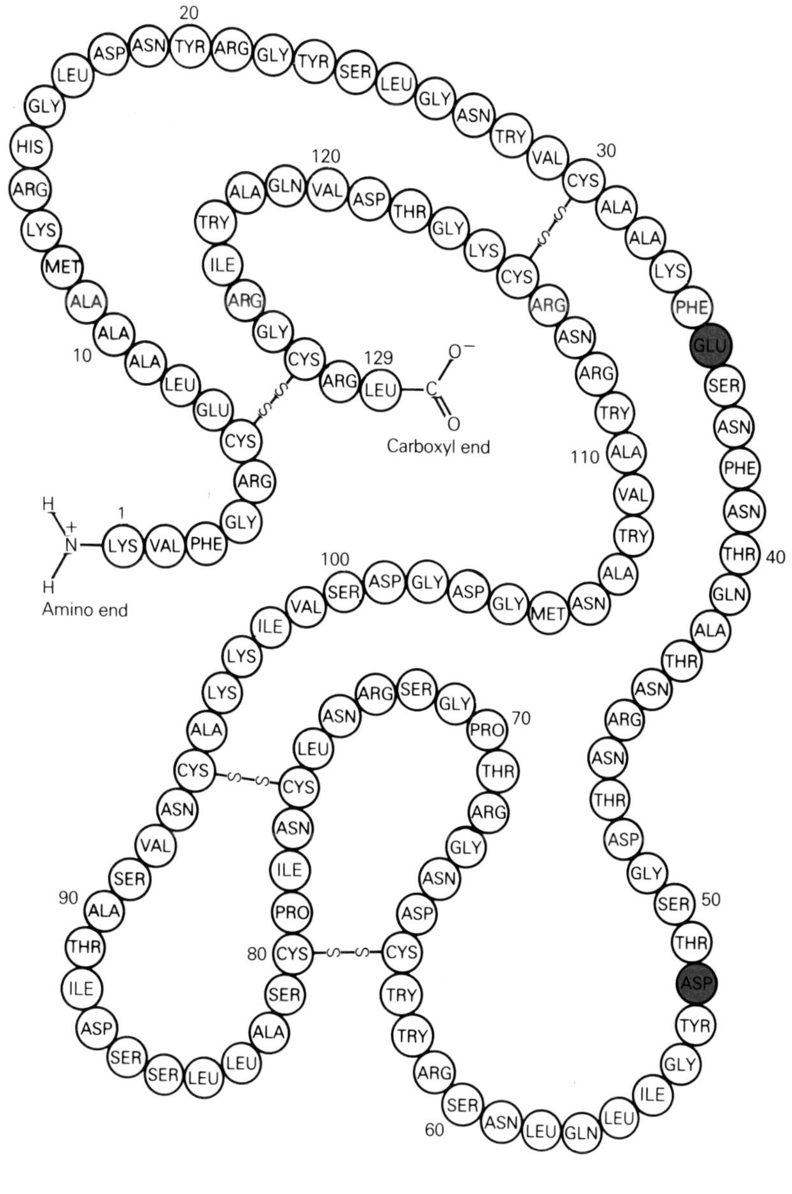

Fig. 5.2 The primary sequence of lysozyme. Amino acid residues are indicated by circles. Those residues associated with the binding of substrate are shaded and the two amino acids directly involved in catalysis are shown in red. Redrawn from Phillips, D.C. (1966) *Scientific American*, **215**(5), 78–90.

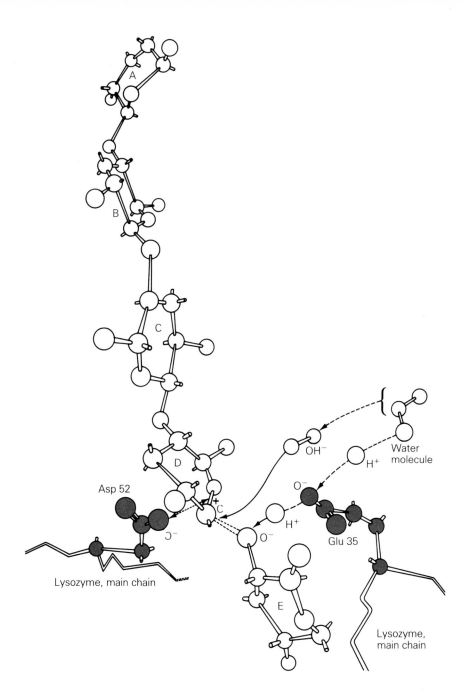

Fig. 5.3 The active site of hen egg white lysozyme. Note the close spatial proximity of the active-site amino acids (shown in red) although they are well separated in terms of primary sequence. The mechanism of catalysis is explained in Section 5.5. Redrawn from Phillips, D.C. (1966) *Scientific American*, **215**(5), 78–90.

bound deep within an enzyme molecule will be able to experience maximal interactions with the active site residues. In addition, the binding properties of groups at the active site may be modified by the surrounding micro-environment. For instance, a hydrophobic environment will stabilize ionic interactions between enzymes and substrates.

Substrates are bound non-covalently to enzymes, although transient covalent linkages may be formed during catalysis. Consequently, the association between substrate and enzyme is relatively weak with the energy stabilizing the interaction ranging from only -10 to -50 kJ mol^{-1}. In comparison the free energy of covalent bonds ranges from -200 to -460 kJ mol^{-1}.

Although enzymes are macromolecules, only a relatively small part of the protein interacts directly with the substrate. The side-chains of as few as five or six amino acid residues within an enzyme may participate in the binding of substrate and in the catalytic reaction (see Table 5.1 for a list of those residues commonly found in active sites). Other amino acids will be involved in a less direct way by creating hydrophobic domains or by having bulky side-chains which may prevent substrate-related molecules from entering the active site. For example, lysozyme from hen egg white is a single polypeptide 129 amino acid residues long in which 19 amino acids are involved in the active site. Yet only two of these, Glu-35 and Asp-52, are directly involved in the cleavage of the glycosidic bond of the substrate (Fig. 5.2). The amino acids Glu-35 and Asp-52 are about 16 residues apart in the primary structure of lysozyme but they are brought into close spacial proximity to each other and to the glycosidic bond of the substrate by the folding of the polypeptide (Fig. 5.3).

The specificity of an enzyme for its substrate is one of the most notable features of biological catalysis and as far back as 1890 Fischer attempted to explain this phenomenon. He proposed that the shape of the active site was complementary to that of the substrate. This was called the **lock and key** model (Fig. 5.4). Although this model has proved to be very useful, however, it suffers from the disadvantage that it implies a rigid enzyme structure. Since that time it has emerged that enzymes are flexible molecules. Therefore, it is more correct to think in terms of an **induced fit** of substrate and enzyme (Fig. 5.5). However, in both cases the structures of active site and substrate are complementary in the transition state.

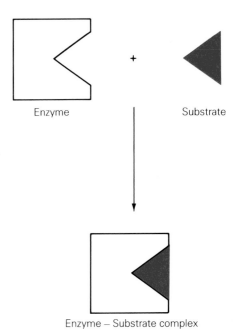

Fig. 5.4 The lock-and-key model of substrate binding. The shapes of the active site of the enzyme (black) and of the substrate (red) are complementary to each other.

5.4 Specificity of enzymes

The remarkable specificity of enzymes and the minimization of side reactions is achieved by a combination of different factors, considered in more detail below.

Binding of substrates

The ability of enzymes to discriminate between substrates and other molecules is a crucial aspect governing specificity. In many cases the differences between substrates and 'non-substrates' are so slight that it is difficult to appreciate how selection occurs. A key to the solution lies in the way that enzymes and substrates are bound to each other. Invariably, non-covalent interactions are involved and arguably the most significant of these, in terms of enzyme specificity, is hydrogen bonding.

Hydrogen bonds are directional and it is this feature that is of importance at the active site. How hydrogen bonding between enzyme and substrate is crucial to specificity is, perhaps, best illustrated by considering the restriction endonuclease *Eco*RI. This enzyme recognizes a specific hexanucleotide sequence of double-stranded DNA and catalyses a single cleavage in each strand (Fig. 5.6). This sequence is palindromic about a point of rotational symmetry and both strands of DNA are cut in identical positions. Perhaps not suprisingly, the enzyme is also symmetrical in that it is comprised of two identical subunits each of M_r 31 000. How is it then that *Eco*RI is able to distinguish this hexanucleotide from all other sequences? It seems that binding of the endonuclease to the DNA molecule causes a perturbation in the normal helical structure of the substrate. In effect the enzyme 'unwinds' the DNA helix by 25 degrees (producing a **neokink**) which causes a widening of the major groove and improves access to the appropriate bases (Fig. 5.7). However, the base pairing is not disrupted. The DNA cleavage site in each

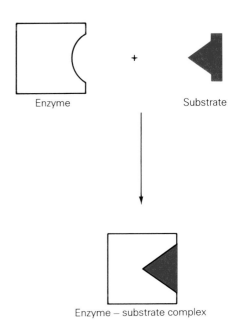

Fig. 5.5 The induced-fit model of substrate binding. The shapes of the enzyme active site (black) and of the substrate (red) alter so that they are complementary only after binding.

See Chapter 4

Neokink: *a localized distortion in the normal structure of B-DNA caused by an alteration in the orientation of the phosphodiester bonds.*

All but the toughest of enzymes are inactivated if exposed to solutions containing more than 50% of a water-miscible organic solvent. However, enzymes are often more stable in the presence of water-immiscible solvents than in completely aqueous solutions. This is because the chemical and structural changes that cause denaturation of an enzyme require the presence of water. Therefore, enzymes suspended in solvent systems where the amount of available water is severely limited (<1%) are often particularly stable. For example, pig pancreatic lipase is inactivated almost immediately in aqueous solution at 100°C. But the same enzyme suspended in heptan-1-ol/tributyrin containing 0.015% water at 100°C still retains 50% of the original activity after 12 hours.

The advantages of using enzymes in organic solvents are two-fold. First, reactions utilizing water-insoluble substrates or products, e.g. steroids, are easier to carry out as higher concentrations of these compounds can be achieved in appropriate solvents. Second, in the absence of water, enzymes may catalyse novel and useful reactions. For example, lipases in organic solvents will catalyse transesterification (i.e. biosynthetic) reactions rather than hydrolyses (i.e. degradative reactions) thus enabling the synthesis of new compounds.

...GAATTC...
...CTTAAG...

Fig. 5.6 The recognition site for *Eco*RI. The restriction endonuclease *Eco*RI recognizes the specific hexanucleotide sequence of DNA shown. The positions of cleavage of the phosphodiester linkages are indicated by arrows.

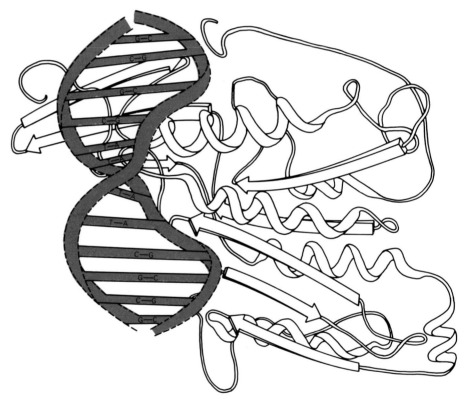

Fig. 5.7 The DNA/*Eco*RI complex. The diagram shows the interaction of one of the two subunits of the restriction endonuclease *Eco*RI with double-stranded DNA (redrawn from McClarin, J.A. *et al.* (1986) *Science*, **234**, 1526–41).

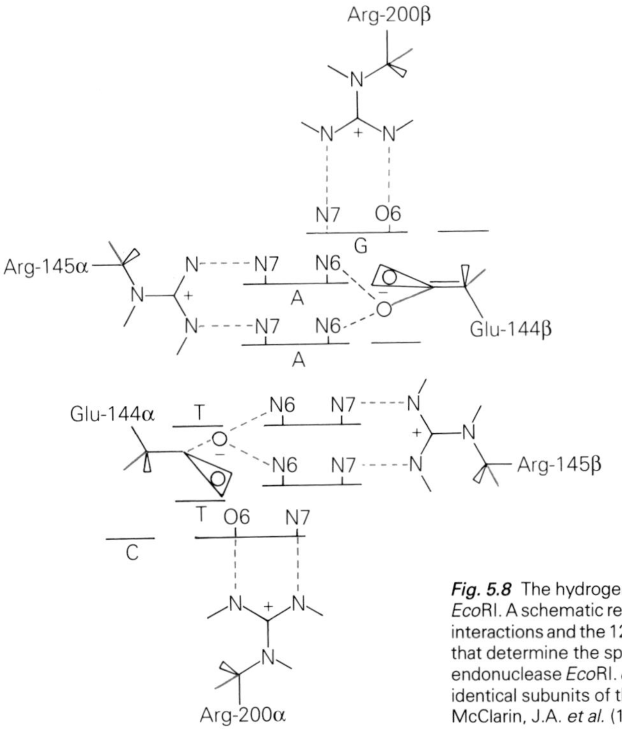

Fig. 5.8 The hydrogen bonding between DNA and *Eco*RI. A schematic representation of the recognition interactions and the 12 hydrogen bonds (dotted lines) that determine the specificity of the restriction endonuclease *Eco*RI. α and β refer to each of the two identical subunits of the enzyme (redrawn from McClarin, J.A. *et al.* (1986) *Science*, **234**, 1526–41).

strand is located within a cleft in each of the two subunits of the enzyme. Here the specific DNA sequence is 'recognized' by the formation of a total of 12 hydrogen bonds. In each subunit of the enzyme the residue Arg-200 forms two hydrogen bonds with guanine whilst residues Glu-144 and Asp-145 each form two hydrogen bonds with adjacent adenine residues (Fig. 5.8). These interactions between six amino acid residues and six bases are sufficient to guarantee the specificity of *Eco*RI.

The substitution of a single base pair would disrupt two or more of these hydrogen bonds between the DNA and the enzyme and prevent productive binding of the substrate to the active site. This specificity is such that even if one of the adenine residues is methylated then cleavage of the DNA will not occur. It may be wondered how the enzyme locates a recognizable site. Clearly, it is unlikely that the enzyme binds directly to the correct hexanucleotide because by chance alone such a sequence would occur only once in every 4^6 base pairs. A more likely scenario is that the enzyme binds anywhere on the DNA and travels along the major groove, causing a transient neokink in the region of binding, until the correct sequence is located. Thus, the kinetically slow step of binding to the DNA molecule only need occur once.

Proof-reading mechanisms

The example of *Eco*RI described above indicated the degree of specificity that can be obtained during the binding of a substrate to an active site. However, sometimes a single binding event is insufficient to produce the required specificity. In these cases a proof-reading mechanism is necessary to ensure adequate selection of substrate. An example of this type of selectivity is the selection of the appropriate amino acid for attachment to tRNA molecules, a reaction essential for protein synthesis. The enzymes that catalyse these reactions are the aminoacyl-tRNA synthetases.

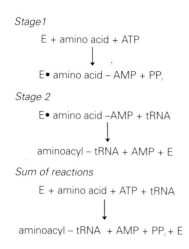

Stage1

E + amino acid + ATP

↓

E• amino acid – AMP + PP$_i$

Stage 2

E• amino acid –AMP + tRNA

↓

aminoacyl – tRNA + AMP + E

Sum of reactions

E + amino acid + ATP + tRNA

↓

aminoacyl – tRNA + AMP + PP$_i$ + E

Fig. 5.9 The reaction catalysed by aminoacyl-tRNA synthetases. This reaction occurs in two discrete stages. If the incorrect amino acid has been used in the first stage of the reaction then the proof-reading mechanism will result in hydrolysis of the aminoacyl-AMP complex.

The reaction catalysed by the aminoacyl-tRNA synthetases (Fig. 5.9) is a two-step process. The enzyme has to select both the correct amino acid and the correct tRNA molecule with considerable accuracy as the overall error rate for the biosynthesis of proteins (of which this is just the first step) is of the order of one mistake per 10^4 amino acids incorporated.

Selection of the appropriate tRNA molecule is not a problem as these are large enough to contain sufficient distinguishing features. Selection of the appropriate amino acid is not so straightforward. The major difficulty is to distinguish between an amino acid and a smaller *homologue*, for example, between the amino acids alanine and glycine. Clearly, the active site of alanyl-tRNA synthetase must be large enough to accommodate alanine, but this may allow the smaller homologue, glycine, to enter the active site. The only form of selection that is available at this stage is the identification of the methylene group on the alanine. This might contribute a 100–200-fold selectivity between alanine and glycine. However, a factor of at least 10^4 is required. Therefore, it is inevitable that the incorrect amino acid glycine will be occasionally linked to AMP in the first stage of the reaction. Thus, a second stage of the reaction is required to 'proof-read' the aminoacyl-AMP complex before transferring the amino acid to tRNA.

Alanyl-tRNA synthetase has a second site with esterase activity. This site is small and unable to accommodate the correctly formed alanyl-AMP complex. However, the smaller glycyl-AMP complex will fit into the site and undergo hydrolysis. Thus the alanyl-AMP complex remains intact and the alanine can be transferred to the tRNA whereas the glycyl-AMP derivative is degraded. This two-stage selection process is necessary to obtain the appropriate degree of fidelity. The price that is paid for this improved accuracy is the hydrolysis of a molecule of ATP. In other words energy is utilized to help prevent loss of information due to entropy. Certain other aminoacyl-tRNA transferases also use proof-reading mechanisms, as do other enzymes, such as DNA polymerase.

5.5 Mechanisms of catalysis

To understand fully the ways in which enzymes are able to enhance the rates of reactions it is important to know something about the chemical reaction itself. Examination of the likely transition states of reactants will often reveal possible mechanisms of catalysis. Catalysis invariably works by lowering the free energy of an intermediate and hence reducing the activation energy of the reaction. To understand the principles more fully it is useful to look at specific enzyme mechanisms in more detail.

Lysozyme

Lysozyme is an enzyme found in a number of different tissues and secretions and is able to destroy certain bacteria. The enzyme degrades the cell wall polysaccharide of the bacterial cell which is then susceptible to osmotic shock, a process refered to as lysis. It is present in milk but is more conveniently purified from egg white. Lysozyme from hen egg white has a M_r of 14 300, with 129 amino acid residues in the chain and contains four disulphide bonds. The amino acid sequence is shown in Fig. 5.2 and the three-dimensional structure of the enzyme and enzyme–inhibition complex are also known. This structural information, in conjuction with mechanistic data have allowed the detailed catalytic mechanism to be elucidated (Fig. 5.3).

THE SUBSTRATE is a polysaccharide consisting of alternate residues of

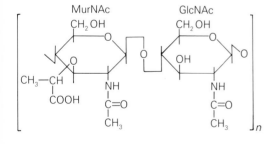

Fig. 5.10 The peptidoglycan substrate of lysozyme. Only the structure of the carbohydrate portion of the complex is illustrated. The polymer comprises alternate units of (β1–4)-linked *N*-acetylmuramic acid (MurNAC) and *N*-acetylglucosamine (GlcNAc). The bond that is hydrolysed is shown in red.

Homologue: *one member of a family of structurally related molecules. This can be illustrated simply by reference to the aliphatic alcohols which form an homologous series:*

n-methanol (CH_3OH), n-ethanol (CH_3CH_2OH), n-propanol ($CH_3CH_2CH_2OH$), etc. *Homologous proteins are ones with a substantial amount of amino acid sequence in common.*

(1–4)-linked β-N-acetylmuramic acid (MurNAc) and β-N-acetylglucosamine (GlcNAc) (Fig. 5.10). Lysozyme hydrolyses the bond between C-1 of β-N-acetylmuramic acid and C-4 of GlcNAc. Chitin (β1–4-linked GlcNAc) is also a substrate and oligosaccharides derived from this polysaccharide have been used extensively to study the mechanism of lysozyme. The disaccharide, trisaccharide and tetrasaccharide derived from chitin are competitive inhibitors of the enzyme. The pentasaccharide is a poor substrate, whereas the hexasaccharide and larger fragments are hydrolysed efficiently (Fig. 5.11).

SUBSTRATE BINDING AT THE ACTIVE SITE has been investigated by X-ray crystallographic studies of the lysozyme–chitin trisaccharide crystalline complex, and this revealed the position of the active-site cleft within the enzyme molecule. The trisaccharide occupies approximately half of the cleft which is consistent with the observation that the hexasaccharide is the minimum size of oligosaccharide that may be hydrolysed rapidly.

Model-building studies indicate that the chitin trisaccharide interacts with the enzyme via six hydrogen bonds. The carboxyl group of Asp-101 forms hydrogen bonds with both the A and B rings (Fig. 5.11). Ring C forms a total of four hydrogen bonds with Trp-62, Trp-63 and the peptide bonds at positions 59 and 107. Van der Waals interactions also stabilize the binding. Attempts to build a model of lysozyme containing the hexasaccharide demonstrated that rings A, B, C, E and F fitted well into the active site cleft. However, ring D had to be distorted from the normal chair conformation into the half-chair or 'sofa' form to bind properly to the enzyme. This correlates well with experimental evidence which shows that ring D makes a negative contribution to the binding of the substrate of about 9 kJ mol^{-1}. The other sugar residues make a positive contribution to binding, that is to say they have a negative free energy of binding. At first sight the need to distort ring D seems to be an unnecessary complication, however, the reasons for this will become clear when the mechanism of catalysis is described (see below).

THE MECHANISM OF CATALYSIS of lysozyme involves two amino acid residues in the enzyme structure. The bond cleaved by lysozyme (when hexasaccharide is used as substrate) is that between C-1 of ring D and the oxygen atom of the glycosidic linkage to ring E. The close proximity of Glu-35 and Asp-52 to this bond (the carboxyl groups are approximately 0.3 nm from the glycosidic bond) implicates the side groups of these two amino acid residues in the reaction mechanism (Fig. 5.12). At the pH optimum for lysozyme (pH 5) Asp-52 is ionized (COO$^-$) as might be predicted from the known pK_a for the side-chain of the free amino acid ($pK_a = 3.9$). However, Glu-35 is in the unionized form (COOH) although the side-chain carboxyl group of the free amino acid has a pK_a of 4.3. This unexpected finding arises because Glu-35 is in a hydrophobic microenvironment, which effectively raises the pK_a of the carboxyl group.

The likely involvement of Glu-35 and Asp-52 plus the distortion of ring D on binding of the substrate to lysozyme suggests an acid-catalysed mechanism of cleavage via a carboxonium ion intermediate. Glu-35 donates a proton to the oxygen atom of the glycosidic bond connecting rings D and E. The cleavage of the glycosidic bond is encouraged because the resultant positive charge (**carboxonium ion**) that forms on C-1 of ring D is stabilized in two ways. Firstly, the distortion of the sugar ring to the 'sofa' form, that occurred during the binding of the substrate, results in a planar structure around C-1 of ring D. This sofa conformation is the preferred structure of the carboxonium ion intermediate. Secondly, the negative charge on Asp-52 will stabilize the positively charged carboxonium ion. The reaction is completed by the addition of a molecule of water to the complex. The hydroxyl group is added

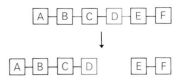

Fig. 5.11 The hydrolysis of the hexasaccharide of chitin by lysozyme. To aid identification, the individual sugar rings are designated A–F. The reducing end is on ring F. The ring that is strained on binding (ring D) is shown in red.

Exercise 1

If the hydrolysis of chitin by the enzyme lysozyme was allowed to proceed in the presence of water labelled with heavy oxygen (^{18}O), where would the label appear in the products? Clue: examine Fig. 5.3 and 5.12.

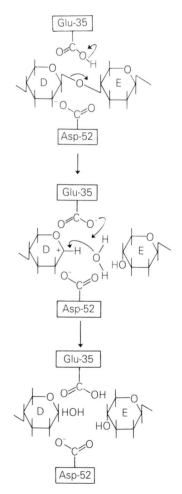

Fig. 5.12 The catalytic mechanism of lysozyme. The proton donated by Glu-35 is shown in red. The substituents on the sugar rings have been omitted to improve the clarity of the diagram.

Carboxonium ion: *a carbon atom that possesses a transiently stable positive charge. Carboxonium ions are often produced as reaction intermediates.*

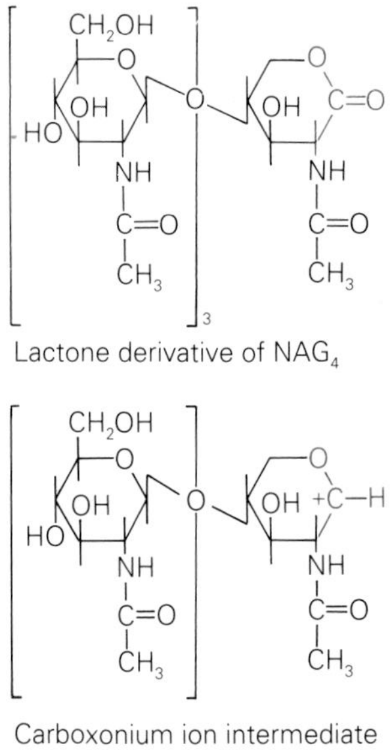

CH_2OH structure

Lactone derivative of NAG_4

Carboxonium ion intermediate

Fig. 5.13 The lactone derivative of the tetra-saccharide of chitin. This compound is a powerful inhibitor of lysozyme as the lactone is an analogue of the carboxonium ion transition-state intermediate that is formed during catalysis of the substrate Both compounds are planar around C-1 of ring D (shown in red).

to the carboxonium ion and the proton regenerates the unionized form of Glu-35 (Fig. 5.12). Finally, the disaccharide and tetrasaccharide products (in the case of hexasaccharide and tetrasaccharide hydrolysis) are released from the enzyme binding site.

The proposed mechanism of catalysis by lysozyme is supported by several observations. A derivative of the tetrasaccharide in which ring D is a **lactone** which is an extremely effective inhibitor of the enzyme (Fig. 5.13). This is because the lactone ring is planar around C-1 and therefore the inhibitor can bind to the active site without further distortion of ring D. This compound is known as a transition-state analogue because it mimics the structure of the reaction intermediate. An alternative approach is to chemically modify selected amino acids and to observe the effects on enzyme activity. Chemical modification of the carboxyl groups of lysozyme results in inactivation of the enzyme under certain circumstances. In the presence of substrate all of the carboxyl groups, with the exception of Glu-35 and Asp-52, may be modified without loss of activity. However, in the absence of substrate, Asp-52 is also modified and activity is lost completely. Therefore, the carboxonium ion transition state stabilized by deformation of ring D and ionic interaction with Asp-52, is the most plausible theory.

Lysozymes from sources other than hen egg white, particularly that from T_4 bacteriophage, have also been studied in some detail. As might be expected, the mechanism of the T_4 enzyme seems to be identical to that proposed for hen egg white lysozyme. However, the location of the two catalytic amino acids within the primary structure of the protein is different. The key glutamate and aspartate residues are at positions 11 and 20, respectively, but the juxtaposition of these amino acids relative to the cleaved glycosidic bond is virtually the same. Site-directed mutagenesis is a procedure for producing protein in which specific amino acid residues have been removed or replaced by others. Such mutagenic experiments have been used to confirm the catalytic role of Glu-11 and Asp-20 in bacteriophage lysozyme. Glu-11 may be replaced by an aspartate residue (this substitution should still leave a proton donor at position 11), but only 16% of the original activity is retained. Any other substitution at either position with other amino acids results in complete loss of activity. Thus, specific mutation of the T_4 bacteriophage lysozyme gene has provided results to support the catalytic mechanism originally proposed on the basis of X-ray crystallography and chemical inactivation studies.

THE pH OPTIMUM of lysozyme (pH 5) may be predicted from the catalytic mechanism. Enzyme activity is dependent on the ionization states of the two key active-site amino acid residues Glu-35 and Asp-52. Consequently, activity decreases rapidly with increasing pH because Glu-35 becomes ionized to the 'inactive' COO^- form. As the pH is decreased below the pH optimum Asp-52 becomes protonated to the 'inactive' COOH form (Fig. 5.14). Therefore the

---

Box 5.5
Abzymes (enzymes from antibodies)

Lerner, R.A. and Tramontano, A. (1987) Antibodies as enzymes. *Trends in Biochemical Sciences*, **12**, 427–30. Sets out the rationale behind the use of monoclonal antibodies to transition-state analogues.

Completely new enzyme-like proteins are being produced using monoclonal antibody technology. Antibodies have been raised against the transition-state analogues of specific reactions. As might be expected, these antibodies will catalyse reactions involving the specific transition state because the substrate(s) will be bound in the correct orientation and proximity for the reaction to occur. Artificial esterases have been produced in this way that are capable of accelerating the reaction by a factor of a million.

Lactone: *formed by the intramolecular condensation of hydroxyl groups to give a cyclic ester with the resultant liberation of a molecule of water.*

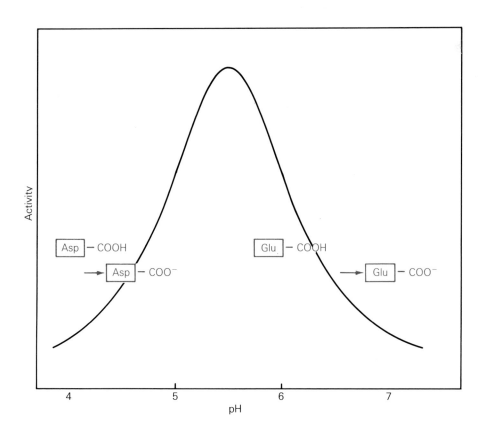

Fig. 5.14 The pH optimum of lysozyme. The contributions of the two active-site amino acid residues (Glu-35 and Asp-52) to the pH optimum are shown.

pH profile of lysozyme is a direct reflection of the ionization states of the two critical amino acids present at the active sites.

This pH phenomenon is common to many enzymes although in some cases it is not necessarily just the reactive amino acids that determine pH optimum. It is also possible for ionizable groups on the substrate or for enzyme instability to have the dominant effects on pH optima.

Serine proteinases

The serine proteinases are a group of related enzymes, all of which utilize an activated serine residue to catalyse the hydrolysis of peptide bonds. The best-known examples are probably the pancreatic enzymes, chymotrypsin and trypsin. The similarities and interrelationships between various serine proteinases are considered in more detail in Section 5.7 on enzyme families. Chymotrypsin will be used as the example for the description of the active site and catalytic mechanism.

☐ **Proteases** (syn. **peptidohydrolases**) are enzymes that catalyse the hydrolysis of peptide bonds. They can show endo- or exopeptidase activity, that is they can cleave internal or terminal peptide bonds in their substrates respectively. **Proteinase** is used synonymously with endo-peptidase. Thus carboxypeptidase is a protease, while chymotrypsin and trypsin can be classified as both proteases and proteinases.

THE SUBSTRATE SPECIFICITY of chymotrypsin is well defined. Chymotrypsin cleaves preferentially those substrates that have an amino acid with an aromatic or large, hydrophobic side chain on the carbonyl side of the peptide bond. The enzyme is also able to catalyse the hydrolysis, of small ester substrates:

Reference Phillips, D.C. (1966) The three-dimensional structure of an enzyme molecule. *Scientific American*, **215**(5), 78–90. A classic article describing the structure and mechanism of lysozyme.

for example *p*-nitrophenyl acetate. Although this is of no particular physiological significance, considerable use has been made of these ester substrates both for mechanistic studies and as a convenient laboratory assay procedure.

THE MECHANISM OF CATALYSIS involves the transient formation of a covalent bond between the hydroxyl group on the side-chain of the active-site serine (Ser-195) and the carbon atom of the peptide bond. In fact an acylenzyme intermediate is formed as a result of the acylation of Ser-195 by the substrate. The hydroxyl group of a serine residue is not normally a particularly reactive species. However, in the presence of substrate the interaction of Ser-195 with His-57 and Asp-102 within the active site of chymotrypsin substantially increases the nucleophilicity of the hydroxyl group. Asp-102 has an abnormally low pK_a of 2 and therefore is negatively charged when the enzyme is active. This negative charge interacts with the imidazole ring of His-57 with the result that one of the ring nitrogen atoms has a tendency to remove the proton from the serine hydroxyl (Fig. 5.15). This specific interaction between the three amino acids of chymotrypsin has been called the **charge-relay system**.

□ A nucleophilic group is an electron-rich chemical group that tends to attack (react with) an electron-deficient centre. An electrophile is an electron-deficient group that can attack electron-rich centres.

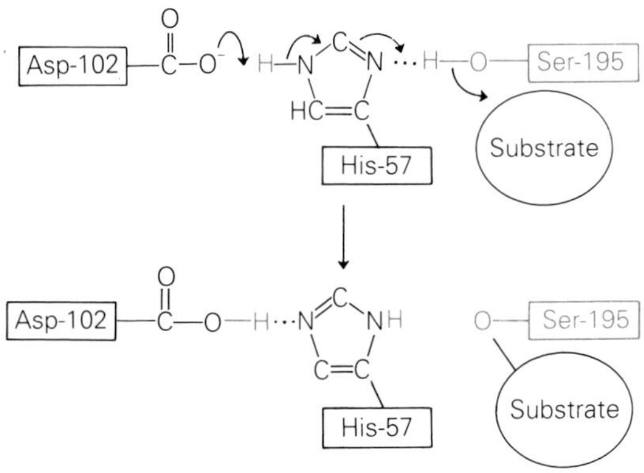

Fig. 5.15 The charge-relay complex of serine proteinases, for example, chymotrypsin. The interrelationships between the active-site serine, histidine and aspartate residues are shown.

The binding of the substrate and the mechanism of hydrolysis of the peptide bond is illustrated in Fig. 5.16 and described below. The hydroxyl group on Ser-195 is acylated by the substrate to produce a tetrahedral transition state. This reaction is facilitated by the transfer of the proton from the hydroxyl group of Ser-195 on to the ring nitrogen of His-57. The tetrahedral transition state is stabilized by the formation of hydrogen bonds between the oxyanion (negatively charged oxygen atom, O⁻) and the –NH– residues of the chymotrypsin backbone (positions 193 and 195). The proton on the nitrogen atom of His-57 (originally derived from Ser-195) is transferred to the nitrogen of the peptide bond of the substrate. This results in hydrolysis of this peptide bond with release of one of the two fragments of substrate.

In the final stage of the reaction chymotrypsin must be deacylated to release the other fragment of substrate and to restore the enzyme to its original form. A molecule of water is activated by the tendency for a proton to be transferred on to a nitrogen atom of His-57 (this is facilitated by the proximity of Arg-102). The hydroxyl group derived from the water molecule can attack the carbonyl carbon of the acylserine intermediate thus deacylating the enzyme. Transfer of the proton from His-57 to Ser-195 completes the reaction and regenerates the enzyme.

Exercise 2

The compound *p*-nitrophenol acetate

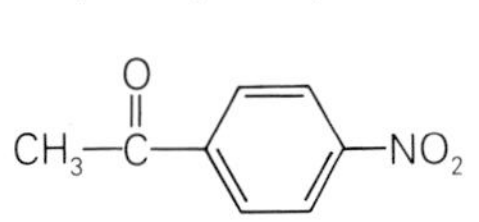

can be hydrolysed to release an acetate group (CH_3COO^-).
(a) Would this reaction be the result of nucleophilic or electrophilic attack?
(b) What types of amino acid residues would you expect to find in the active sites of enzymes that catalyse this hydrolytic reaction?

Box 5.6
Non-protein enzymes

A knowledge of the mechanism of certain enzymes, e.g. chymotrypsin, has allowed chemists to synthesize non-protein analogues that have catalytic activity. An example of a chymotrypsin analogue is β-benzyme in which *O*-[4(5)-mercaptomethyl-4(5)-methylimidazol-2-yl]benzoate is covalently attached to the rim of a β-cyclodextrin molecule. This derivative contains an hydroxyl group, an imidazole ring and a carboxylate group and mimics the active site of chymotrypsin (Ser-195, His-57 and Asp-102). This analogue displays Michaelis–Menten kinetics and is able to hydrolyse certain esters as efficiently as chymotrypsin.

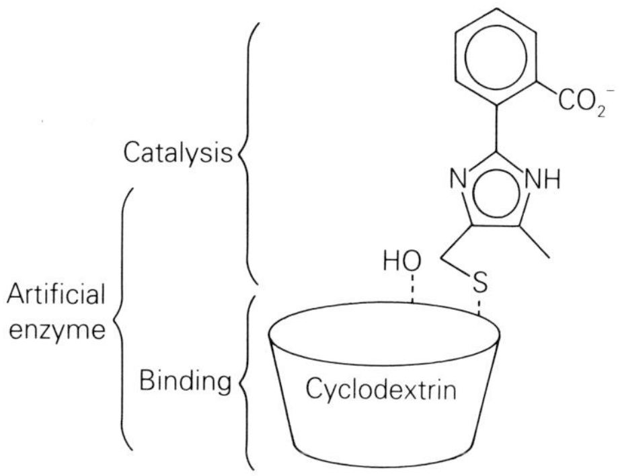

Bender, M.L., D'Souza, V.T. and Lu, X. (1986) Miniature organic models of chymotrypsin based on α-, β- and γ-cyclodextrins. *Trends in Biotechnology*, **4**, 132–5. A readable account of the synthesis of a range of non-protein chymotrypsin analogues.

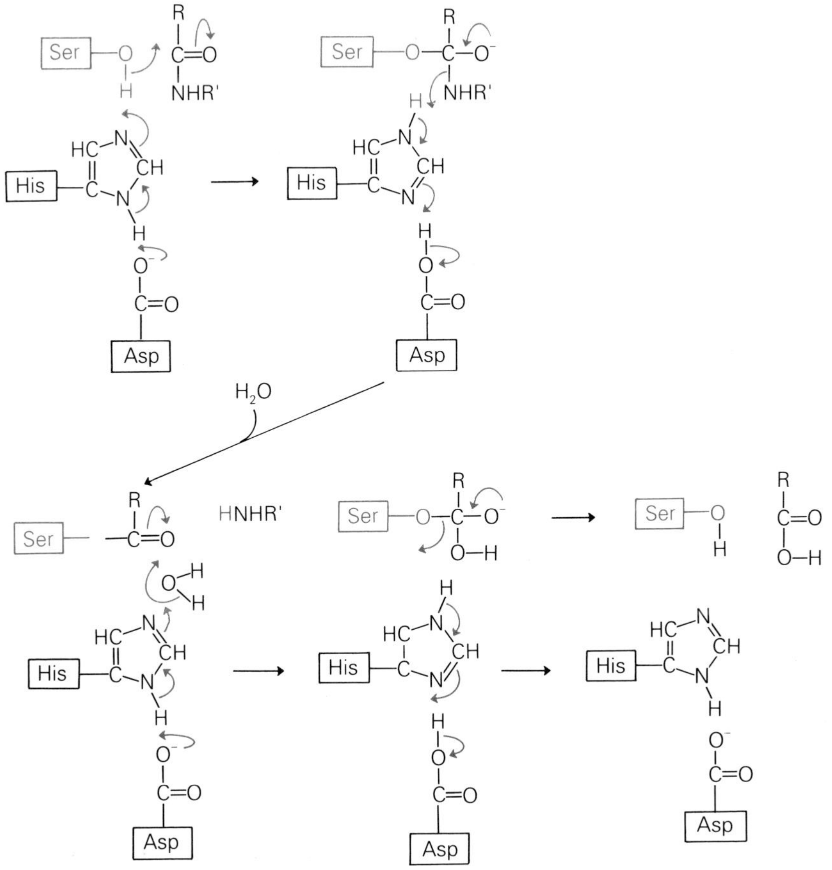

Fig. 5.16 The catalytic mechanism of chymotrypsin.

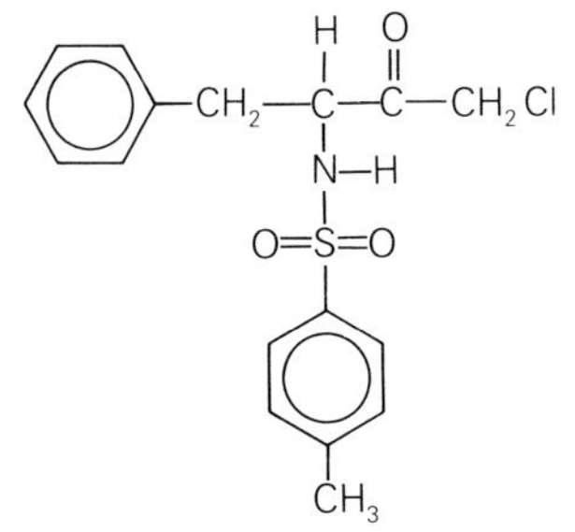

Fig. 5.17 The structure of the substrate analogue, tosyl-L-phenylalanine chloromethylketone.

EVIDENCE FOR THE MECHANISM of chymotrypsin comes from four main types of experimental approach.

1. X-ray diffraction analysis of the enzyme has demonstrated the close spatial arrangement of the three critical amino acid residues, Ser-195, His-57 and Asp-102.
2. The enhanced reactivity of Ser-195 has been demonstrated using the inhibitor diisopropylphosphofluoridate, a compound that has tetrahedral geometry similar to the transition state. This inhibitor reacts very much faster with Ser-195 than any of the other serine residues in the enzyme.
3. A substrate analogue, tosyl-L-phenylalanine chloromethylketone (Fig. 5.17) specifically alkylates one of the ring nitrogens of His-57, thus implicating this residue in the mechanism.
4. Site-directed mutagenesis of the enzyme to replace Asp-102 with an asparagine residue results in a dramatic loss of activity.

Therefore, evidence from structural studies, chemical modifications and alterations to the amino acid sequence of the enzyme all support the charge-relay network mechanism for chymotrypsin. However, there is still considerable controversy about the exact role of the three amino acids at the active site.

5.6 General comments about enzyme mechanisms

The elucidation of an enzyme mechanism requires information from a variety of different sources. Often it is the chemical inhibition studies that are performed first and this gives some idea about the type of amino acid side-chain groups that may be critical to the catalytic mechanism. X-ray diffraction patterns of enzyme–inhibitor complexes provide information on the spatial arrangement of possible catalytic groups. Clearly if a putative catalytic group is not closely associated with the substrate or other catalytic residues in the enzyme then it is unlikely to be of mechanistic importance. The advent of site-directed mutagenesis will do much to solve certain mechanistic problems. However, this technique is only really useful if used to make rational changes based on the likely chemistry of the catalytic mechanism. Ultimately, any proposed mechanism for an enzyme must accommodate the data obtained using these various techniques and, most importantly, must make sense from a chemical point of view.

5.7 Enzyme families

So far in this chapter various individual enzyme mechanisms have been looked at in detail. However, many groups of enzymes utilize similar or identical mechanisms. These groups of enzymes are known as enzyme families, of which the serine proteinases are probably the most extensively studied example (Table 5.3).

Serine proteinases do show overall homology in amino acid sequence, although this does not appear to be that great. However, this is rather a gross method of comparison. A more detailed analysis reveals several important features. First, some amino acid residues are always conserved. This is particularly true of those that are involved in the catalytic mechanism. For example, all of the enzymes have a charge-relay system comprising residues of aspartate, histidine and serine. In addition, the amino acid sequence adjacent to the active-site serine is conserved (Table 5.4). Second, the overall

---

Exercise 3

There are four groups of proteolytic enzymes. In addition to the serine proteinases there are the acidic, the thiol and the metal-containing enzymes. Which amino acid residues are likely to be involved in catalysis by acid and thiol proteinases? Can you make predictions about the likely pH ranges over which these enzymes might operate?

Reference Shaw, W.V. (1987) Protein engineering: the design, synthesis and characterization of factitious proteins. *Biochemical Journal,* **246**, 1–17. A review of applications of site-directed mutagenesis to the study of enzyme mechanisms.

Table 5.3 *Structural homologies in serine proteinases*

Enzyme	Source	Species	% Homology
Trypsin	Pancreas	Cow	100
		Dogfish	69
	Fungal	*Streptomyces griseus*	43
Chymotrypsin	Pancreas	Cow	53
Elastase	Pancreas	Pig	48
	Bacterial	*Myxobacter sorangium*	26
	Fungal	*Streptomyces griseus*	20
Thrombin	Plasma	Cow	38
Factor Xa	Plasma	Cow	50
Subtilisin	Bacterial	*Bacillus* spp.	0

Data from Hartley, B.S. (1974) *Symp. Soc. Gen. Microbiol.*, **24**, 152, and Delbaere, L.T.J. *et al.* (1975) *Nature*, **257**, 758.

Table 5.4 *Amino acid sequences of serine proteinases. The amino acid sequence around the active-site serine is shown for a number of different serine proteinases. Only subtilisin has a completely different sequence. All of the other enzymes, regardless of the extent of homology in other regions of the molecule, have highly conserved sequences around the serine residue*

Enzyme	Sequence					
Chymotrypsin	Gly	Asp	Ser	Gly	Gly	Pro
Trypsin	Gly	Asp	Ser	Gly	Gly	Pro
Elastase	Gly	Asp	Ser	Gly	Gly	Pro
Thrombin	Gly	Asp	Ser	Gly	Gly	Pro
Plasmin	Gly	Asp	Ser	Gly	Gly	Pro
Subtilisin	Gly	Thr	Ser	Met	Ala	Ser

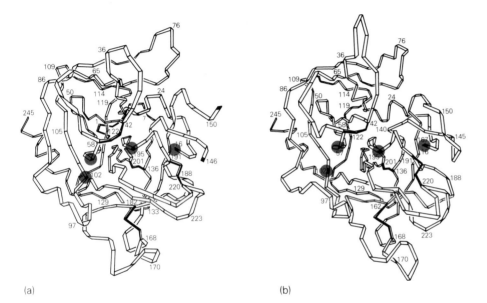

(a) (b)

Fig. 5.18 The three-dimensional structure of serine proteinases. Note the similarity in overall structure between (a) α-chymotrypsin and (b) elastase. The active-site amino acid residues, Ile-16, His-57, Asp-102 and Ser-195 are shown in red. Redrawn from Hartley, B.S. and Shotten, D.M. (1971) in *The Enzymes*, Vol. III (ed. P. Boyer) Academic Press, New York, USA, p. 323.

three-dimensional structure of the proteins is very similar with modifications largely confined to the surface (Fig. 5.18). This is so even for enzymes with as low a degree of homology as 20%. The most likely explanation for the similarity in these enzymes is that they evolved from a common ancestor protein.

Enzymes with a high degree of homology will have diverged relatively recently whereas those with considerably less homology will have diverged earlier during the course of evolution.

The presence of different amino acid residues at certain sites in the structure of enzymes with an otherwise high degree of homology indicates a specific purpose for these residues (Fig. 5.19). For example, the active site of chymotrypsin contains a large pocket to accommodate the aromatic or large hydrophobic amino acid side chains of the preferred substrates. The amino acid residue at the base of the pocket is Ser-189. In comparison, the residue at the base of the binding pocket in trypsin is Asp-189. This alters the substrate specificity such that trypsin binds preferentially substrates with a positively

☐ Divergent evolution is not the only way in which enzyme families are generated. It is also possible for similar mechanisms to arise by convergent evolution: here two unrelated enzyme proteins evolve towards a common mechanism. This has been the case for certain of the serine proteinases. For example, subtilisin a serine proteinase derived from *Bacillus* spp. has an active-site serine and a charge-relay system, and yet there is no other homology with chymotrypsin. Therefore, it seems likely that subtilisin has evolved independently of the mammalian enzymes but uses an identical catalytic mechanism.

Hydrophobic surfaces — Chymotrypsin

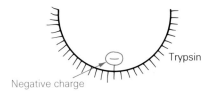

Negative charge — Trypsin

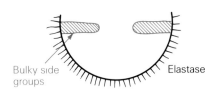

Bulky side groups — Elastase

Fig. 5.19 Binding sites of various serine proteinases. The substrate specificity of the different serine proteinases is determined by the nature of the amino acid residues which line the substrate binding site. Chymotrypsin binds aromatic or bulky non-polar substrates whereas trypsin binds positively charged and elastase binds small amino acid side chains in the substrate molecules.

charged residue, for example, Arg, Lys, adjacent to the peptide bond that is to be cleaved. Similarly, the specificity of elastase for amino acids with small side chains, especially alanine, is achieved by substitution of two glycine residues at the mouth of the chymotrypsin pocket with the relatively bulky valine (Val-216) and threonine (Thr-226) residues. Thus, three very similar enzymes, chymotrypsin, trypsin and elastase, have different substrate specificities simply because of the alteration of one or two amino acid residues at the binding site.

5.8 Overview

Elucidation of enzyme mechanisms requires the application of several different techniques.

☐ A detailed knowledge of the three-dimensional structure of the enzyme and enzyme–substrate/inhibitor complex is required. Often it is possible to identify amino acid side-chains that are in a suitable spacial orientation, relative to the substrate, to promote catalysis. Sometimes it may be possible to postulate a likely transition-state complex.
☐ Active-site amino acid residues may be identified by the action of specific enzyme inhibitors or by affinity labelling. Often an amino acid residue at the active site is unusually reactive.
☐ Alteration of single amino acids by side-directed mutagenesis can provide invaluable data on active site residues and possible mechanisms.
☐ A proposed enzyme mechanism must be chemically feasible.

Certain groups of enzymes have common mechanisms and form discrete families. Some enzymes are produced in precursor form and require activation by proteolytic cleavage before they are able to catalyse the appropriate reaction.

Answers to Exercises

1. On carbon atom 1 of the D ring.
2. (a) Nucleophilic attack, since the carboxyl carbon atom would have a partial

$$CH_3 \overset{\delta^+}{-} \overset{\overset{\displaystyle \overset{\delta^-}{O}}{\|}}{C} - R$$

positive charge ($CH_3 \overset{\delta^+}{-}\overset{O}{\overset{\|}{C}}-R$), and therefore be deficient in electrons.

(b) For example, residues His, Cys, Ser, Tyr.
3. Acidic – aspartate or glutamate; thiol – cysteine.

The acid proteinases will act at low pH because of the mechanistic require-ment for a protonated carboxyl group. The thiol proteinases will work over a fairly broad range of pH although at pH greater than 8 there is an increasing chance of the thiol group becoming ionized and hence the enzyme will become inactive.

FILL IN THE BLANKS

1. The original theory to explain the binding of substrate to enzyme based on the _____ of structure between these two components was called the _____ _____ _____ model. The concept of complementarity in the transition state is called the _____ _____ model. Enzyme catalysis occurs by a _____ of the _____ energy of the _____ state.

Hen egg white lysozyme can catalyse the hydrolysis of _____ which is a polysaccharide comprised of _____ _____ residues. The hexasaccharide is cleaved to produce the _____ and _____ products. The two amino acid residues directly involved in the hydrolysis are _____ and _____ .

Choose from: activation, Asp-52, β1–4-linked N-acetylglucosamine, chitin, complementarity, disaccharide, Glu-35, induced fit, lock and key, lowering, tetrasaccharide, transition.

MULTIPLE-CHOICE QUESTIONS

In the following questions, select those options (any number may be correct) that correctly complete the initial statement, or correctly answer the question posed. Indicate which of the following statements are true.

2. The following enzymes are serine proteinases:
A. chymotrypsin
B. pepsin
C. elastase
D. trypsin

3. Hen egg white lysozyme has:
A. a single polypeptide chain
B. a metal ion at the active site
C. the ability to hydrolyse peptidoglycan
D. a mechanism that produces a carbonium ion containing intermediate
E. an active site which can bind several small oligosaccharides

SHORT-ANSWER QUESTIONS

4. What modifications should be made to chymotrypsin to produce an enzyme with 'trypsin-like' specificity?

5. Suggest enzymes other than the serine proteinases that may be grouped into 'families'?

6. Why are enzyme active sites usually situated within clefts or pockets?

7. Outline how you might produce an 'abzyme' that has esterase activity.

ESSAY-TYPE QUESTIONS

8. Describe the charge-relay mechanism that operates in the serine proteinases.

9. Describe the experiments that might be carried out to elucidate the mechanism of lysozyme.

6

Polysaccharides

Objectives

After reading this chapter you should be able to:

☐ explain the circumstances that influence polysaccharide chains to adopt particular patterns of folding and so produce particular shapes of macromolecule;

☐ describe with specific examples, how the shape of a polysaccharide may be related to its biological functions;

☐ summarize the contributions of polysaccharides to the viability of all types of cell, whether microbial, plant or mammalian.

6.1 Introduction

Do polysaccharides possess properties necessary for cell viability that are absent in other molecules? This is not an easy question to answer. Cells invest a lot of energy and metabolites in producing macromolecules. As much as 80–90% of the dry weight of a cell is made up of proteins, nucleic acids, and polysaccharides. Each addition of an amino acid, nucleotide or sugar requires the hydrolysis of at least one, and usually more, phosphoanhydride bonds during the synthesis of the relevant polymer. Most cells (microbial, plant or animal) include polysaccharides as part of their composition. This chapter mainly concerns the structures and functions of these polysaccharides.

See *Cell Biology*, Chapter 2

☐ For most cells, the variety of exogeneous organic compounds that they absorb to sustain growth is wide. Using the intracellular pathways of intermediate metabolism, the assimilated carbon sources are altered to produce the precursors for polysaccharide, protein, nucleic acid and lipid biosyntheses. Gluconeogenesis is the metabolic process whereby sugars, such as glucose 1-phosphate and glucose 6-phosphate, precursors for polysaccharide assembly, are produced from metabolites, such as lactate, glycerol and oxaloacetate, that bear little or no structural resemblance to carbohydrates. The reactions that comprise gluconeogenesis are described in detail elsewhere (*Biosynthesis*, Chapter 3).

6.2 Distinguishing features of polysaccharides

It might be asked why cells should produce polysaccharide even to the extent that some cells perform gluconeogenesis.

If polysaccharides fulfil unique functions then they would be expected to possess structures that are different from other cellular polymers such as the proteins. An obvious difference is that proteins contain nitrogen whereas polysaccharides frequently lack it. That is probably why plants, which often grow in nitrogen-limited environments, produce polysaccharides to support and protect their cells rather than a nitrogen-rich fibrous protein. Apart from chemical composition (Tables 6.1 and 6.2) there are other distinguishing features, all to do with shape and size. Firstly, a particular protein always has a specific number and sequence of amino acid residues. This number might alter for a variety of reasons, such as the modification of preproteins, but changes are always precise. Individual polysaccharides, in contrast, rarely have a specific M_r but rather show a range or distribution of M_r. Nevertheless, they can be thought of as large or small, depending on the **average** M_r.

The second difference is in the variety of components. Each type of

Table 6.1 *Structures of monosaccharides incorporated into polysaccharides*

α-D-glucose (Glc) α-D-galactose (Gal) α-D-mannose (Man)

β-D-fructose (Fru) α-L-fucose (Fuc) α-L-rhamnose (Rha)

α-abequose (Abe) α-L-arabinose (Ara) β-D-xylose (Xyl)

α-D-N-acetylglucosamine (GlcNAc) α-D-N-acetylgalactosamine (GalNAc) α-D-glucuronate (GlcUA)

α-D-galacturonate (GalUA) β-D-mannuronate (ManUA) α-L-guluronate (GluUA)

α-L-iduronate (IdUA) N-acetylneuraminate (NeuNAc)

Structures are depicted in two ways: as Haworth perspective formulae devised by Professor Sir W.N. Haworth, and as conformational structures that provide a more realistic presentation of molecular shape. Most sugars, both free or incorporated into polymers, adopt the C1 rather than the 1C chair conformation.

Table 6.2 Periodic polysaccharides

Name	Repeating sequence	Conformation
Amylose	-Glcα1–4-Glcα1–4-Glcα1–4-Glc-	Helix
Cellulose	-Glcβ1–4-Glcβ1–4-Glcβ1–4-Glc-	Ribbon
Chitin	-GlcNAcβ1–4-GlcNAcβ1–4-GlcNAcβ1–4-Glc-NAc	Ribbon
Xylan	-Xylβ1–4-Xylβ1–4-Xylβ1–4-Xyl-	Extended helix
Pectate	-GalUAα1–4-GalUAα1–4-GalUAα1–4-GalUA-	Puckered ribbon
Hyaluronate	-GlcNAcβ1–4-GlcUAβ1–3-GlcNAcβ1–4-GlcUA-	Helix
Lipopolysaccharide of *Salmonella* sp.	Abeα1–3 Abeα1–3 | | -Manα1–4-Rhaα1–3-Galα1–2-Man-	Helix

The repeating residues and linkages are highlighted. See Table 6.1 for abbreviations.

polysaccharide comprises very few types of sugars (Table 6.2). Frequently there is but one. Thus, a polysaccharide may consist of only glucose residues (a **glucan**) or mannose residues (a **mannan**), etc. Many polysaccharides have repeating disaccharide units. Occasionally there are larger blocks, e.g. the tetra-, penta- and hexasaccharides of lipopolysaccharides found on the surface of Gram-negative bacteria. Examples of these **periodic** or **repeating sequences** are shown in Table 6.2. The small number of sugar types in any one polysaccharide means that their structures are more uniform in shape than those of proteins. A sequence of repeating blocks of mono-, di-, tri-saccharides, etc. produce polysaccharides of simple structures, such as ribbons or helices. This means that there is a limited array of groups on the polysaccharide surface capable of interaction with other molecules. Contrast this with the much greater range of chemical or physical interactions displayed by amino acid side-chains on the surface of a protein. This is probably why polysaccharides cannot act as catalysts. They simply do not have the subtle array of chemical groups oriented in particular, specific ways, as found at the active site of an enzyme. However, their very simplicity, especially in extended linear conformations, allows them to interact with other polymers to form large molecular networks with important and desirable properties.

Thirdly, although there is a limited range of sugar types, the glycosidic bond between adjacent residues is not confined to just one group in the sugar ring. In contrast to proteins, where identical peptide bonds join amino acid residues, a number of **glycosidic bond** arrangements are found between sugar residues (Fig. 6.1). A carbon atom in the sugar ring, at position 1, often may be linked through the oxygen of the glycosidic bond to any of the other carbon atoms in the adjoining sugar residue.

The availability of different linkages means that the preferred **conformation** (or shape) of the polysaccharide is dependent on the glycosidic bond as well as the nature of the residues themselves. It will be seen that glucose residues

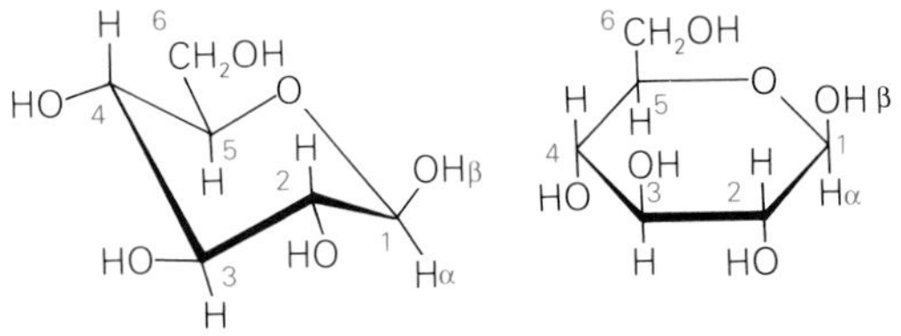

See Chapter 5

□ Until comparatively recently it was also assumed that only enzyme proteins could function as biological catalysts. However, it is now known that some types of RNA show all the characteristic properties associated with catalysis by protein enzymes (see Chapter 4 and *Molecular Biology and Biotechnology*, Chapter 2).

Fig. 6.1 The nomenclature used to describe the configuration of glycosidic bonds between adjacent sugar residues is derived from that describing the position of carbon atoms in monosaccharides and is illustrated with reference to β-D-glucose. α and β describe the configuration of the –OH at C-1. In the L-series of monosaccharides where, for instance, the –CH₂OH at C-5 is below the plane of the ring the orientation of α and β is reversed (see examples in Table 6.1).

Residue: *a monosaccharide linked to others in a polysaccharide. Glycosyl describes a sugar residue of any composition. Glucosyl, mannosyl, etc. denote the specific sugar, i.e. glucose, mannose, etc.*

Box 6.1
Glycoproteins and glycolipids

See *Molecular Biology and Biotechnology*, Chapter 5

The term *glycoprotein* describes a wide range of biopolymers. In general, glycoproteins are proteins to which sugar residues have been covalently attached (Fig. 6.2) after assembly of the peptide chain. The size, number of attachment points and variety of sugar residues varies enormously. However, only a limited number of types of amino acid residues are utilized: Ser, Thr and Asn are common sites of carbohydrate attachment. The structure of mucins consists of about 200 disaccharides attached to a single protein of some 1200 amino acid residues through serine residues. Viscous solutions of mucins lubricate and protect epithelial surfaces of the digestive and urinogenital tracts. Mucins from other sources carry additional larger oligosaccharide side-chains. Later in this section the complex of macromolecules constituting articular cartilage is described. The principal components of this complex are proteoglycans, the structure of which contains up to 150 polysaccharide chains attached to serine residues of a single core protein. Some biopolymers that were thought to be pure polysaccharides are now believed to have protein covalently attached to them. Glycogen is such an example.

Mucins, proteoglycans and glycogen are biopolymers that contain carbohydrate as the principal structural component. The term glycoprotein, however, is more often used to describe proteins, such as glycophorin referred to in this section, that are mainly protein but contain between 5 and 10% by weight of carbohydrate. There is a range of structures. The saccharide side-chains vary in number between 1 and about 20 and are linked to the protein at Asn, Ser and Thr residues (Fig. 6.2). The saccharide moiety contains anything up to about 20 sugar residues and may be branched in structure (Fig. 6.6). The range of oligosaccharyl moieties bound to protein can also be found bound to lipid (Fig. 6.18), giving rise to glycolipids.

A more comprehensive term, describing proteins and lipids that contain covalently bound carbohydrate, is **glycoconjugate**.

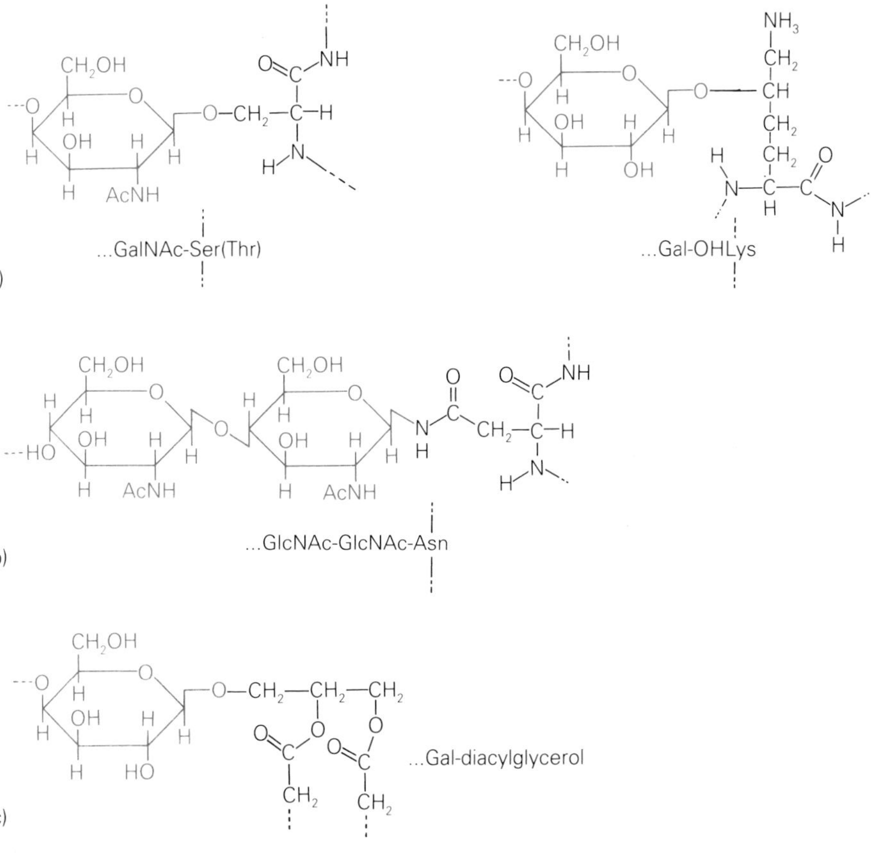

(a) ...GalNAc-Ser(Thr)

...Gal-OHLys

(b) ...GlcNAc-GlcNAc-Asn

(c) ...Gal-diacylglycerol

Fig. 6.2 Structure of the covalent links between oligosaccharides (coloured) and protein or lipid: (a) *O*-glycoprotein where the serine residue may be replaced by a threonine residue; (b) *N*-glyco-protein; (c) glycolipid where acylglycerol may be replaced by ceramide and the galactose residue by a glucose residue. Abbreviations: Asn, asparagine; OHLys, hydroxylysine; Thr, threonine; Ser, serine.

linked α1–4- produce a polymer with a shape and solubility quite different from one with identical residues, linked 1–4, but with a β-**configuration**.

There is also the possibility of producing branched structures. Small but extensively branched chains, say 10–20 sugar residues, create unique molecular shapes that can be recognized specifically by proteins. In this manner, attached to proteins and lipids (Fig. 6.2), carbohydrates aid in **cellular recognition** processes. In other examples, short branches or side-chains are added to regularly shaped polysaccharides, thus producing local modifications to the contour. Such alterations in the shape of polysaccharides change the way in which they interact with other molecules. For example, some polysaccharides, because of their complementary surfaces, align with each other, strengthen the association through hydrogen, hydrophobic and ionic bonds, and form polysaccharide aggregates. Local changes to the shape of the polysaccharide hinder the association. These changes produce polysaccharide networks that trap clusters of water molecules to form gels, and bridging polysaccharides that act as cross-links between polysaccharide microfibrils.

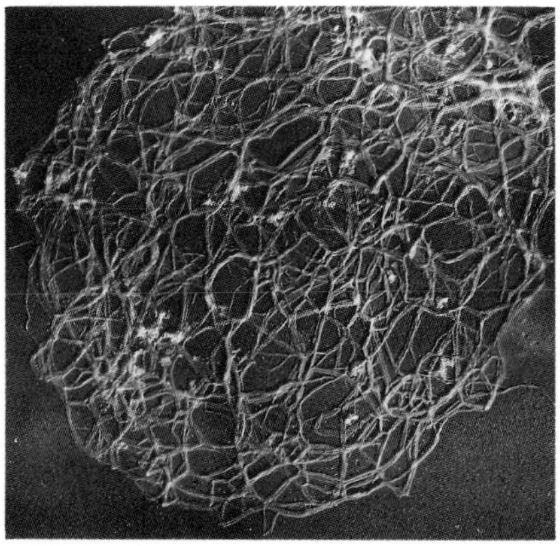

The rest of this chapter will describe examples of polysaccharides and concentrate on their structures and functions. Like all macromolecules, polysaccharides have been tailored by evolutionary pressures to interact with other molecules, large and small. The questions are what do they do and how do they do it?

Functions

Sugar-based structures have a variety of biological functions. These functions can usually be described as structural, energy storage and recognition.

STRUCTURAL POLYSACCHARIDES are molecules that protect and support biological structures. An important function of polysaccharides, especially in cells of microbes and plants, is to provide support. To accomplish this a **molecular net** is thrown around the fragile plasma membrane containing the cytoplasm (Fig. 6.3). A consequence of this is that the extracellular skeleton then largely dictates the shape and size of the cell. Typical constituents of this array are **insoluble fibrous polysaccharides** (Table 6.3) **cross-linked** to one another to form a net (Fig. 6.4). Polysaccharides that participate in cross-linking in this way are the other major class of skeletal cell-wall polymer.

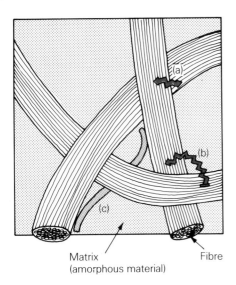

Fig. 6.3 The fibrillar molecular bag surrounding *Saccharomyces cerevisiae* which comprises the main component of the cell wall (×4800). Courtesy of H. Kopecka.

Fig. 6.4 Ways of cross-linking fibrillar components of cell walls to produce an encapsulating framework: (a) short (e.g. –S–S–) covalent bonds; (b) longer series of covalent bonds; (c) polymers comprising a number of sites binding to fibres through physical bonds.

Matrix (amorphous material)

Fibre

Table 6.3 *Structural polysaccharides*

Source	Polysaccharide	
	Fibrous	**Matrix (gel and cross-link)**
Bacteria	Peptidoglycan	
Yeast and fungi	β1–3-Glucan Chitin Cellulose	β1–3/1–6-D-Glucan Mannoprotein
Plants	Cellulose Callose (β1–3-glucan)	Xyloglucan Arabinoxylan Rhamnogalacturonan ⎱ Pectic Arabinogalacturonan ⎰ substances Agar Carrageenan
Animals	Chitin	Glycosaminoglycans

This table is not intended to be comprehensive.

Plant cell walls contain a number of these polysaccharides, which are known as **hemicelluloses** and which contain a variety of sugars. Cross-linking of the fibrous polysaccharides **β1–3-glucan**, **chitin** and **cellulose** that support yeast and fungal cell walls is less well understood but is probably through hydrogen and hydrophobic bonds with branched β1–3/1–6-D-glucans or formation of covalent bonds with proteins. The structure of the molecular net surrounding the cell is not permanent but is continually undergoing change to accommodate growth or morphological change.

The formation of extracellular support around both Gram-positive and Gram-negative bacterial cytoplasmic membranes is achieved in a different fashion to that in plant and yeast cell walls but with the same result. Fibrous polysaccharides are structurally modified to accommodate peptides that covalently cross-link them. The bacterial cell is thus surrounded by an enormous glycopeptide bag-shaped molecule that contains unique amino acid and sugar residues linked in a unique fashion. This giant biopolymer is called a peptidoglycan.

Polysaccharide or peptidoglycan nets, however, are not the universal solution to extracellular support. A number of mammalian cell types are surrounded, albeit with a much looser structure, by cross-linked fibrous proteins such as collagen and elastin. The requirements for support of these cells are the same, the response is similar, but the materials used are different.

Another important function of polysaccharides found in the cell walls of yeast, fungal, plant and some mammalian cells is to act as in-fill between the supportive skeletal fibres. This is **matrix** or **amorphous material** (Fig. 6.4). Unlike the insoluble skeletal polymers, these polysaccharides provide protection rather than support. They produce a compact **molecular barrier** protecting the fragile plasma membrane from a potentially harmful environment, yet are sufficiently porous to allow small nutrient molecules access for binding to transport proteins lodged in the membrane. The matrix is a layer of partially **immobilized water**, a **gel**, surrounding the cell. Mannoproteins and branched β1–3/1–6-D-glucans in yeast and fungal cell walls and pectins in plant cell walls are good examples of these types of carbohydrates (Table 6.3).

☐ Bacteria differ from one another in many ways, one of them being in the structure of their cell surfaces. The terms Gram-positive and Gram-negative refer to the ability of the bacterial surface to take up (+ve) or reject (−ve) the Gram stain (see *Cell Biology*, Chapter 2).

☐ Some of the amino acid residues, for example the alanine and glutamine residues, that are part of the peptide structure of the peptidoglycan are in the D configuration rather than the usual L form. Of the two carboxyl groups present in glutamate it is the one in the side-chain, the γ-COOH, rather than the usual α-COOH bonded to the α-carbon atom that is used in the formation of an amide link to the neighbouring residue. Diaminopimelate, a dibasic amino acid, may replace the more common lysine.

Box 6.2
Polysaccharide
nomenclature

Polysaccharides are named in two ways. They can be described by their traditional trivial name, e.g. amylose, glycogen, agarose, pectin, but they are often given names that describe their composition. Thus, glucans contain only glucose residues, mannans are formed from mannose, and arabinoxylans from arabinose and xylose. The names can be preceded by a description of the bonds between the sugar residues. Thus, amylose is described as an α1–4-D-glucan because neighbouring D-glucose residues are united by a glucosidic bond between the C-1 position (α-configuration) of one residue to the C-4 position of the next residue. Amylopectin, which has a branched structure (see later), can be described as an α1–4/α1–6-D-glucan.

Descriptions of polysaccharide structure employ a number of terms. Chains of sugar residues, like sequences of amino acids in proteins or nucleotides in nucleic acids, are directional. For example, just as proteins have amino and carboxy terminal amino acid residues, so polysaccharides have *reducing* and *non-reducing* terminal residues or ends. The term reducing refers to the chemistry of those sugar residues possessing a ring structure that can open to produce an aldehyde or a ketone group. Among other properties, these groups can reduce Cu^{2+} to Cu^+.

In addition to listing relative molecular weight values, M_r, a description of polysaccharide size is conveyed by quoting the number of sugar residues that comprise the polymer. This is the *degree of polymerization* (D.P.). *Chain length* (CL) refers to the number of sugar residues in a linear sequence.

Reference Candy, D.J. (1980) *Biological Functions of Carbohydrates*, Blackie, Glasgow, UK. A readable and well-illustrated account of simple sugar and polysaccharide chemistry and biochemistry.

Box 6.3
Cartilage

Cartilage consists of cells, chondrocytes, embedded in an extensive extracellular network, or matrix, of their own making. The matrix contains collagen fibres that provide the tensile strength of the tissue, and proteoglycans that are responsible for its resilience. Morphological cartilage provides shape to certain parts of the body, e.g. the nose and ears. Fibrocartilage has great tensile strength and is found in intervertebral discs. Elastic cartilage forms part of the rib cage and articular cartilage is found in joints between bone surfaces. Differences in the proportions of collagen and proteoglycan, together with variations in the fine details of their structure, account for the different properties of the cartilages.

See *Cell Biology*, Chapter 7

☐ Loose hydrated coats of polysaccharide surround the cells of some bacteria and yeasts, and are often thicker than the cell wall. Different species and strains produce different capsular polysaccharides, thus providing ways of distinguishing between cell types by immunological techniques. The pathogenicity of microorganisms can often be linked to the absence, presence, or structure of the capsule. Examples of the structure of some of these polysaccharides found on the surface of *Klebsiella* spp. are shown in Figure 6.5. They illustrate the variety of composition found within a species. In all cases, the structures consist of repeating sequences of residues.

See *Cell Biology*, Chapter 2

Fig. 6.5 Structures of the repeating units of capsular polysaccharides surrounding some *Klebsiella* strains.

See *Energy in Biological Systems*, Chapters 3 and 5

Surrounding some mammalian cells, particularly those comprising articular cartilage tissue between bone surfaces, is a gel layer which has the specialized and curious property of altering viscosity in response to forces applied to it. Cartilage itself has the property of absorbing shock. Energy is absorbed through deformation of the cartilage which then returns to its original shape. These properties of lubrication and resilience are provided' by a combination of **proteoglycans** (specialized forms of glycoproteins), glycosaminoglycans (polysaccharides), and proteins (such as collagen). The ability of proteoglycan solutions to become less viscous when shaken, stirred or forced through a nozzle is found in other solutions of biopolymers. Solutions of compounds possessing this **pseudoplastic** or **thixotropic** behaviour have many industrial uses.

The cell walls of bacteria do not appear to contain gel-forming polymers, though there are other major components besides peptidoglycan , e.g. the teichoic acids of Gram-positive cells, that contribute to the extracellular architecture. However, some bacteria and yeasts are surrounded by highly hydrated **capsules** of polysaccharide gel that are bound, often loosely, to the outer surface of the wall (Fig. 6.5).

STORAGE POLYSACCHARIDES are molecules rich in easily released chemical potential energy. As well as forming components of the extracellular framework, polysaccharide structures are found *within* the cell, in the cytoplasm and organelles. With few exceptions most cells are capable of producing and storing large ($M_r > 10^6$) branched α-D-glucans. Bacteria, fungi and mammalian cells produce **glycogen**. Plant cells contain **starch**, a mixture of **amylopectin** and **amylose**. These polymers are stores of potential metabolic energy and are therefore referred to as **storage** or **reserve polysaccharides**. In most cases this means that the glucosyl residues comprising the polymer may be converted to glucose 6-phosphate, which can enter the glycolytic pathway. Some ATP is produced by substrate-level phosphorylation and considerably more is produced through oxidative phosphorylation under aerobic conditions. Conversely, in times of plenty, whether carbohydrate is taken up, or manufactured within the cell (by photosynthesis or by gluconeogenesis), glucose 6-phosphate can be converted to glycogen or starch and the carbohydrate stored.

Glycogen and starch, though by far the most common, are not the only storage polymers. Some plants have unusual reserve polysaccharides. Jerusalem artichokes and dahlia tubers contain inulin, a fructan, while some grasses contain levans (Table 6.4). All of these polysaccharides are smaller than glycogen or starch, with M_r of 3500–8000 (about 20–50 sugar residues per molecule). The considerable advantages of storing sugar residues as polysaccharides are described later, but they are not the only stores of cellular

---

Teichoic acids: *from the Greek* teichos, *wall.*

energy. Fats, which are more efficient energy stores than polysaccharides, and some proteins perform the same function.

COMPLEX OLIGOSACCHARIDES are carbohydrates, participating in recognition between molecules. Later in the chapter it will be seen how polysaccharides form extended fibres for cellular support, water-trapping gels, and highly branched polymers for storage of metabolic energy. A common feature of all of these is that, for the most part, **structures are simple and repetitive**. There are, however, smaller carbohydrate polymers termed

See Chapter 7

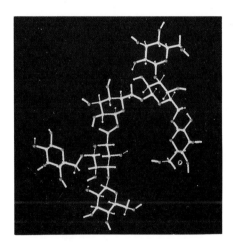
Reference Albersheim, P. and Darvill, A.G. (1985) Oligosaccharins. *Scientific American*, **253**(3), 44–50. The account of yet another unexpected way in which cells regulate their activities and the commercial possibilities this discovery could bring.

Box 6.6
Polysaccharide storage diseases

See Chapter 7

Beneficial as polysaccharides are, the pathways of synthesis and breakdown must be carefully balanced and controlled. Some pathological states are caused by genetic defects giving rise to enzymes, usually hydrolases, that are impaired in their catalytic activity or response to control. The glycogen storage diseases, which are sometimes fatal, are examples of these inborn errors of metabolism. McArdle's disease produces muscle cramps on exercise and is caused by accumulation of glycogen in muscle tissue because of the absence of glycogen phosphorylase. Pompe's disease arises from a deficiency of α-glucosidase in the lysosomes. There are also neurological disturbances and enlargement of the heart.

Mucopolysaccharidoses are characterized by the accumulation of various glycosaminoglycans. Hurler's disease starts in the first year of life and kills before ten years. Sufferers have enlarged elongated heads, flattened upturned noses and are dwarfs. The lack of lysosomal α-L-iduronidase is the cause. In Sanfilippo's disease the absence of N-acetyl α-glucosaminidase or heparan sulphamidase produces the same effects.

Glycolipidoses are a complex group of disorders described by a misleading term since there can be an accumulation of glycopeptides as well as glycolipids. They affect the nervous system. Tay–Sachs disease is caused by a deficiency of N-acetyl-β-hexosaminidase. Other disorders in this group are Gaucher's disease and Fabry's disease.

Table 6.4 *Storage polysaccharides*

Source	Polysaccharide
Bacteria	Glycogen
Yeast and fungi	Glycogen Galactomannan
Plants	Amylose } Starch Amylopectin Phytoglycogen Galactomannan Inulin } Fructans Laevan Laminarin (β1–3/1–6-glucan*)
Animals	Glycogen Galactan Mannan

* The principal link is β1–3: the few β1–6 links present inhibit association and account for the solubility of glucans.
This table is not intended to be comprehensive.

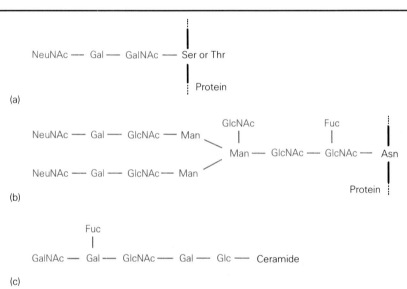

(a)

(b)

(c)

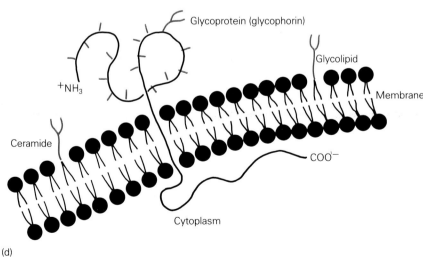

(d)

Fig. 6.6 Example of carbohydrate structures covalently bound to protein: oligosaccharides attached to the polypeptide portion of **glycophorin**, a principal glycoprotein of the red blood cell membrane. Fifteen serine or threonine residues carry the shorter chains shown in (a). There are some variations on this structure which are shown as —— in (c). The larger oligosaccharide (b) is carried on just one asparagine residue and is shown as —— in (c). A combination of small changes in amino acid sequence and the smaller oligosaccharides account for changes in antigenicity in the expression of the MN blood group. The carbohydrate portion is also the binding site of the influenza virus. The disposition of carbohydrates on the membrane penetrating glycophorin is shown in (d)

Oligosaccharides: *carbohydrate polymers comprising 3–20 sugar residues. A Greek affix denotes the number of residues; tetrasaccharide (4), octasaccharide (8), etc. From the Greek oligos, small.*

oligosaccharides, that are covalently attached to lipid and protein (Fig. 6.2) and where the structure is anything but simple. There may be many such carbohydrate groupings attached to a single protein molecule. The structure of these carbohydrate or oligosaccharide units is often **branched** and lacks repeating sequences (Fig. 6.6). What is the function of these side-chains?

Earlier in the chapter it was seen that one of the structural features distinguishing polysaccharides from proteins is the great variation possible in the configuration of the glycosidic bond between adjacent sugar residues. The many possibilites for linking allows a variety of shapes and surfaces to be formed from just a few sugar residues. Using a limited range of monosaccharides, such as mannose, galactose, *N*-acetylglucosamine, fucose and *sialic acid*, which are some of the commonest, a range of unique molecular shapes may be generated. These complex oligosaccharides act as **tailored keys** that assist a glycoprotein circulating in some vascular system, or a glycoprotein or glycolipid located on a cell surface, specifically to adhere to another cell surface. The oligosaccharide side-chain is the key, the cell-surface receptor protein is the lock. The combination of the two entities is very similar to the formation of an enzyme–substrate complex, but without the subsequent transformation of the substrate. In this way some circulating hormones are targetted to specific cells and cell–cell adhesion is made

---

Box 6.7
Glycolipids, glycoproteins and cell–cell recognition

The term blood group antigen or blood group substance is misleading for it suggests that these molecules, glycoproteins or glycolipids, are only associated with cells found in the bloodstream; such as the glycoprotein **glycophorin** (Fig. 6.6) which is on the surface of erythrocytes. This was, indeed, where they were first discovered but they are abundant elsewhere in organisms. They occur in relatively high concentrations on the surface of many cells. It is now clear that glycolipids and glycoproteins play major roles in many cell–cell recognition processes.

Specific interactions can occur between the surface proteins of one cell and the carbohydrate moieties of glycolipids or glycoproteins of the other. The cell density presumably influences the frequency of contact and leads to the promotion or inhibition of cell growth. In other words, glycolipids and glycoproteins participate in transcytoplasmic membrane-signalling mechanisms.

Glycolipids occur in high concentrations on the surface of epithelial cells that constitute the mucous lining of many organs. Over 80% of human cancers are epithelial in origin, and it is perhaps the malfunction of the cell density sensing mediated by glycolipids that is one of the causes contributing to the excessive cell proliferation that is characteristic of cancerous tissues.

During the early stages of embryo development changes occur in cell surface glycolipids. This process enables the dividing and differentiating cells to be sorted into the correct three-dimensional cellular patterns, so that the embryo grows and develops appropriately.

An example of binding between cells of different species is seen in N_2 fixation. Specific strains of the N_2-fixing bacteria *Rhizobium* adhere to the root-hair cells of selected plants because a lectin (that is, a glycoprotein that binds to specific sugars) on the plant cell surface binds to carbohydrate on the bacterial cell surface. The binding is specific; *R. trifolii* bind to clover and *R. phaseolus* to beans.

See *Biosynthesis*, Chapter 5

Glycolipid structure

See Chapter 8

There are over 100 different types of glycolipid structure, reflecting variations in the fatty acyl chains of the hydrophobic ceramide portion and the different combinations and types of sugar residue in the hydrophilic carbohydrate portion. The latter fall into five groups depending on the structure of the outer portion. They are ganglio, globo, lacto, type I and type II, and a remainder of disparate structure. A galactosyl–glucosyl disaccharide bridges these outer structures to the ceramide.

Sialic acids: a number of related acidic sugars: N-acetyl, N-glycolyl, 4,7,8,9-O-acetyl neuraminic acid.

Reference Paulson, J.C. (1989) Glycoproteins: what are the sugars for? *Trends in Biochemical Sciences*, **14**, 272–76. A useful survey of the role of carbohydrate in glycoproteins, e.g. protein stabilization and mechanisms of cell sorting and differentiation.

□ Lectins are carbohydrate-binding glycoproteins. They were first discovered in plants, and are now recognized to be present in bacteria, fungi, slime moulds, molluscs and body fluids of vertebrates and invertebrates. Different types of lectins bind to different sugars. This specificity is used in a number of biochemical purification and cytological procedures. When attached to insoluble, inert supports, for example, lectins provide a convenient means of fractionating glycoproteins. Derivitized with fluorescent compounds they are used for selectively binding to, and visualizing, cell-surface carbohydrates. The multivalency of lectins is used in blood group typing where they act as agglutinating agents.

See Chapter 2 and 3

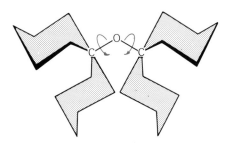

Fig. 6.7 Examining rotation of sugar residues, here shown in chair conformation, about the C–O–C bonds of the glycosidic bond show that ring collision can occur and that there is restricted freedom of movement.

selective. Another role is in the control of selective cleavage of proteins. The same glycoprotein can be cut at different points by different proteases to produce a range of peptides with different activities and targets. Proteins that bind to specific sugar residues within specific glycosidic linkages are termed **lectins**.

Structures

The size of polysaccharides varies enormously. In contrasting the structure of proteins and polysaccharides, attention has already been drawn to the enigma of polysaccharide size: its variability compared with the preciseness of proteins. Apart from the seemingly carefully controlled structure of oligo-saccharides of glycoproteins and glycolipids, the sizes of polysaccharides show a distribution of M_r within the limits: 10^3–10^4 for inulin and 10^7–10^9 for glycogen, for example. With some exceptions, the functions of poly-saccharide structure lie in the uniform and repeating shapes and surfaces rather than size itself. In contrast, the result of protein folding is to produce a unique but irregular surface, a feature that requires a precise sequence and fixed number of amino acids.

Knowing the shapes of polysaccharides is of great importance if the way they carry out their function is to be understood. Whilst little is known about the termination of polysaccharide extension, more is known about the influences controlling their folding. These are, of course, exactly the same forces that control the folding of other molecules. In constructing a scheme of **polysaccharide folding**, and to understand why they tend to remain in a particular conformation, it is necessary to examine interactions occurring between neighbouring residues, for instance those within a disaccharide located somewhere in the polymer. The individual sugars are relatively rigid in shape and have a more or less fixed conformation. However, there is some **freedom of movement** about the C–O–C bonds of the glycosidic link (Fig. 6.7). This freedom can be illustrated by arbitrarily considering one residue to be stationary and rotating the other about one of these bonds. It is found that in some circumstances the two **residues could potentially collide**. In terms of molecular electronic structure the van der Waals spheres of individual atoms approach too closely and are repulsed. Thus, the freedom of movement about the glycosidic bond is restricted. X-ray diffraction studies show that, at least in the solid state, disaccharides adopt a particular conformation. So why, out of the many possible conformations remaining, is one preferred and the others rejected? The answer lies in considering what happens when individual sugars approach one another.

In only a few conformations is proximity and linearity of some C, H and O atoms in the individual residues appropriate for **inter-residue H-bonds** to form. The restriction in movement is compensated by the formation of bonds, albeit weak ones. Hydrogen bonds that constrain maltose, an α1–4-linked glucosyl disaccharide and cellobiose, a β1–4-linked glucosyl dissacharide to take up particular conformations are shown in Fig. 6.8.

Forces that challenge these bonds are increases in bond vibration with raised temperature and competition for H-bonding from neighbouring molecules, typically water. Other bonds also contribute to influencing and stabilizing polysaccharide shapes. Just as the stacking of bases in double-helical DNA is assisted by hydrophobic bonds between the planes of purine and pyrimidine rings so hydrophobic bonds assist in the stacking of sugar chains on top of one another in, for example, cellulose and chitin microfibrils. The mutual repulsion of negative ion charges prevents anionic poly-saccharides from coiling up in, for example, the glycosaminoglycans and pectins.

Reference Hakomori, S. (1986) Glycosphingolipids. *Scientific American*, **254**(5), 32–41. A well-illustrated account of glycolipid structures and their relationships to functions, especially in cancerous tissues.

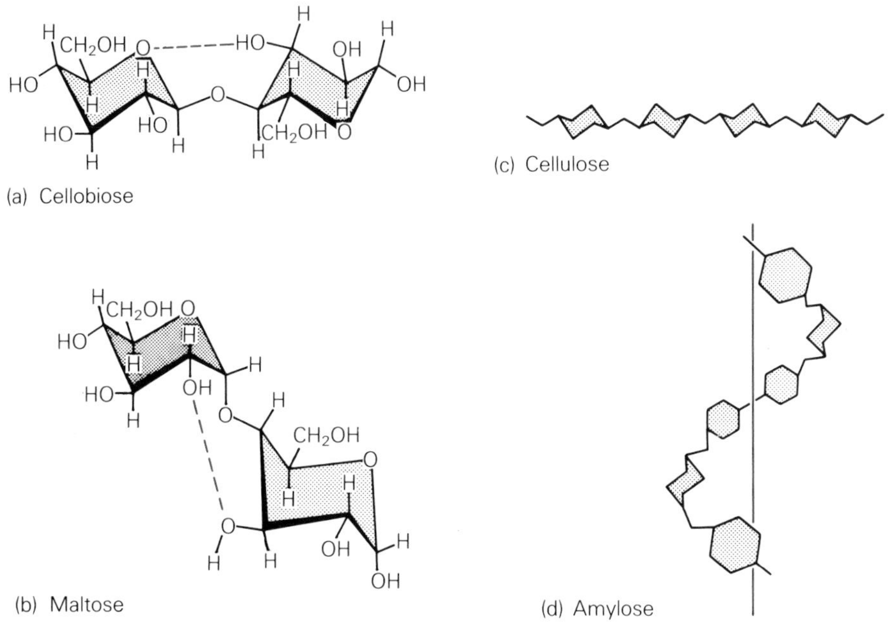

(a) Cellobiose

(c) Cellulose

(b) Maltose

(d) Amylose

Exercise 1

(a) Of all the residues in a polysaccharide only one has the reducing reactions of a monosaccharide. Explain why this should be? (b) If you were to introduce a tritium (^{3}H) label into this residue, which would contain the greater amount of radioactivity: 10 mg of amylopectin or 10 mg of amylopectin that had been exposed to an enzyme that hydrolyses α1–6-glucosidic bonds?

Fig. 6.8 Examples of physical bonds, in this case hydrogen bonds O...H–O, between neighbouring sugar residues that help to stabilize them in a particular conformation: (a) cellobiose (Glc,β1–4,Glc), (b) maltose (Glc,α1–4,Glc). Conformation obtained by extending these disaccharides with the same linked residues: (c) a portion of cellulose; (d) a portion of amylose which has been extended to illustrate the conformation. In reality it is more compressed.

If maltose or cellobiose is now extended by glucose residues linked in the same fashion (α for maltose and β for cellobiose), then two different patterns begin to appear. The α1,4-linked residues adopt a **helical conformation** and the β1,4-linked take up an **extended ribbon** or **linear conformation** (Fig. 6.8). Nevertheless, these conformations are maintained by rather weak bonds and they require stabilization for permanence. In the case of the β-1,4-linked ribbons, this occurs when they align to form bundles or **microfibrils** (Fig. 6.16). Further examples of combinations of residues and glycosidic bonds that produce ribbons or helices are shown in Table 6.2.

It is not unexpected to find that regular repeating sequences of residues and linkages form regular shapes. This presents a problem if polysaccharides are to produce net-like structures for the containment of cells. There must be molecular devices for cross-linking the fibrous molecules (Fig. 6.4). For example, the bacterial cell-wall peptidoglycan comprises a series of glycan ribbons that are covalently attached to one another through short peptides to form a bag-shaped molecule.

In the walls of plant cells the cross-linking is through **hemicelluloses** that bind via non-covalent bonds to the cellulose microfibrils. Polysaccharides that both bridge and bind must be adhesive in some parts but not in others. As might be expected, polymers with alternating properties have alternating sequences, producing alternating conformations. The regularity of periodic structure is occasionally altered by replacement of one sugar residue by another, covalent attachment of sugars producing side-chains, or covalent modification that changes the conformation of a residue (Fig. 6.9). Alterations tend to occur in groups and so produce **domains**. By these means, variations in folding patterns are introduced into the polymer at selected points. The consequence is a change in the ability of polymers to associate with one another. Helical sequences that were interweaving can no longer associate. Alternatively, lengths of helical polymers that become modified now extend into ribbon conformations and can associate with other linear polymers.

These structural changes, which need only be minor in occurrence, give rise to the term **interrupted** to describe this important group of polysaccharides.

Other, minor, interruptions in regular sequences can have a profound influence on structure. Thus, the occasional branching of polysaccharide

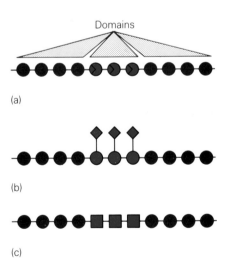

Fig. 6.9 Occasional changes to a regular sequence (●—●—●—●—) produce interrupted polysaccharides with domains that have different shapes and properties. Examples are: (a) structural changes (●) to the regular residue; (b) attachment of sugar residues (◆) to form side-chains; (c) replacement of the regular residues by another sugar (■).

Box 6.8
Multifunctionality

It is instructive to compare multifunctionality in proteins and polysaccharides. The idea that proteins have domains or specially folded regions of peptide within a single protein molecule that are responsible for specific functions has been established, for instance, in the dehydrogenase and immunoglobulin families. Some single polypeptide chains have more than one catalytic function. This idea of multifunctionality can be extended to polysaccharides in terms of their binding. At some points they associate with neighbouring polymers, at others with water. Moreover, in most polysaccharides the structural variations can be altered after initial synthesis. Since these changes are enzyme-catalysed, a hierarchy of controls exists to assist the changing requirements of the cell.

chains allows linear molecules to become spherical. The cross-linking of polymers through mutual dissociation and reassociation induced by structural interruptions can trap and immobilize water molecules forming gels.

From this general introduction to polysaccharide function and structure it can be seen that, like proteins, they adopt helical, linear and extended coil conformations. Specific polysaccharides will now be described in more detail to illustrate how the structure of the molecule accounts for the role it plays in the protection and survival of the cell.

6.3 Specific polysaccharides

The structure–function relationship of particular polysaccharides with a range of biological roles will now be discussed. This is not a comprehensive list, but rather examples have been chosen to illustrate the versatility of polysaccharides.

Glycogen

□ Interconversion of glucose 1-phosphate (G1P) and glucose 6-phosphate (G6P) is rapid. The enzyme phosphoglucomutase ensures that whenever G1P is produced, for example in the phosphorolysis of glycogen, or G6P from assimilation of glucose into the cell, photosynthesis or gluconeogenesis, then a supply of the other glucose phosphate is maintained.

Maintaining a supply of cytosolic glucose 1-phosphate and glucose 6-phosphate is essential to most cells. Glucose 1-phosphate is a source of carbohydrate residues for cell wall and extracellular polysaccharides. As a reduced molecule it is a potential source of energy because oxidation of glucose 6-phosphate can provide both NADPH and ATP. To store large amounts of a small molecule would cause an unacceptably high osmotic pressure to build up within the cell with a consequent entry of water. Continually pumping water out of the cell would require the expenditure of a great deal of energy, and the solution to this potentially explosive problem in almost every cell type is to polymerize the glucose residues to form large biopolymers. This avoids excessive cytosolic osmotic pressure building up. In plant cells the polysaccharide is starch, which consists of two types of polymer, amylose and amylopectin. In bacterial, fungal and animal cells it is usually glycogen.

Glycogen consists of chains of α1–4-linked glucose residues, the size of chain varying but usually being 8–12 units long. These chains are linked to one another in an extensive tree-like structure through α1–6-glucosidic bonds forming branch points. Experimentally, it is quite difficult to establish the details of branching patterns. One description is that, apart from chains at the very centre and on the outermost surface, the great majority carry two branch points (Fig. 6.10c).

Is there any reason why, during synthesis, this enlarging and continually branching molecule should be interrupted in its growth? Electron microscopy

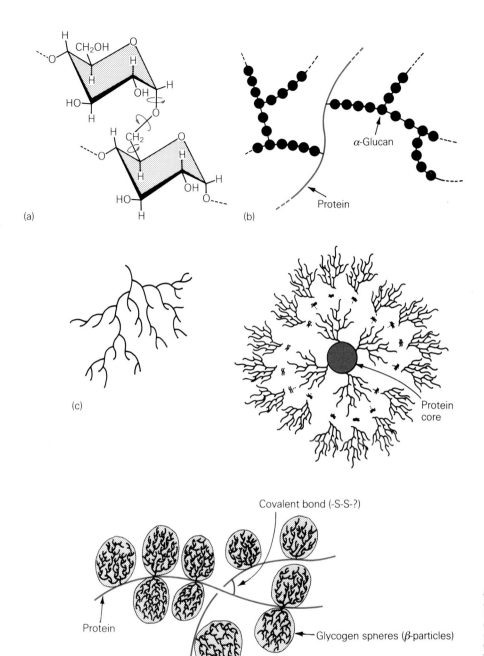

(a)

(b)
α-Glucan
Protein

(c)

Protein
core

Covalent bond (-S-S-?)

Protein

Glycogen spheres (β-particles)

Fig. 6.10 Aspects of glycogen structure: (a) the α1–6-glucosidic bond where extra flexibility is introduced into the molecule; (b) structure of glycogen in the centre of the molecule where glucan chains (—●—●—●—●—) are bound to protein (~); (c) branching pattern of glucan chains (λ) throughout the molecule; (d) cross-section through a glycogen particle; (e) possible arrangement of proteins uniting glycogen spheres.

of glycogen that has been carefully extracted or left undisturbed in tissue, shows it to be present as spheres, the so-called β-**particles** (Figs 6.10e and 6.11). These are of M_r 10^7 (degree of polymerization, D.P. 62 000). Computer modelling has predicted this size and shown the reason for limitation. Starting from a small chain, and after a number of branchings in all directions, the density of outer chains becomes so great that continuation of extension and branching is impossible (Fig. 6.10d). This point is reached after about 60 000 residues have been added. The final sphere has a radius of 20 nm.

Construction of this sphere depends on regular branching of the α1–4-glucan chains. But why should the branch point be sited at C-6? Why not use the presumably equally available OH-groups at C-2 or C-3? Model building

Reference Whelan, W.J. (1986) The initiation of glycogen synthesis. *BioEssays*, **5**, 136–40. An informative account chronicling changing opinions of the proposal that glycogen is a glycoprotein.

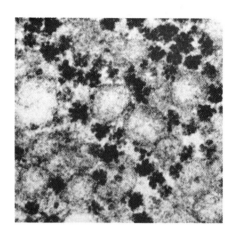

Fig. 6.11 Transmission electron micrograph of glycogen β-particles. Magnification × 50 000. Reproduced from Biava, C. (1963) *Laboratory Investigation*, **12**, 1179–97.

□ Glycogen spheres, or particles, bind a range of enzymes that are related to glycogen metabolism. Glycogen phosphorylase, for instance, has a specific binding site for glycogen, whch is quite separate from the catalytic site.

Fig. 6.12 Corncob (fruit of *Zea mays*) showing smooth and wrinkled seeds.

shows that addition of chains at these sites and at C-6 would distort the folding. In addition, there is a distinct advantage in using the 6-position that is not available when using C-2 or C-3. Carbon-6 is part of the hydroxymethyl group outside the sugar ring and is more mobile than the ring carbons. The linkage has an extra C–C bond in its structure and this gives it *one more degree of rotational freedom* (Fig. 6.10a) than an α1–4-bond (Fig. 6.8) and thus 1–4 more flexibility. Neighbouring chains can pack more easily in the region of this branch point and a more compact polymer is produced. Model building also shows that the concentration of glucose residues both within the glycogen sphere and on the surface is $3\,\mathrm{mol\,dm^{-3}}$. Since the M_r of a glucose residue is 162, this corresponds to 3×162 or $486\,\mathrm{g}$ glucose per $\mathrm{dm^3}$. Thus, enzymes that attack the non-reducing termini, such as phosphorylase which releases glucose 1-phosphate, have a high concentration of substrate readily available to them as they diffuse on to the surface of the glycogen molecule. The structure of glycogen is more complex than a simple sphere. The spheres themselves are organized, a fact that emerged when extraction techniques were modified.

An ever-present experimental hazard in the analysis of cells is that methods used for their disruption may distort or modify the very component under examination. If a glycogen-containing organ, e.g. rat liver, is rapidly isolated, stored and the glycogen extracted from it in conditions that minimize changes to the polysaccharide through enzyme or chemical action, then glycogen of M_r larger than 10^7 is obtained. This observation, together with evidence showing that glycogen always seems to contain small amounts of protein, has led to the belief that the self-limiting spheres are covalently attached to a protein backbone or core (Fig. 6.10b, e), and that the backbones themselves are bound to one another through disulphide bridges. The M_r of this molecule is about 10^9 (D.P. 6.2×10^6). Indeed, the protein backbone may serve as the initial primer, or foundation, molecule on which the glycogen spheres are constructed. The structural organization of glycogen is complex and, in mammalian cells at least, is probably clusters of β-particles (Fig. 6.10e). Possibly this is a method of controlling the number of glucose residues on the surface available for interaction with enzymes of glycogen synthesis and degradation.

Starch

A common reserve of carbohydrate in plants is starch, which is composed of the two α-D-glucans, amylose and amylopectin. Most starches contain 20–30% amylose and 70–80% amylopectin but there are a few plants where the proportions are quite different. Wrinkle-seeded peas and some maizes and barleys for example, contain more than 60% amylose (Fig. 6.12). A few cereal starches, notably those from waxy varieties of maize and sorghum, consist almost entirely of amylopectin.

AMYLOSE consists of a series of chains of α1–4-linked glucosyl residues of D.P. greater than 100. The chains are linked together at each end, i.e. at their reducing and non-reducing termini, through α1–4-glucosidic linkages to produce a linear, unbranched polymer. Most molecules are of D.P. 2000–3000 that is with M_r in the region of $3–5 \times 10^5$, though heterogeneity is such that sizes probably vary from D.P. 2000 to 6000.

A combination of restricted rotation and stabilizing bonds between residues produces repeating shapes from periodic polysaccharides. The α1–4-bonded glucose residues of amylose tend to form a **left-handed helix**. A particularly stable form occurs when there are about six residues per turn, for not only are neighbouring covalently bound residues H-bonded to one

Reference Rees, D.A. (1977) *Polysaccharide Shapes* (Outline Studies in Biology series). Chapman and Hall, London, UK. An excellent description of the stereochemistry of sugars and polysaccharides.

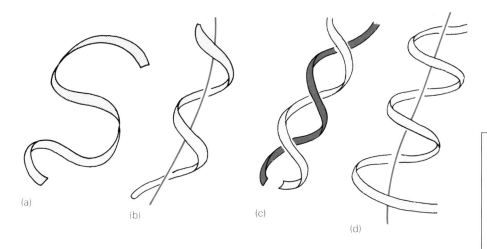

(a)

(b)

(c)

(d)

Fig. 6.13 Aspects of amylose (Glc,α1–4,Glc)$_n$ conformation. Tendency of the linear polymer (∼) to form a left-handed helix (a → b) which may combine with a neighbouring molecule to form a double helix (c). In some circumstances, not occurring *in vivo*, the helix widens and compresses (d) to accommodate guest molecules such as I$_2$.

another but so are residues separated by 5 units, which therefore sit on top of one another at the helix surface (Figs 6.8 and 6.13). The deep blue colour produced when iodine is added to solutions of amylose is thought to be caused by I$_2$ molecules lining up in the hollow centre of the spiral, thus stabilizing the coiled amylose. Similarly, molecules of polar organic substances, such as thymol or *n*-butanol can penetrate the spiral and precipitate amylose from solution. This property is used to separate amylose from amylopectin.

It is suggested on the basis of X-ray studies that two amylose helices can merge to form a double helix. One possible arrangement is shown in Fig. 6.13. This is the compact, naturally crystalline, form of amylose that is present in starch granules.

AMYLOPECTIN, like amylose, consists of glucose residues linked by α1–4- and α1–6-glucosidic bonds. However, whereas amylose has very few α1–6-bonds (less than 1% of total bonds), amylopectin has many more, usually about 5% and they act as branch points; chain lengths of amylopectin are 20–25 residues long. The D.P. of most amylopectins is in the region 10^4 to 10^5, giving an M_r of about 10^7, which is considerably greater than amylose and similar to that of glycogen. Within the cell, amylopectin appears as a series of clusters located in granules.

In green plants, starch is almost always found in granules. Here, contrary to what might be expected from its structure, it is virtually water insoluble. Granules vary in shape and may be from 1 to 100 μm in diameter. They are found in chloroplasts where starch is synthesized and temporarily held, or in amyloplasts, which are long-term storage sites. Cross-sections of granules reveal a layered structure of concentric rings 100–400 nm thick (Fig. 6.14). Each ring (Fig. 6.15) contains radial arrays of amylopectin molecules; each polymer stretching the width of the ring and comprising a series of domains or clusters every 5–10 nm. Where chains are aligned, there is a strong possibility for double-helix formation. Helices like these are more likely to form with amylopectin than with glycogen because of the longer stretches of uninterrupted α1–4-linked residues. Interactions like these produce a very compact molecule and reduce the extent to which water can penetrate the amylopectin array. Obviously, storing 'glucose' in this insoluble form reduces still further the osmotic pressure that it exerts in the cell.

☐ There are some exceptions to this portrayal of amylopectin structure. Phytoglycogen, the branched component of starch in sweet corn (*Zea mays*) has shorter chains. It is, as its name implies, more like glycogen than amylopectin.

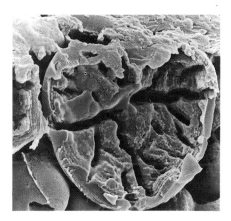

Fig. 6.14 Scanning electron micrograph of a cross-section of a barley starch granule showing erosion during germination. The lamellae (layered) structure of the granule is apparent. Granules are found in pollen, leaves, stems, woody tissues, roots, tubers, bulbs, rhizomes, flowers, fruits and the pericarp, cotyledons and endosperm of seeds. Magnification × 10 000. Courtesy of Dr G.H. Palmer, Department of Biological Sciences, Heriot-Watt University, UK.

Main sources of starch are the grains, tubers and roots of cereals. Extraction of starch from granules is called gelatinization. Slurries of granules are heated until disruption and hydration occurs. The temperature at which this takes place is characteristic for each cereal. Starch, often chemically modified, is used for coating papers and fabrics to improve their binding properties, as a thickening and gelling agent in the food industry, drilling muds, for the preparation of adhesives and printing inks, and as a binding agent in coal bricks. Starch is also used as a support medium in electrophoresis.

Products of starch hydrolysis are used in the food, beverage and fermentation industries. The action of acid and carbohydrases, followed by controlled evaporation, converts starch to syrups containing glucose, maltose and oligosaccharides. The extent of hydrolysis is described by a dextrose equivalent (D.E.) number. Starch has a D.E. of 0 and dextrose, another name for glucose, has a D.E. of 100. Maltodextrins have a D.E. < 20. Corn syrup, also known as glucose syrup and liquid glucose, has a D.E. > 20. High-fructose corn syrup (HFCS) is a corn syrup produced with the added step of the enzymic conversion of some glucose to fructose.

Maltodextrins, because of their viscosity, are used to control the texture of foods. Corn syrups are used to prevent the crystallization of sucrose in the preparation of toffees and boiled sweets, or to produce very fine sucrose crystals in the manufacture of fondants. An additive like this is called a 'doctor' in the trade. HFCSs, because of their fructose content, are used as sweeteners, and syrups of high D.E. are a source of metabolizable sugar in the fermentation industries.

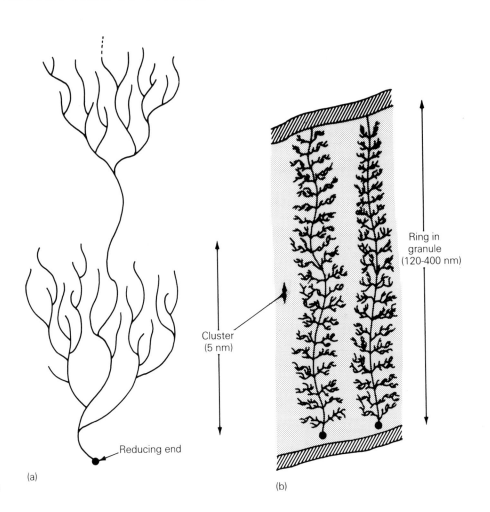

Fig. 6.15 Aspects of amylopectin structure: (a) the cluster structure. A number of variations have been proposed, the differences being in the extent and regularity of branching; (b) ordered alignments of amylopectin cluster structures are thought to account for annular rings seen as internal structures of starch granules.

(a)

(b)

Reducing end

Cluster
(5 nm)

Ring in granule
(120-400 nm)

Amylose and amylopectin are but two of many polysaccharides that form helical structures. However, it must not be assumed that helical structures are found only in reserve polysaccharides. Another commonly occurring helix is that of the β1–3-D-glucan found in the cell walls of fungi and yeast. It is, however, considerably less soluble in water than amylose, is more extended, and associates to form a triple helix that is part of the cell wall fibrillar structure.

Exercise 3

What happens to a viable yeast cell when it is suspended in a mixture of active β-glucanase and protease enzymes? Why should this be and how might the final outcome be prevented?

Cellulose

Cellulose is a major constituent of plant, and some fungal, cell walls. It is also secreted by a few bacterial species. Thus, it is one of the most widely occurring of all biopolymers. Although both cellulose and amylose have identical chemical compositions and are both glucans possessing 1–4-linkages, they are very dissimilar in function. A difference in role suggests a difference in polymer shape. This difference is seemingly simple, but the effects are far reaching. The 1–4 linkage has a β configuration in cellulose, whereas it is α in amylose. In other words, the configuration at C-1 has been reversed.

The function of cellulose is to provide a supportive and protective framework outside the plant cell membrane. Whether that skeleton is sturdy and rigid, as in wood fibre, or thin and flexible, as in seedling shoots, depends on the amount of cellulose laid down and its interaction with other cell-wall polysaccharides, the hemicelluloses and pectins, non-carbohydrate polymers, such as lignin, and proteins, for example extensin. Compared with starch, cellulose is merely a component molecule contributing to a function. It

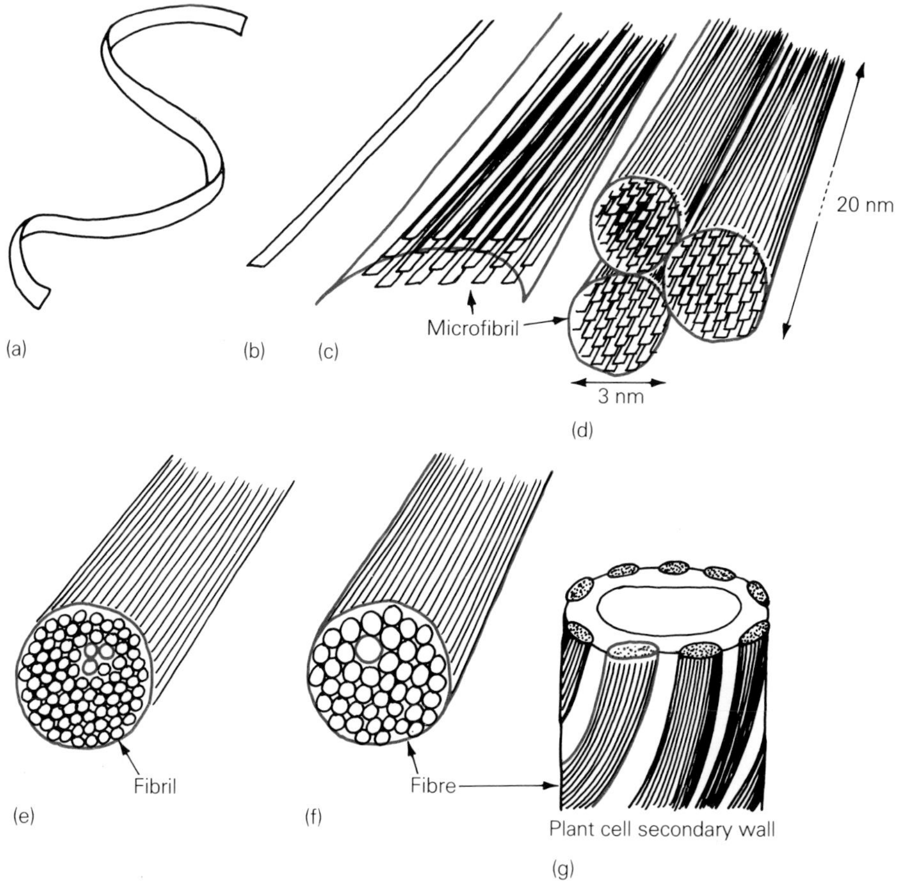

Exercise 4

It is important to 'translate' the nomenclature that describes bonds between sugars and to remember the shapes of residues. Without referring to the text, draw the conformational structures of (a) gentiobiose, a disaccharide consisting of two D-glucose residues linked through a β1–6-glucosidic bond, and (b) two D-mannose residues linked through an α1–2-mannosidic bond. Both these structures are found in the polymers of the yeast cell wall.

Fig. 6.16 Aspects of cellulose (Glcβ1–4,Glc)$_n$ conformation and structure. Tendency of the linear polymer ($\sim$) to fully extend (a → b), followed by association to form a microfibril (c) which in turn aligns with others (d → e) to eventually form a cellulose fibre (f). (g) Orientation of fibres in a layer of secondary wall. Other layers of secondary wall have different orientations. The extending and association of chitin (GlcNAcβ1–4GlcNAc)$_n$ molecules follows the same pattern.

Configurational change: *rearrangement of covalent bond structure, e.g. mannose and glucose have different configurations of –H and –OH about C-2.*

Fig. 6.17 Scanning electron micrograph of cellulose fibres. Magnification × 30 000. Reprinted from Preston, R.D. (1964) *Endeavour*, **23**, 153–9.

See Chapter 3

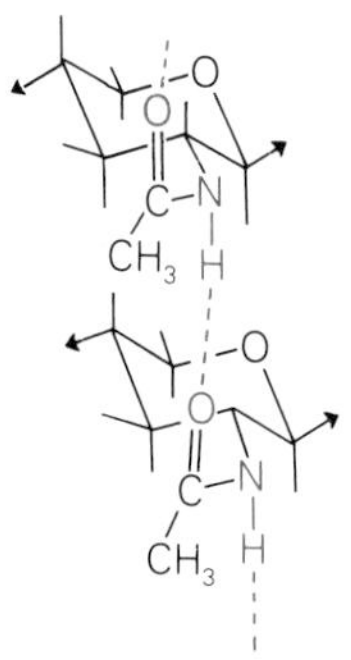

Fig. 6.18 Hydrogen bonding (dotted lines) between acetamido groups of adjacent chitin chains that is in addition to the inter- and intra-chain hydrogen bonding similar to that in cellulose microfibrils.

might be asked why such a trivial difference as changing an α1–4- to a β1–4 linkage should produce a polymer with different properties. It has already been mentioned that α1–4 linked glucose residues tend to form a helix and β1–4 linked residues tend to form an extended ribbon (Fig. 6.8), and that it is possible for double helices to form, but that is the end of helix association.

The ribbons of cellulose, on the other hand, can pack together in huge arrays, by establishing non-covalent bonds between themselves rather than with water. Hydrogen bonds and hydrophobic bonds form between individual polymers, and association is not random. A series of molecular bundles associate on an ever-increasing scale. (Comparison of this scheme with that of collagen assembly reveals a remarkable similarity.) The stages of cellulose fibre assembly are shown in Fig. 6.16, commencing with microfibrils containing about 40 molecules packed together in a layered and staggered fashion. This produces an inextensible, fibrous water-insoluble array (Fig. 6.17) suited to the encasement and protection of the plant cell. Not only is there an ordered regime of molecular associations but also their alignment produces a series of concentric cylinders. The outermost of these, the primary cell wall, comprises randomly orientated bundles. The inner or secondary walls, of which there may be many, display aligned bundles which differ in orientation from one layer to the next (Fig. 6.17).

The size of individual cellulose molecules depends on their location in the cell wall. For primary wall polymers the D.P. is in the region of 2000 (M_r 320 000). It is considerably higher in the inner secondary walls with D.P.s of 15 000 (M_r 2.4×10^6) or greater.

Chitin

Chitin is a commonly occurring polysaccharide, similar to cellulose in both structure and function. It is the main constituent of the arthropod (crustaceans, insects, centipedes, millipedes and spiders) exoskeleton and is a major component of fungal cell walls. In place of glucose residues are *N*-acetylglucosamine residues, but the linkage is still β1–4. Perhaps surprisingly, the much bulkier acetamido group appears to make little difference to the packing of the ribbon-like polymers. The individual polymers remain fully extended, as in cellulose. In addition, there is further enhancement of inter-ribbon interactions through H-bonding between acetamido groups in the ribbons stacked above and below each other, thus producing an even more rigid and inert association than that found in cellulose (Fig. 6.18).

Box 6.10
Commercial use for chitin

The amount of chitin biosynthesized may be similar to the 10^{11} tonnes of cellulose that are produced globally each year. Attempts have been made to harvest, extract and find commercial uses for chitin or its derivatives. One such example is the treatment of crab shell to extract and convert chitin to chitosan, the deacetylated form of the polymer. Unlike chitin, chitosan is water soluble and can be used to coat fruits, producing a thin film that provides protection against their deterioration.

Reference Albersheim, P. (1975) The walls of growing plants. *Scientific American*, **232**(4), 81–95. A lucid description of plant wall assembly with useful descriptions of experimental procedures.

6.4 Overview

The contributions of polysaccharides to microbial, plant and most animal cells are vital. Their functions may be categorized under three headings. They form *extracellular support and protection* through a combination of fibrous and matrix polymers. In addition they are major determinants of cellular shape and size. They provide *intracellular storage* of carbohydrate, which is subsequently oxidized to produce essential ATP and NAD(P)H. Covalently bound to protein and lipid they act as *molecular tags* that participate in intercellular communication and recognition.

In common with all naturally occurring polymers it is presumed that evolutionary pressures have selected polysaccharide structures to perform particular functions. Two major conformations appear: the fully extended ribbon and the less-extended helix. Explanations for these shapes are found when the covalent and non-covalent bonds between neighbouring groups are examined. Studying the shape of these molecules and how they interact with each other and smaller molecules in the environment provides an explanation of their functions.

Answers to Exercises

1. (a) It is the only residue that does not participate in linkage from the (usually) C-1 carbon to the next residue. The ring can therefore open to produce the reducing aldehyde group. (b) Amylopectin exposed to a debranching enzyme will produce many more reducing groups and will be more extensively labelled. The large D.P. of amylopectin results in very little label being introduced.

2. The blue coloration will disappear because the products of digestion cannot form helices to contain the iodine. α-amylase will destroy the coloration faster because of its random attack. With 50% hydrolysis by β-amylase there is still half an amylose molecule left that can readily complex with iodine.

3. These enzymes will digest the yeast cell wall. The cell will then burst through its internal hydrostatic pressure. This can be prevented by suspending the cell in a hypertonic medium (e.g. 0.7 mol dm^{-3} KCl) that abolishes pressure difference and allows the cell wall to be removed without destroying the cell. This is the technique of producing stable protoplasts.

4.

Glc,β1–4,Glc

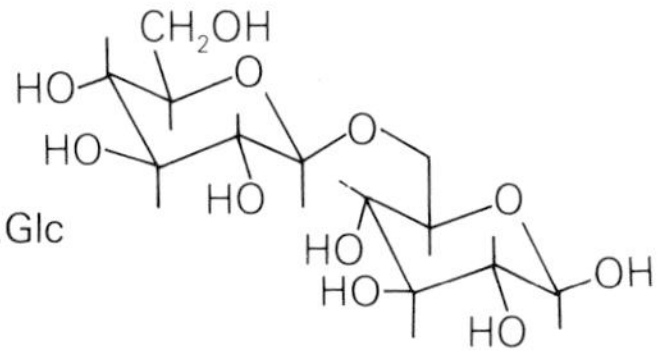

(a)

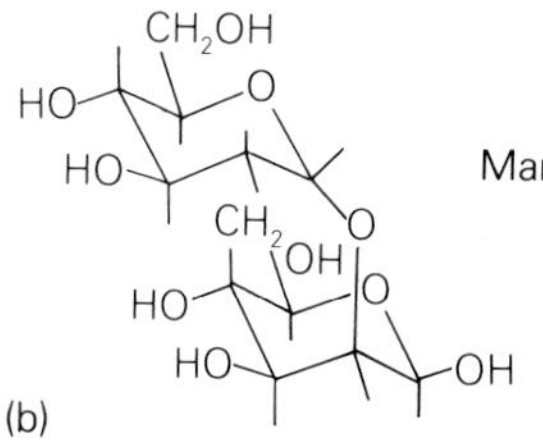

Man,α1–2,Man

(b)

FILL IN THE BLANKS

1. _____ differs from _____ in having less branching. However, _____ has even more _____ branch points than _____ , although all three polysaccharides are composed of _____ residues. _____ consists of β-D-glucose residues, while the similar carbohydrate, _____ contains N-acetylglucosamine.

The conformations of differing polysaccharides vary. Amylose forms a _____ structure. Cellulose and chitin both consist of extended _____ . Pectate has a _____ _____ conformation which is stabilized by _____ binding to adjacent _____ residues giving an _____ structure.

Peptidoglycans of bacteria occur in a number of types although all are _____ in some manner forming a _____ 'bag-shaped' molecule which gives extracellular support. Proteoglycans of _____ cells also occur _____ forming part of the cell _____ . A proteoglycan complex consists of polysaccharides called _____ linked to proteins all bound to a _____ molecule of _____ .

Complex oligosaccharides usually consist of _____ structures which lack _____ sequences. These are therefore suitable for participating in recognition phenomena. Such oligosaccharides generally occur as part of _____ and _____ .

Choose from: $\alpha1$–6, amylopectin (2 occurrences), amylose, animal, branched, buckled, Ca^{2+}, cellulose, chitin, cross-linked, egg-box, extracellularly, galactouronate, α-D-glucose, glycogen, glycolipids, glycoproteins, glycosaminoglycans, helical, hyaluronate, matrix, repeating, ribbon, ribbons, single (2 occurrences).

MULTIPLE-CHOICE QUESTIONS

In all the following questions, select those options (any number may be correct) that correctly complete the initial statement, or correctly answer the question posed.

2. Match the letter and the number(s) that best describe the following polysaccharides:

A. amylose
B. amylopectin
C. cellulose
D. glycosaminoglycan

as structures that are:

1. branched
2. linear
3. glucans
4. glycans
5. composed of more than one type of sugar residue
6. anionic

3. Match the letter and the number that best describe the following polysaccharides:

A. chondroitin 4-sulphate
B. glycogen
C. chitin
D. β1–3-glucan
E. inulin
F. galactomannan

as used for:

1. metabolic energy reserve
2. cellular exoskeletal support
3. extracellular gel formation

4. Match the letter and the number(s) that best describe the following polysaccharides:

A. cellulose
B. hyaluronate
C. rhamnogalacturonan
D. callose
E. chitin

as being found in:

1. bacterial cells
2. fungal cells
3. plant cells
4. mammalian cells

SHORT-ANSWER QUESTIONS

5. Describe a method for the separation of amylose from starch granules. What criterion of purity could be used?

6. What solvents might be used to solubilize a polysaccharide that is insoluble in water?

7. Describe a method for determining whether the hydrolysis of a polysaccharide by a carbohydrase occurs at random sites or sequentially from one end.

8. What chemical analyses might be used to distinguish between an α1–4 and an α1–3-glucan? What other approaches might be used?

9. How might the extent and permanence of immobilized water in a gel be investigated?

10. How could the extent to which polysaccharides associate with one another be studied?

11. Suggest ways in which the following carbohydrate structures:

–A–A–B–B–B–B–A–A–A–A–A–A–

and   –A–A–B–A–A–B–A–A–B–A–A–B–

might be distinguished.

7

Lipids: structures and functions

Objectives

After reading this chapter you should be able to:

☐ summarize the structural diversity exhibited by lipid molecules;

☐ describe the classification of lipids;

☐ explain the importance of lipids as structural and functional units of biological membranes;

☐ explain the importance of lipids as specific metabolites, and as modulators of biochemical activities.

7.1 Introduction

The term **lipid** is used to describe the oily, greasy and waxy materials that can be extracted from organisms using organic solvents. Lipids have many biological functions, and this is reflected in the variety of individual structures encountered. Lipids are important as fuel molecules whose oxidation is described elsewhere (see *Energy in Biological Systems*). The structural diversity of lipids is also used by organisms to produce a variety of specific metabolites. These metabolites function as hormones, receptors and modulators of metabolic activities. In addition lipids have structural roles as major components of membranes, separating the cell contents from the external environment and dividing it into discrete structural and functional compartments. Lipids also form important protective and insulating barriers. For example, plant leaves are covered with a waxy cuticle to minimize dehydration, and to help protect the plant from microorganisms. Mammals, typically have a subcutaneous layer of fat that acts as a thermal insulator, and also protects against physical damage. The lipid-rich myelin sheath of some nerve fibres is important in the transmission of nerve impulses.

7.2 Classification of lipids

Although lipids show a wide variety of structures, they possess a number of common features, which may be used in classifying them. Classification schemes are often useful in highlighting common features, but they are usually arbitrary. Since lipids show a wide range of structures, it is not surprising that they are classified in several ways. None is completely satisfactory, but the scheme outlined in Table 7.1 is good as any and highlights some of the problems of nomenclature. Thus, many lipids have several names: for example, phosphoacylglycerols and sphingomyelins are also called phospholipids, since they contain phosphate, while sphingomyelins, cerebrosides, and gangliosides are sometimes referred to as

Lipids: *from the Greek* lipos, *fat*.

Table 7.1 *A common method of classifying lipids*

Group	Example	Products of hydrolysis (molecular constituents)
Simple	Waxes	Monohydroxyalcohols and fatty acids
	Acylglycerols (neutral fats)	Glycerol and fatty acid(s)
Complex — Phospholipids ┬ Phosphoacylglycerols		Glycerol, fatty acids, phosphate and a variety of other molecules, e.g. choline
└ Sphingomyelins		Sphingosine, fatty acid, phosphate and a variety of other molecules, e.g. choline
└ Glycolipids ┬ Cerebrosides		Sphingosine, fatty acid and simple sugar(s)
└ Gangliosides		Sphingosine, fatty acid and simple sugars, including sialic acid
Polyprenyl lipids or isoprenoid compounds	Steroids Carotenoids Lipid vitamins	Cannot be hydrolysed

sphingolipids, because they contain sphingosine. Since cerebrosides and gangliosides contain sugars they are also called glycolipids.

7.3 Simple lipids

Simple lipids are esters: their hydrolysis yields an alcohol and fatty acid(s) only. The fatty acids generally have long hydrocarbon portions.

WAXES are esters formed from a fatty acid and an alcohol. Both acid and alcohol components contain long hydrocarbon chains. The general formula of a wax is given in Fig. 7.1, which also gives the structure of myricyl palmitate, a major component of beeswax. Waxes are totally insoluble in water, a consequence of their hydrocarbon, non-polar, nature. Their major biological function is to act as coatings of surfaces, preventing dehydration and reducing the risk of infection. Plant leaves have a waxy cuticle and birds, particularly aquatic types, produce a waxy secretion from the preen gland, to be spread over the feathers to waterproof them.

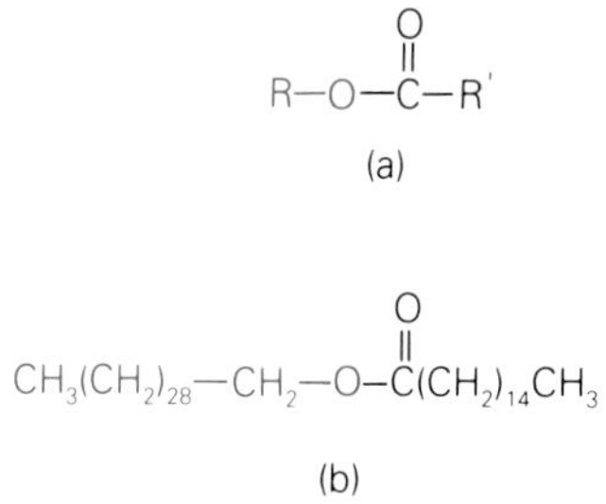

ACYLGLYCEROLS or neutral fats are formed from the trihydric alcohol glycerol and up to three fatty acids (see later) (Fig. 7.2). Each of the fatty acids

---- Exercise 1 ----

Try to devise an alternative, simple, classification of lipid molecules.

Fig. 7.1 (a) The generalized structure of a wax, and (b) the structure of the wax, myricyl palmitate.

☐ The term *acyl* refers to the fatty acid residue in a lipid joined to an alcohol by either an ester or an amide linkage.

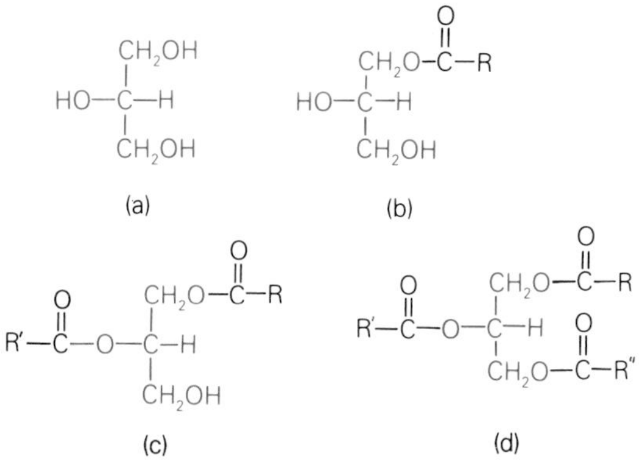

Fig. 7.2 Structures of (a) glycerol, (b) 1-mono-acylglycerol, (c) 1,2-diacylglycerol and (d) 1,2,3-triacylglycerol.

Reference Gurr, M.I. and Harwood, J.T. (1991) *Lipid Biochemistry*, 4th edn, Chapman and Hall, London, UK. A very good, in-depth coverage of lipid biochemistry.

Reference Gunstone, F.D., Harwood, J.L. and Padley, F.B. (eds) (1986) *The Lipid Handbook*, Chapman and Hall, London, UK. Splendid reference book covering most aspects of the chemistry, biochemistry and industrial and medical uses of lipids.

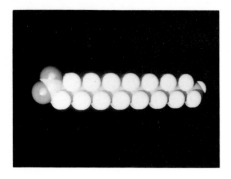

Fig. 7.3 Computer-drawn model of palmitate. The terminal carboxyl oxygen atoms are shown as darker spheres. Courtesy of Dr C. Freeman, Polygen, University of York.

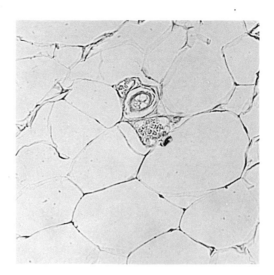

Fig. 7.4 Light micrograph of white adipose tissue (magnification × 900). Note the large lipid(L)-filled cells, with only a small amount of peripheral cytoplasm (C). Courtesy of M.J. Hoult, Department of Biological Sciences, Manchester Polytechnic, UK.

Exercise 2

Draw the structure of 1-palmitoyl, 3-oleoyl-diacylglycerol.

See *Energy in Biological Systems*, Chapter 7

$$R-\overset{O}{\underset{OH}{C}} \quad R-\overset{O}{\underset{O}{C}}H$$

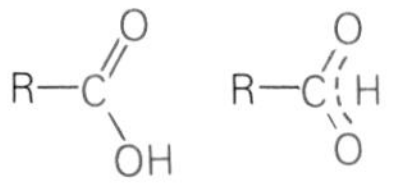

$$R-COOH \quad R-CO_2H$$

Fig. 7.5 Four equivalent representations of a fatty acid: R is generally an extended hydrocarbon chain.

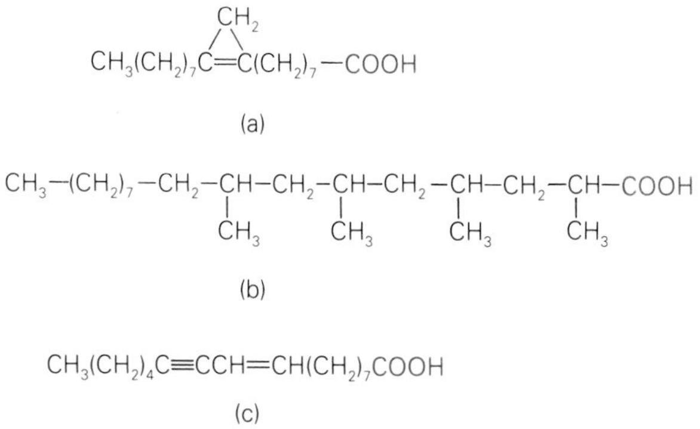

in a triacylglycerol may be different, giving a variety of potential structures. Typically the fatty acid at C-2 is unsaturated, while C-1 and C-3 have a high proportion of palmitate (Fig. 7.3).

Neutral fats are stored as large droplets in the cells of white adipose tissue. These lipid stores occupy much of the volume of the fat cell (Fig. 7.4), providing an energy-rich reserve. The energy reserves of adipose tissues are mobilized by hydrolysing the acylglycerols to release free fatty acids. These can then be oxidized to give metabolic energy. The most significant biological process for oxidizing fatty acids is the β-oxidation pathway although other oxidative pathways are known. White adipose tissue acts as a protective cushion around certain organs, such as the kidneys. The layer of fat beneath the skin of mammals forms an effective thermal insulating layer, as well as providing mechanical protection.

Triacylglycerols are transported around the body combined with protein to form **lipoprotein** particles. Many of the biologically relevant properties and reactions of triacylglycerols are those of their constituent fatty acids.

FATTY ACID is a common and useful a term meaning an organic acid of the aliphatic series (see Fig. 7.5 for structure) potentially to be found in fats. A

$$CH_3(CH_2)_7 \overset{CH_2}{\overset{/\backslash}{C}=C}(CH_2)_7-COOH$$

(a)

$$CH_3-(CH_2)_7-CH_2-\underset{CH_3}{\overset{|}{CH}}-CH_2-\underset{CH_3}{\overset{|}{CH}}-CH_2-\underset{CH_3}{\overset{|}{CH}}-CH_2-\underset{CH_3}{\overset{|}{CH}}-COOH$$

(b)

$$CH_3(CH_2)_4C\equiv CCH=CH(CH_2)_7COOH$$

(c)

Fig. 7.6 Structures of three unusual fatty acids:
(a) sterculic acid from higher plants; (b) mycocerosic acid, which can be isolated from *Mycobacteria* species, and (c) crepenynic acid from *Crepis foetida* seeds.

Table 7.2 *Structure and nomenclature of common fatty acids*

No. of carbons	Common name	Systematic name	Structure	Comments
Common saturated fatty acids				
4	Butyric	Tetranoic	$CH_3(CH_2)_2COOH$	Major fuel molecule for ruminants; high concentrations found in ruminant milk
6	Caproic	Hexanoic	$CH_3(CH_2)_4COOH$	
8	Caprylic	Octanoic	$CH_3(CH_2)_6COOH$	
10	Capric	Decanoic	$CH_3(CH_2)_8COOH$	Effective limit of water solubility
12	Lauric	Dodecanoic	$CH_3(CH_2)_{10}COOH$	Widely distributed; a major component of some seed fats
14	Myristic	Tetradecanoic	$CH_3(CH_2)_{12}COOH$	
16	Palmitic	Hexadecanoic	$CH_3(CH_2)_{14}COOH$	Widespread; one of the commonest fatty acids in plants and animals
18	Stearic	Octadecanoic	$CH_3(CH_2)_{16}COOH$	
20	Arachidic	Eicosanoic	$CH_3(CH_2)_{18}COOH$	
Common unsaturated fatty acids				
One double-bond (C_{18})	Oleic	9-Octadecenoic	$CH_3(CH_2)_7CH{=}CH(CH_2)_7COOH$	Most abundant unsaturated fatty acid in plants and animals
Two double-bonds (C_{18})	Linoleic	9,12-Octadecadienoic	$CH_3(CH_2)_4CH{=}CHCH_2CH{=}CH(CH_2)_7COOH$	Major component of plant lipids; in animals it is derived from plant sources
Three double-bonds (C_{18})	Linolenic	9,12,15-Octadecatrienoic	$CH_3CH_2CH{=}CHCH_2CH{=}CHCH_2CH{=}CH(CH_2)_7COOH$	
Four double-bonds (C_{20})	Arachidonic	5,8,11,14-Eicosatetranoic	$CH_3(CH_2)_4CH{=}CHCH_2CH{=}CHCH_2CH{=}CHCH_2CH{=}CH(CH_2)_3COOH$	Major component of animal lipids; precursor of prostaglandins and related compounds

considerable variety of fatty acids occur in natural fats. The commonest are shown in Table 7.2.

Generally, the acids are monocarboxylic, unbranched and contain an even number of carbon atoms. At pH values near neutrality fatty acids will be in their ionized form. The commonest fatty acids contain 12–20 carbon atoms. However, fatty acids that are branched or have odd numbers of carbon atoms are found and these types are more common in plants and bacteria. The structures of some unusual fatty acids are given in Fig. 7.6.

The numbering system of the carbon atoms of straight-chain fatty acids is shown in Fig. 7.7. The carbon atoms are also designated by Greek letters, but when this is done the α-carbon atom is the one having a carboxyl group attached to it and corresponds to C-2. Unsaturation is common, particularly in the C_{18-20} fatty acids (Table 7.2), but animals lack the enzymes necessary to insert double bonds beyond C-9. Since such fatty acids are biologically necessary for mammals, they must be supplied in the diet and are referred to as **essential fatty acids**. In these respects they resemble vitamins but in general are required in much larger amounts than vitamins.

$$\overset{\omega}{CH_3}\text{---}\overset{\gamma}{CH_2}\text{---}\overset{\beta}{CH_2}\text{---}\overset{\alpha}{CH_2}\text{---}COOH$$

$$\quad n \quad\quad 4 \quad\quad 3 \quad\quad 2 \quad\quad 1$$

Fig. 7.7 Numbering and naming of carbon atoms in a fatty acid.

Table 7.3 *The effects of unsaturation upon the melting point of a fatty acid. Generally fatty acids greater than ten carbon atoms long are solid at room temperature. The examples of stearic and oleic acids emphasize the effects of unsaturation: in this case the presence of a single double-bond reduces the melting point by over 50°C*

Fatty acid	Numerical symbol*	Melting point (°C)
Stearic	$C_{18:0}$	69.6
Oleic	$C_{18:1}$	16.3
Linoleic	$C_{18:2}$	-5.0
Linolenic	$C_{18:3}$	-11.0
Arachidonic	$C_{20:4}$	-49.5

* The first subscript number refers to the number of carbon atoms, the second to the number of double bonds.

Polyunsaturated: *usually used in the context of fat or fatty acid, having multiple double bonds. From the Greek* polys, *much or many.*

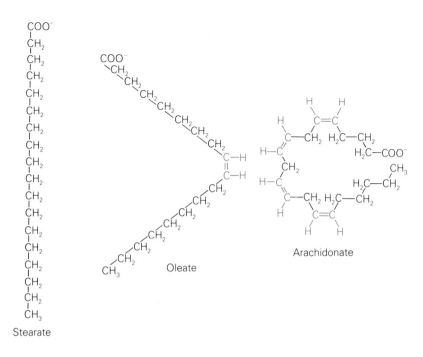

Fig. 7.9 Effects of unsaturation on fatty acid conformation.

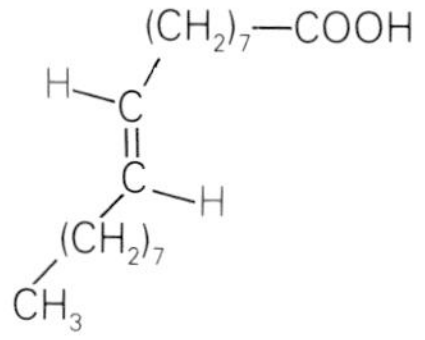

Arachidonate

Oleate

Stearate

(CH₂)₇—COOH

trans-9-octadecenoic acid
(elaidic acid)

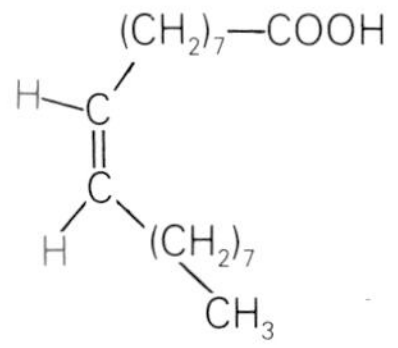

(CH₂)₇—COOH

cis-9-octadecenoic acid
(oleic acid)

Fig. 7.8 *Cis–trans* isomers of a fatty acid.

Carbon–carbon single bonds permit free rotation *around* the bond but this cannot occur with double bonds. This gives rise to *cis–trans* isomerism of fatty acids (Fig. 7.8) and in naturally occurring fatty acids, the more unstable *cis* isomers predominate. The presence of double bonds in the extended hydrocarbon chains introduces kinks or bends along its length (Fig. 7.9).

The absence of free rotation has important consequences for the fluidity of biomembranes because the insertion of double bonds reduces the melting point and increases the fluidity of the fatty acid. The effects of unsaturation on melting point are illustrated in Table 7.3. Unsaturated fatty acids occupy a larger volume than saturated fatty acids of the same chain length and this increased volume affects their packing in biomembranes.

Box 7.1
Saponification

Soaps are the Na⁺ and K⁺ salts of a fatty acid. Hydrolysis of triacylglycerol is called saponification and is thought to proceed as follows:

$$\text{Triacylglycerol} \xrightarrow{\text{NaOH}} \text{'soap'} + \text{diacylglycerol} \xrightarrow{\text{NaOH}} \text{'soap'} +$$

$$\text{monoacylglycerol} \xrightarrow{\text{NaOH}} \text{'soap'} + \text{glycerol}$$

although mono- and diacylglycerides are not easily detected during saponification. Following saponification, the products are washed with water and brine to recover the commercially important glycerol, and remove impurities. The washed product is then adjusted ('fitted') to give desirable properties, e.g. high fatty content, controlled electrolyte content and homogeneity.

The soap is dried to reduce its water content from 30–35% to 10–15% and is 'finished' by the addition of minor compounds such as colouring and perfumes. It is then mixed uniformly and the particulate mass is compressed to give a compact product.

Reference Pond, C. (1990) The natural way to be both fat and fit. *Biological Sciences Review*, **3**, 36–40. Fascinating article written at a basic level on the role of subcutaneous adipose tissue in a variety of animals.

Margarine was invented by the Frenchman Mege-Mouries in 1869. His initial recipe consisted of churning cream prepared from water and/or milk with oleomargarine obtained from beef tallow (this is largely the glyceride of oleate with some palmitate). The development of hydrogenation using hydrogen under pressure with a finely divided nickel catalyst by the German Normann in 1902, marked a significant advance in margarine manufacture. Unsaturated fatty acids could be hydrogenated, increasing the melting point and stability to oxidation. The degree of hydrogenation may be stopped at any point up to complete saturation, to give the final margarine the desired physical properties of plasticity and spreadability at ambient temperatures.

Margarine is a water-in-oil emulsion. The water content is about 20%, the fat about 80% and there are minor amounts of flavourings added to give a butter-like taste. Typical flavourings are diacetyl, fatty acids and ketones. Colouring agents, generally carotenes, are added, as are vitamins A and D. Finally, antioxidants are added to retard the oxidation of fatty acids, which would lead to rancidity.

Chemical properties of triacylglycerols

Hydrolysis of acylglycerols is catalysed by the presence of bases or by specific enzymes called **lipases**.

The hydrocarbon chains of the fatty acid components of lipids are relatively inert when saturated, but unsaturated fatty acids are more reactive because they may participate in addition reactions involving both hydrogen and halogens:

$$\text{Hydrogenation } -CH=CH- + H_2 \xrightarrow{\text{Ni,Pa}} -CH_2-CH_2-$$

$$\text{Halogenation } -CH=CH- + I_2 \xrightarrow{\text{catalyst}} -CHI-CHI-$$

Hydrogenation is commercially important in margarine manufacture. Halogenation by iodine and saponification using potassium hydroxide are used to characterize fats by determining their **iodine** and **saponification numbers** respectively. However, these relatively crude measures of the degree of unsaturation and proportion of smaller fatty acids in a fat sample have now been largely superseded by modern chromatographic methods. The relationship between the fatty acid compositions of some naturally occurring fats and their iodine and saponification numbers is given in Table 7.4.

Table 7.4 *Percentage of major fatty acids and saponification and iodine numbers of some common fats and oils. The general trend is towards more unsaturation in plant oils than animal fats, but this does not apply to fish oils, where the colder environment has selected for unsaturation and therefore lower melting points*

Source	Percentage of total fatty acids								Saponification no.	Iodine no. (g)
	Saturated acids			Unsaturated acids						
	$<C_{16}$	C_{18}	$>C_{18}$	$<C_{18}$	$C_{18:1}$	$C_{18:2}$	$C_{18:3}$	$>C_{18:n}$*		
Coconut oil	92	2.0	0	–	5	1	–	–	257	9
Mutton fat	33	32	–	2	32	2	–	–	193	40
Beef fat	30	7	–	14	48	2	1	–	198	42
Lard	29	11	–	4	44	11	–	–	199	58
Olive oil	10	2	0.4	1	78	7	1	–	190	84
Groundnut oil	13	3	4	–	32	4	–	2	192	93
Corn oil	13	3	0.5	–	31	52	1	–	191	116
Sunflower oil	6	6	1	–	18	69	–	–	191	127
Herring oil	21	19	–	7	–	–	–	46	186	138
Sardine oil	24	4	–	–	11	1	1	41	194	179

* n = 1 to 6.

Box 7.3
Iodine and saponification numbers

The iodine number of a fat is defined as the number of grams of iodine taken up by 100 g of the fat. A high iodine number is associated with fats containing a high proportion of unsaturated fatty acids. The saponification number is the number of milligrams of KOH needed to hydrolyse (saponify) 1 g of the fat. Acylglycerols containing high proportions of shorter chain fatty acids have relatively higher saponification numbers than those with high proportions of longer chain fatty acids.

Note, the approximate mean molecular weight of the fatty acids in a triacylglycerol can be obtained from the saponification number using:

$$M_r = \frac{3 \times 56 \times 1000}{\text{saponification number}}$$

Box 7.4
Auto-oxidation of unsaturated fatty acids

The polyunsaturated fatty acids of lipids are prone to auto-oxidation. This has important physiological and pathological effects and is the cause of rancidity in foods. Auto-oxidation is due to the susceptibility of the methylene group between two double bonds ($-CH=CH-CH_2-CH=CH-$) to removal (abstraction) of one hydrogen atom, forming a **free radical**. The initial cause of the hydrogen abstraction is the action of oxygen-derived free radicals. Light and other radiations, metal ions (notably iron), elevated concentrations of oxygen, toxins (such as paraquat), cigarette smoke, carbon tetrachloride and some drugs are all thought to lead to superoxide radical (O_2^-) and hydrogen peroxide (H_2O_2) formation within cells. These may cause damage directly, or interact to give the very reactive hydroxyl radical, $OH^.$. The effects of $OH^.$ radicals on unsaturated fatty acids are as follows:

$$\underbrace{H_2O_2 + O_2^-} \longleftarrow \text{light, metal ions, etc.}$$

$-CH=CH-CH_2-CH=CH- + OH^.$
polyunsaturated fatty acid $\downarrow \to H_2O$

$-CH=CH-\overline{C}H-CH=CH-$
lipid-derived free radical

$\downarrow O_2$

$\overset{O-O^-}{-CH=CH-CH-CH=CH-}$ peroxyl radical $\quad [H]$

$-CH=CH-CH-CH=CH-$
polunsaturated fatty acid

$-CH=CH-\overline{C}H-CH=CH- \longleftarrow$

$\overset{O-O[H]}{-CH=CH-CH-CH=CH-} \longrightarrow$ breakdown products, e.g. ketones ketoacids, dialdehydes

Abstraction of hydrogen from the methylene group by the $OH^.$ radical leads to a lipid-derived free radical. This can react with molecular oxygen to give a peroxy radical, which, in turn, can attack other lipids. This attack generates a further lipid radical, leading to a self-perpetuating, autocatalytic chain. The lipid peroxides formed fragment to give a variety of products, including ketoacids, ketones and dialdehydes,

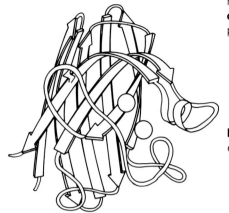

See *Cell Biology*, Chapter 1

the accumulation of which in fatty foods causes 'off flavours' (rancidity). Within the cell, these products are highly toxic. The hydroxyl radical can also attack other molecules such as proteins or the bases of DNA leading to peroxy radical formation and similar autocatalytic cycles.

Organisms protect themselves against the excessive accumulation of free radicals and the protective devices include several enzymes. The **superoxide dismutases** and **catalase** effectively inactivate superoxide radicals and hydrogen peroxide:

$$O_2^- + O_2^- + 2H^+ \longrightarrow H_2O_2 \longrightarrow H_2O$$
$$\searrow O_2 \qquad \searrow \tfrac{1}{2}O_2$$

Peroxidases are able to neutralize peroxide radicals using reducing agents, for example, dihydroascorbic acid (reduced vitamin C).

$$R-\overset{|}{\underset{|}{C}}-OOH + AH_2 \longrightarrow A + R-\overset{|}{\underset{|}{C}}-OH + H_2O$$

Further protection is afforded by scavenging molecules which effectively trap free radicals, interrupting the self-propagating chain reactions. Scavengers include **ascorbic acid** (vitamin C), **vitamin E**, (α-tocopherol) and **glutathione**.

Superoxide dismutase. Circles denote positions of Cu^{2+} and Zn^{2+}. Redrawn from Richardson, J. (1981) *Adv. Prot. Chem.*, **34**, 167–339.

7.4 Complex lipids

Complex lipids are subdivided into two groups, the **phospholipids** and the **glycolipids** (see Table 7.1) and occur mainly in biomembranes.

PHOSPHOLIPIDS are in turn classified into two main groups based on glycerol (**phosphoacylglycerols**) or on the alcohol, **sphinogosine** (see Figs 7.2 and 7.10) (**sphingomyelins**).

Phosphoacylglycerols, or glycerol phospholipids, are the most abundant of the complex lipids. They occur in all types of cells, forming about 40–50% of the lipid content of the biomembranes. The three major types present in membranes of eukaryotic cells are phosphatidyl choline (lecithin), phosphatidyl ethanolamine (cephalin) and phosphatidyl serine (see Fig. 7.11). Minor amounts of other phosphoacylglycerols are also present.

Phosphoacylglycerols, and complex lipids in general, are **amphipathic** molecules, having a *hydrophobic* portion that is water insoluble, and a *hydrophilic* region that is water soluble. Amphipathic molecules have a very low true solubility in water, but nevertheless form stable structures in water. Two such structures are micelles and bilayers. If the concentration of an amphipathic lipid in water is raised, then at a specific concentration, the *critical micelle concentration* (cmc), the lipid molecules associate to form spheres or micelles. Within a micelle the hydrophobic hydrocarbon tails interact with one another to exclude water, while the hydrophilic head groups, interact with the aqueous exterior (Fig. 7.12a). Micelles are stable because of the hydrophobic interactions of their hydrocarbon chains. Under somewhat different conditions, amphipathic lipids may associate to form a bilayer (Fig. 7.12b) that could be considered an extended, flat micelle. The interior of the bilayer is formed by the association of the hydrocarbon chains and is therefore hydrophobic in character. The hydrophilic polar groups are on the outside in

☐ Reagents that supply only one electron during covalent bond formation are called free radicals. Free radicals are normally highly reactive and only appear as transient intermediates during chemical reactions.

☐ The tripeptide γ-glutamylcysteinylglycine is called glutathione. It contains a metabolically active sulphydryl group

$$\gamma\text{-glu}-\text{cys}-\text{gly}$$
$$\underset{SH}{|}$$

and can reduce hydrogen peroxide and organic peroxides in the cell. Glutathione also participates in many bioreductive reactions.

Exercise 3

Calculate the *theoretical* iodine number of oleic acid.

$$\begin{array}{c} CH_3 \\ | \\ (CH_2)_{12} \\ | \\ H-C \\ \| \\ C-H \\ | \\ H-C-OH \\ | \\ H-C-NH_2 \\ | \\ CH_2OH \end{array}$$

Fig. 7.10 Sphingosine.

Hydrophobic: from the Greek hydor, *water, and* phobos, *fear, i.e. water-hating.*
Hydrophilic: from the Greek hydor, *water, and* philos, *love, i.e. water-loving.*

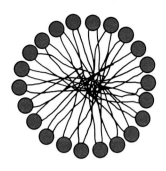

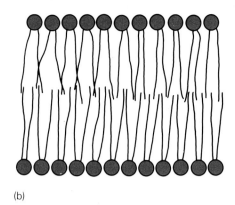

(a)

(b)

Fig. 7.12 Diagrammatic structures of (a) a lipid micelle and (b) a lipid bilayer.

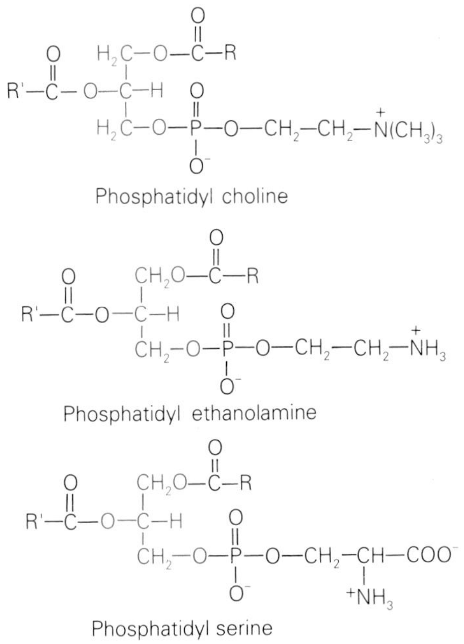

Phosphatidyl choline

Phosphatidyl ethanolamine

Phosphatidyl serine

Fig. 7.11 Structures of major phosphoacyl glycerols from membranes of eukaryotic cells.

contact with the watery environment. Simple spherical micelles are limited in size, having diameters of less than 20 nm, but bilayers can be extensive and indeed they form the basic structural unit of biological membranes.

SPHINGOMYELINS are based on the alcohol sphingosine. In glycerol-based lipids, the fatty acids are joined to the glycerol via ester linkages. In contrast, in sphingolipids the fatty acid is attached by an **amide linkage** to produce a **ceramide**:

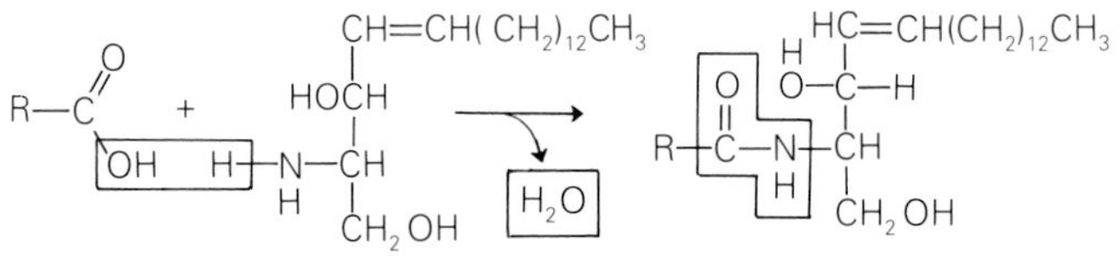

fatty acid   sphingosine   ceramide

This formal equation does not describe the biosynthesis of ceramides, which occurs by a rather more complex route. The attachment of a polar group (such as phosphoryl choline) to the terminal hydroxyl groups produces a sphingomyelin (Fig. 7.13).

Sphingomyelins are amphipathic and are major constituents of biomembranes, particularly those of brain and nerve tissue. Sphingomyelins and phosphoacylglycerols have similar gross conformations: both possess polar head groups and have two hydrophobic tails. However, in the case of sphingomyelins, one tail is part of the alcohol sphingosine, rather than a fatty acid (Fig. 7.14).

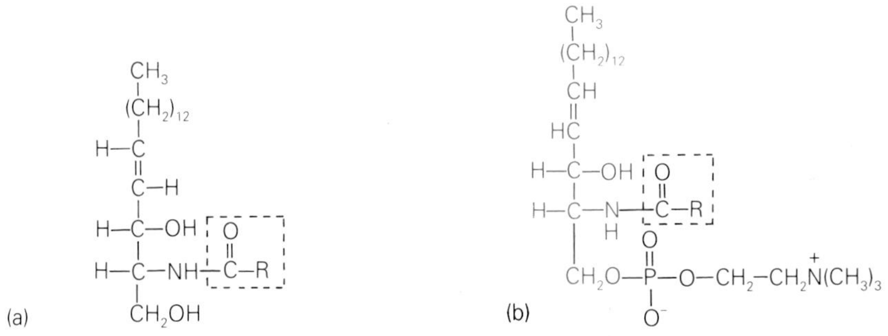

(a)

(b)

Fig. 7.13 Structures of (a) a generalized ceramide and (b) a complete sphingomyelin (phosphorylcholine head group).

Reference Kuksis, A. (ed.) (1978) *Fatty Acids and Glycerides*, Plenum Press, New York. A rather old, but very informative account of many aspects of the structure, metabolism, synthesis and separation of fatty acids and glycerides.

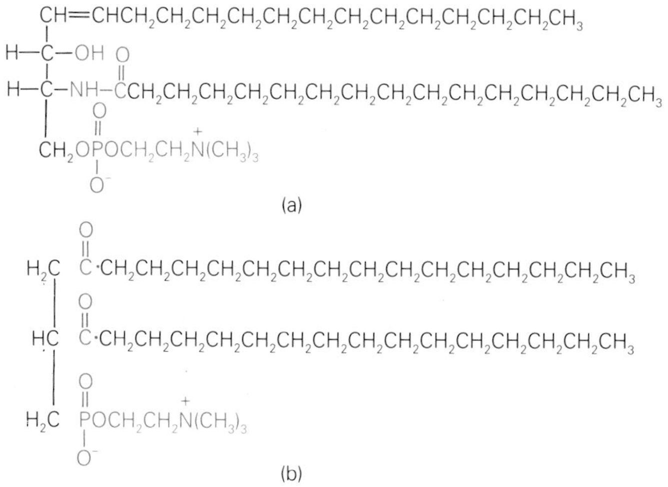

Fig. 7.14 Diagrammatic representations showing the gross structural similarity between (a) a sphingomyelin (palmitoyl sphingomyelin) and (b) a phosphoacylglycerol (1,2-dipalmitylglycerol, 3-phosphocholine).

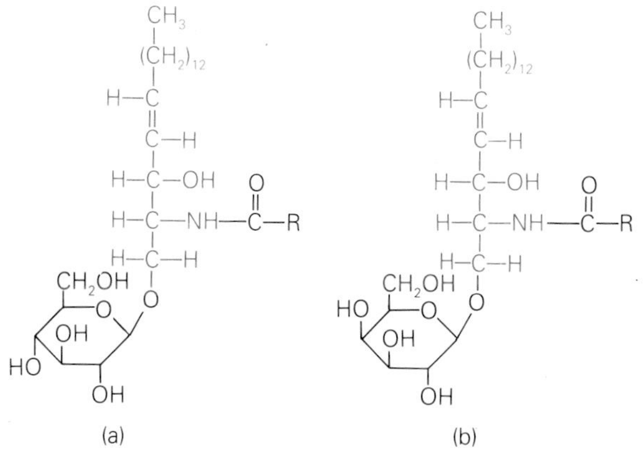

Fig. 7.15 Structures of (a) a glucocerebroside and (b) a galactocerebroside.

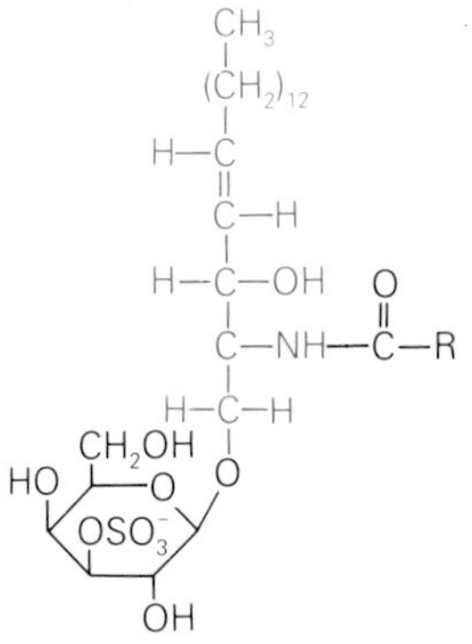

Fig. 7.16 A sulphatide.

GLYCOLIPIDS are formed by the attachment of carbohydrates to a ceramide (see earlier) and are also based on sphingosine. The two major classes of glycolipids are the **cerebrosides** and the **gangliosides**, which differ in the sugars attached.

Cerebrosides contain neutral sugars such as glucose or galactose and the simplest contain a monosaccharide attached to the terminal hydroxyl group of the ceramide. If the sugar is glucose or galactose (Fig. 7.15) the compound is a glucocerebroside or a galactocerebroside, respectively. *Sulphatides* are sulphated galactocerebrosides (Fig. 7.16), which form a large proportion of brain cerebrosides. Other cerebrosides have chains of two to ten sugar residues rather than a single monosaccharide.

Gangliosides differ from cerebrosides in that their oligosaccharide head-group contains one to four *sialic acid* residues (Fig. 7.17). Since sialic acid residues are negatively charged at pH 7, gangliosides are markedly amphipathic. Examples of two common gangliosides are shown in Fig. 7.18.

Gangliosides should not be thought of as mere structural units of biomembranes. The wide variety of types of oligosaccharides carried by gangliosides embedded in the cell membrane project away from the cell surfaces (Fig. 7.18) forming **antigenic determinants**. For example, the

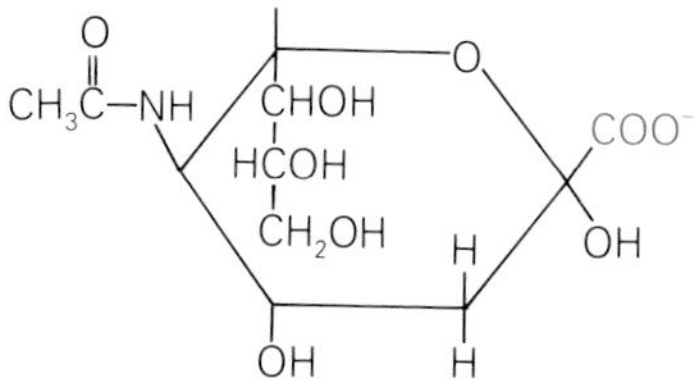

Fig. 7.17 A sialic acid (N-acetylneuraminate).

Sialic acids: derivatives of the nine-carbon sugar neuraminate.

Reference Robertson, R.N. (1983) *The Lively Membranes*, Cambridge University Press, UK). A short, readable book on membranes. Particularly good on selected aspects of lipids in membranes.

The ABO blood group system of human beings and many other animals is an interesting example of oligosaccharide-mediated recognition. The system is based on three common antigenic determinants called A, B and H carried on gangliosides and glycoproteins of erythrocyte membranes, and by glycoproteins in a variety of body secretions. The antigens are short oligosaccharides that differ from each other by only a single monosaccharide:

H

Fucα1–2Galβ 1–3 Glc NAc · · · ceramide

A   GalNAcα1–3 Galβ 1–3GlcNAc· · · ceramide
      Fucα1–2

B   Galα 1–3 Galβ–3GlcNAc· · · ceramide
      Fuc α 1–2

The determinant H is a precursor of A and B. The high specificity of recognition is generated not only by the nature and sequence of the monosaccharides, but also by the types of glycosidic bonds joining them together.

□ Interferons are a group of proteins secreted by virus-infected cells, which diffuse to neighbouring cells and stimulate them to synthesize proteins that give some protection against viral attack.

Cell membrane

Galβ 1-3 Gal NAc β,1-4
               Galβ,1-4 Glc —ceramide
     Siaα,2-3

GM$_1$

Gal NAcβ1-4
           Galβ1-4 Glc—ceramide
   Siaα2-3

GM$_2$

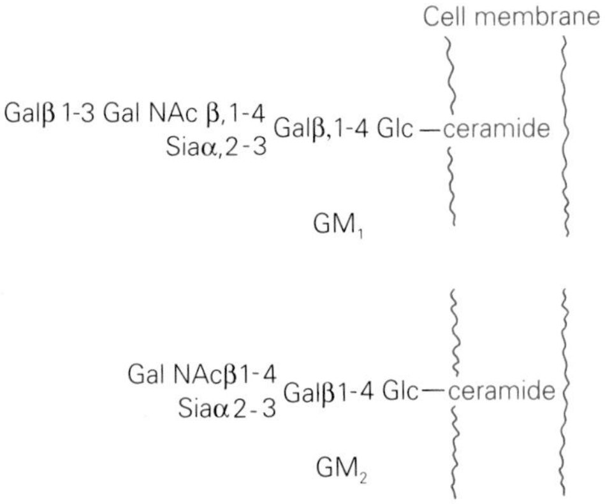

Fig. 7.18 The structures of the GM$_1$ and GM$_2$ gangliosides. The ceramide portion is embedded in the cell membrane.

CH$_3$

CH$_2$=C—C=CH$_2$
        |
        H

Fig. 7.19 Isoprene.

common ABO blood group system is characterized by the presence or absence of three different oligosaccharides.

The oligosaccharides carried on the cell surface act as biochemical markers identifying stages in cellular differentiation. This is necessary for the normal growth and development of the cell or tissue. Gangliosides in tumour cells show a simplified and individual composition compared with their normal cellular counterparts.

Glycolipids function as receptors for bioactive substances, mediating the specific binding of a variety of agents to the cell surface. Thus glycoprotein hormones and interferons bind to specific gangliosides. Unfortunately, the specific recognition sites provided by oligosaccharides have been exploited by several parasites. Thus, the common bacteria, *Escherichia coli* and *Salmonella typhimurium*, bind to epithelial cells via mannose-rich cell surface oligosaccharides of glycolipids. This is a first step in infection by these organisms.

The recognition phenomena just described are mediated by sugars carried by glycoproteins as well as gangliosides. A specific example of cell-to-cell recognition is illustrated by erythrocytes. Each erythrocyte carries a coating of about 2×10^7 sialic acid residues on gangliosides and glycoproteins, giving a net negative charge to the cell. Removal of sialic acid residues with neuraminidase exposes galactose sugars, which are recognized by liver cells. Thus, desialysed erythrocytes are identified by the liver and removed from the circulation.

7.5 Polyprenyl lipids

The **polyprenyl lipids**, often referred to as **isoprenoid compounds**, are a heterogeneous group of components. All can be considered to be derived, at least in part, from **isoprene** (Fig. 7.19). The group includes a number of compounds with biological roles in membrane structure, bone deposition, photoreception and reproduction. Some of the more important examples are described below.

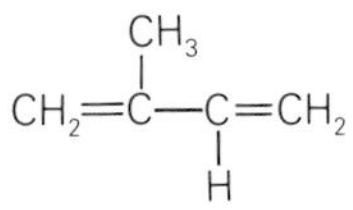

Box 7.6
Lysosomal storage diseases

Gangliosides are of medical interest since defects in their metabolism have serious physiological consequences. Among these are the inherited disorders of metabolism characterized by the accumulation of sphingolipids in various organs and tissues. These **sphingolipidoses** are a group of lysosomal storage diseases. Lysosomes are membrane-limited sacs within the cytoplasm containing about 60 types of **hydrolytic enzymes** whose concerted action is to degrade all biological material. They are involved in the normal digestion of both extra- and intracellular materials. Lysosomal storage diseases result from a marked reduction in the level of a lysosomal enzyme, which results in the accumulation of the substrate. Lysosomes in the afflicted cell become enlarged with undigested material, resulting in gross interference with the normal processes of the cell. The table summarizes the main effects and metabolic defects associated with the sphingolipidoses. These disorders are transmitted as recessive, autosomal abnormalities, with the exception of **Fabry's disease** which is carried on the X-chromosome.

Sphingolipidoses (from Brady, R.O. (1978) *Annual Review of Biochemistry*, **47**, 689–92.)

Disorder	Main symptoms	Major lipid accumulated	Enzyme deficiency
Ceramide lactoside lipidosis	Slow progressive brain damage; liver and spleen enlargement	Ceramide lactoside	Neural β-galactosidase
Fabry's disease	Skin disorder; kidney failure; pain in lower extremities	Ceramide trihexose	Ceramide-trihexosa-α-galactosidase
Farber's disease	Hoarseness; dermatitis; skeletal deformation; mental retardation	Ceramide	Ceramidase
Fucosidosis	Cerebral degeneration; muscular spasticity; thick skin	Pentahexosyl-fucoglycolipid	α-Fucosidase
Gaucher's disease	Spleen and liver enlargement; erosion of long bones and pelvis; mental retardation only in infant form	Glucocerebroside	Glucocerebrosidase
Generalized gangliosidosis	Mental retardation; liver enlarged; skeletal deformities	Ganglioside G_{M1}	β-Galactosidase
Krabbe's disease (globoid leukodystrophy)	Mental retardation; virtual total absence of myelin	Galacto-cerebroside	Galacto-cerebrosidase
Metachromic leukodystrophy	Mental retardation; nerves stain yellow-brown with cresyl violet dye	Sulphatide	Sulphatidase
Niemann–Pick disease	Enlargement of spleen and liver; mental retardation	Sphingomyelin	Sphingomyelinase
Tay–Sachs disease	Mental retardation; red spot in retina; blindness; muscular weakness; death at 2 to 3 years	Ganglioside G_{M2}	Hexaminidase A
Tay–Sachs variant	Same as Tay–Sachs, but progressing more rapidly	Ganglioside G_{M2} and globoside	Hexaminidases A and B

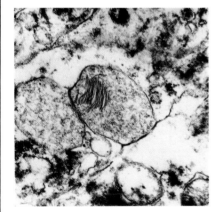

Electron micrograph of enlarged lysosomes from the cerebral cortex of a 19-week-old fetus. One lysosome contains accumulated lipid (a 'Zebra' body). Diagnosis of Tay–Sachs disease was made by demonstrating lack of hexaminidase A activity in cultured amniotic fluid cells obtained by amniocentesis, following which the fetus was aborted. Courtesy of Dr A. Cooper, Willink Biochemical Genetics Unit, Royal Manchester Children's Hospital. (Magnification × 20 000).

The diagnosis of a specific sphingolipidosis is possible from biopsies or histological examination since the stored lipid gives the lysosomes a distinctive appearance (see Figure).

Biochemical methods are available for the assay of a particular enzyme to confirm a diagnosis.

Reference Stanbury, J.B., Wyndgaarden, D.S., Fredrickson, D.S., Goldstein, J.L. and Brown, M.S. (eds) (1983) *The Metabolic Basis of Inherited Disease*, 5th edn, McGraw-Hill, New York. Covers many aspects of individual inherited disorders. Many chapters are devoted to genetic diseases involving lipids. Very good reference text.

Tissue samples can be used as the source of the enzyme to diagnose the specific enzyme deficiency and identify the particular sphingolipidosis. In many cases artificial substrates are available that produce products that fluoresce at, or absorb light of, a specific wavelength, simplifying assay procedures. For example, **Tay–Sachs disease** can be investigated using 4-methylumbelliferyl-β-D-N-acetyl-glucosamine as substrate for hexasaminidase A. Hydrolysis of this substrate releases 4-methylumbelliferone which fluoresces strongly in alkaline conditions allowing measurement of enzymic activity.

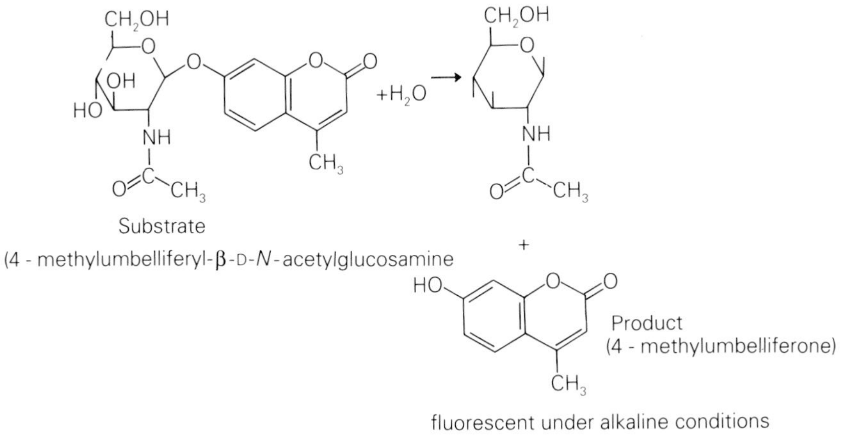

Substrate
(4 - methylumbelliferyl-β-D-N-acetylglucosamine

Product
(4 - methylumbelliferone)
fluorescent under alkaline conditions

There is no effective therapy for sphingolipidoses. Most effort is concentrated on prevention by genetic counselling of potential parents. Detection of sufferers at an early stage of pregnancy is possible following **amniocentesis**, allowing the possibility of abortion.

STEROIDS are based on a common structural unit called **perhydrocyclopentanophenanthrene**. This consists of four fused rings, A to D, with the carbon atoms numbered as shown in Fig. 7.20. In this versatile structural unit varying levels of unsaturation and addition of different chemical groups at specific positions produce a variety of different molecules with different biological properties.

The **sterols**, characterized by the possession of a hydroxyl group at C-3 and a variety of groups at C-17, are widespread. *Cholesterol* (Fig. 7.21) is the most abundant sterol in animals. It is present in biological membranes, although the amounts vary. For example, prokaryotic membranes contain negligible amounts, while the myelin sheath of nerve cells contains about 20% cholesterol. Since the fused rings of cholesterol form a rigid structure, the presence of high concentrations of cholesterol in a biomembrane greatly reduces its fluidity.

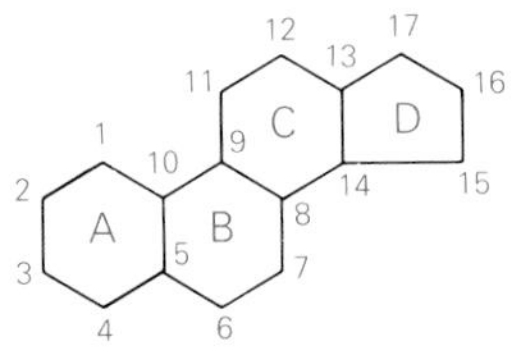

Fig. 7.20 Perhydrocyclopentanophenanthrene.

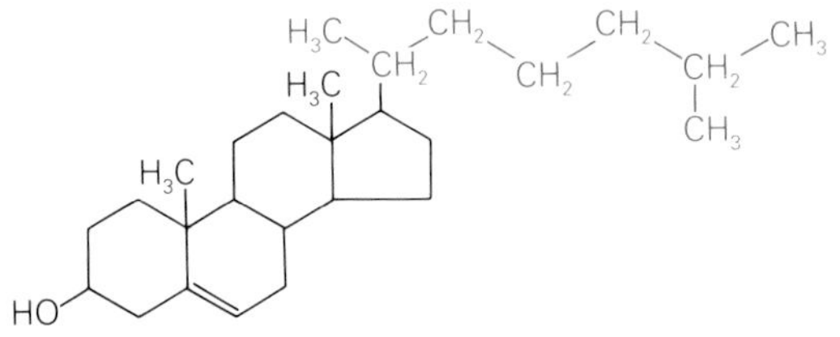

Fig. 7.21 Cholesterol.

Exercise 5

Is cholesterol an amphipathic molecule?

Cholesterol: from the Greek chole, bile *(appreciable amounts of cholesterol are found in bile),* and stereos, *solid.*

Table 7.5 *Major types of steroid hormones: their structures and principal physiological effects. Note that only the oestrogens have an aromatic ring A*

Hormone	Class	Structure	Site of synthesis	Physiological effects
Oestradiol	Oestrogen		Ovary	Development and maintenance of female sex characteristics
Progesterone			Corpus luteum, placenta	Prepares uterus for implantation; suppresses ovulation during pregnancy
Testosterone	Androgen		Testis	Development and maintenance of male sex characteristics
Aldosterone	Mineralocorticoid		Adrenal cortex	Promotes retention of Na^+ by renal tubules
Cortisol	Glucocorticoid		Adrenal cortex	Promotes gluconeogenesis; suppresses inflammatory responses

Cholesterol is transported around the body in blood plasma bound in lipoprotein particles. Cholesterol is infamous because elevated levels of plasma cholesterol are correlated with increased risk of cardiovascular disease. Cholesterol is also a precursor of the **steroid hormones**, the **D vitamins** and the **bile salts**. The major sites of biosynthesis of the steroid hormones are the gonads (sex hormones) and adrenal cortex (mineralocorticoids and glucocorticoids) In pregnant females the placenta also produces steroid hormones. The structures and physiological functions of some of the major steroid hormones are given in Table 7.5.

VITAMIN D is now considered to be a hormone rather than a vitamin, since **vitamin D_3 (cholecalciferol)** is produced by ultraviolet irradiation of 7-dehydrocholesterol, a normal metabolite of cholesterol in the skin (Fig. 7.22). Vitamin D_3 absorbed from dietary sources or formed in the skin is hydroxylated to form **1,25-dihydroxycholecalciferol** in two steps by specific enzymes in the liver and kidney. This hormone controls the metabolism of Ca^{2+} and phosphate in target tissues.

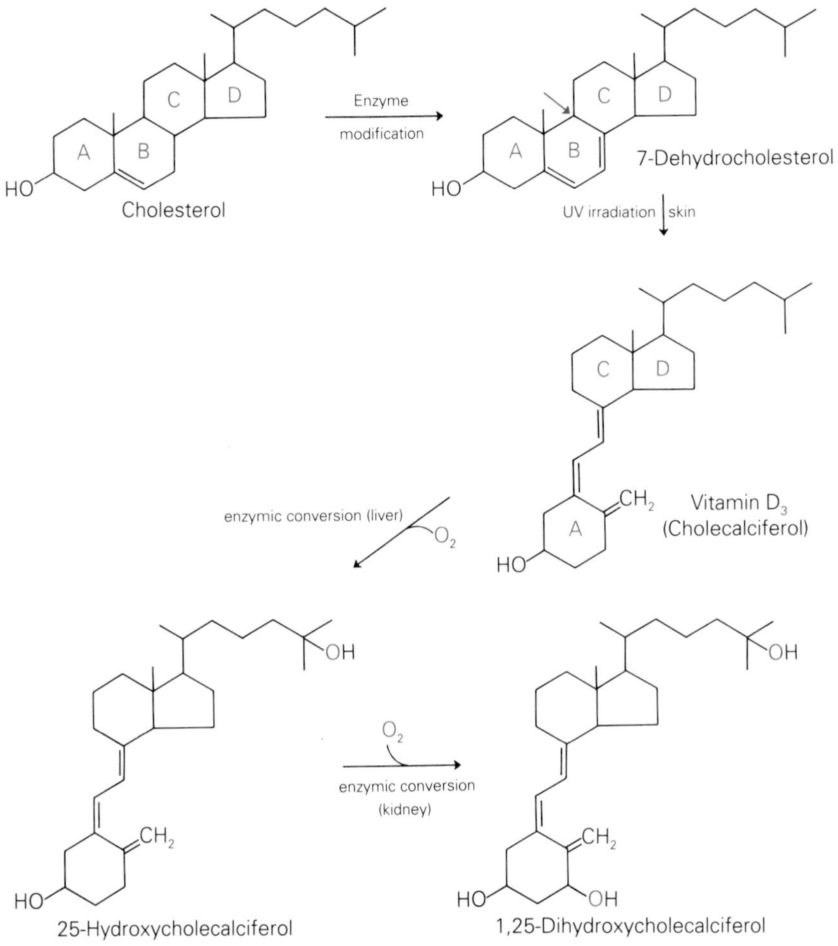

Fig. 7.22 Conversion of cholesterol to the biologically active forms of vitamin D.

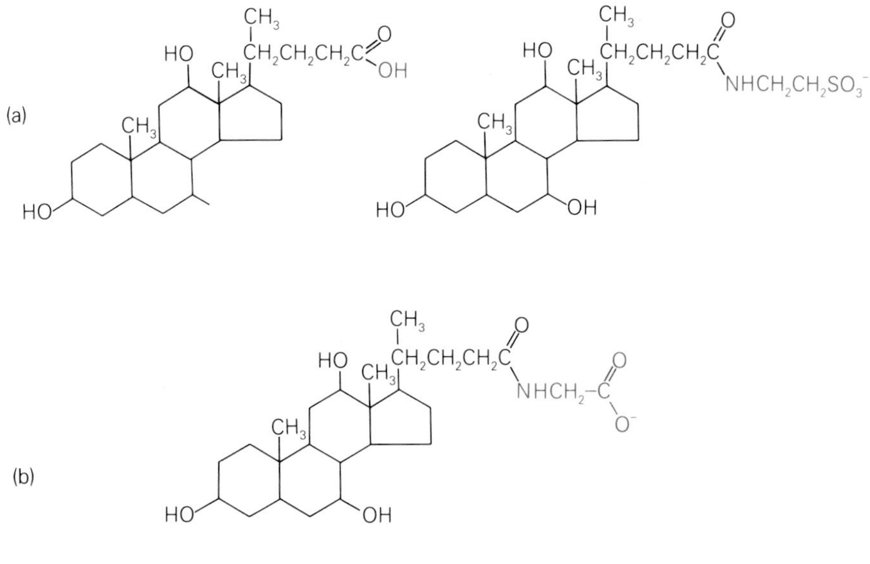

Fig. 7.23 The bile acid, cholic acid and its salts (a) taurocholate and (b) glycocholate.

BILE ACIDS are catabolic products of cholesterol. In humans the major bile acids are **cholic** and **chenodeoxycholic** acids, although **deoxycholic** acid is also functional in other mammals. The acids usually occur as conjugates with **taurine** and **glycine**, and are called **bile salts** (Fig. 7.23). Bile salts are synthesized in the liver. Bile is stored in the gall bladder, from which it travels, via the bile duct, to the small intestine. Here the bile salts act as emulsifiers of dietary fats, aiding their hydrolysis by digestive lipases.

Lipid-soluble vitamins

Isoprenoid (lipid-soluble) vitamins include **vitamins A, E** and **K**.

The active forms of vitamin A are **retinol, retinal** and **retinoic acid** which may be formed from the dietary provitamin, β-**carotene**. Carotenes are yellow to red pigments widely distributed in nature, especially in plants, and β-carotene is one of the commonest. Carotenes function as accessory pigments in **photosynthesis** trapping light energy in the 450–500 nm wavelength region, where **chlorophylls** are ineffective in absorbing radiant energy. However, the primary function of β-carotene may be in preventing the photo-oxidation of chlorophylls.

See *Cell Biology*, Chapter 12

VITAMIN A (all-*trans* retinol) is formed by the enzymic splitting of β-carotene, to retinal followed by its reduction to retinol (Fig. 7.24). Retinal may also be oxidized to retinoic acid. Both retinol and retinoic acid function in a hormone-like fashion, affecting cell growth and differentiation. They appear to be particularly important in promoting healthy epithelia by preventing keratinization (Fig. 7.25).

If keratinocytes, the cells of the epidermis, are grown in cell culture medium in which the vitamin A is absent there is increased synthesis of some of the high M_r keratins that combine to form the intermediate filaments of the cytoskeleton. Synthesis of high M_r keratins is one of the first indications that a keratinocyte has undergone a commitment to terminally differentiate. Simultaneously there is a much reduced synthesis of the lower M_r keratin proteins.

☐ Keratins are major structural proteins in skin, hair, horn, etc. (see Chapter 3). Keratinization is the process by which epidermal cells become packed with keratin polypeptides, lose most of their cellular organelles, and develop an inert, cross-linked cell envelope. The cells die and form dry, scaly structures.

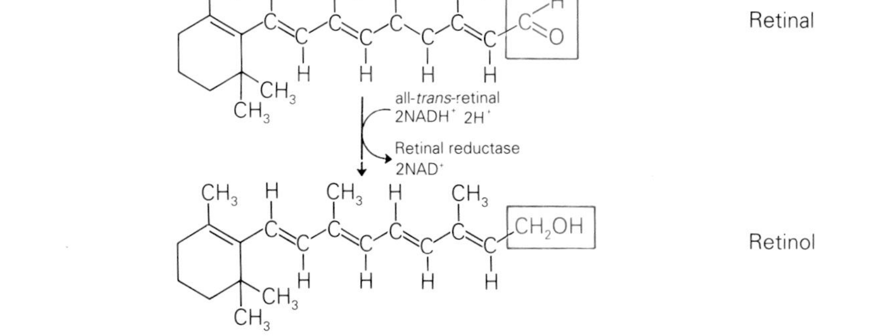

Fig. 7.24 Production of retinol, via all-*trans*-retinal, from β-carotene.

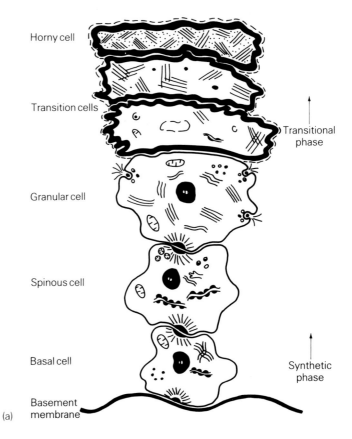

Horny cell

Transition cells

→ Transitional phase

Granular cell

Spinous cell

Basal cell

→ Synthetic phase

Basement membrane

(a)

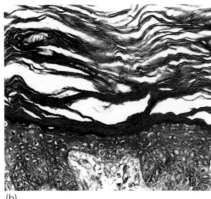

(b)

Fig. 7.25 The cell types in human epidermis (not to scale). From the basement membrane moving 'outwards', the cell layers are (1) *stratum basale* or *stratum germinativum* (basal or germinative layer); (2) *stratum spinosum* (spinous layer or spinous cells); (3) *stratum granulosum* (granular layer); (4) *stratum lucidum* (transitional layer); (5) *stratum corneum* (horny layer of dead cells). Note that there are several layers (3–6) of each cell type, except for the basal layer which is only 1–2 cells thick. The thickness of the stratum corneum obviously depends partly on the rate of sloughing off of dead cells. (b) Photomicrograph of human skin to show epidermal region (magnification × 120).

See *Cell Biology*, Chapter 9

Retinal is the form of vitamin A involved in vision. The molecular events involved in vision are illustrated in Fig. 7.26. Retinol is transported to the eye bound to the retinol-binding protein in plasma and in the retina is oxidized to all-*trans*-retinal. This is isomerized to 11-*cis*-retinal which complexes with a protein, **opsin**, to form **rhodopsin** or visual purple. Rhodopsin is the primary photoreceptor of the rod cells of the retina. Light causes dissociation of rhodopsin to opsin and all-*trans*-retinal, which initiates the production of a nerve impulse informing the brain of a light detection event. Regeneration of the visual pigment requires isomerization of the all-*trans*-retinal back to the 11-*cis* isomer.

VITAMIN E is a mixture of several closely related compounds called **tocopherols**. α-Tocopherol (Fig. 7.27) has aromatic rings and long

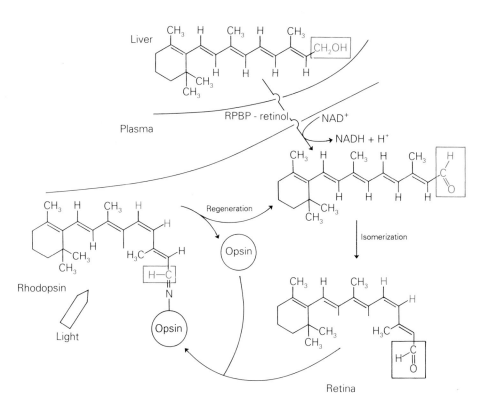

Fig. 7.26 Overview of the molecular events involved in light detection. RPBP, retinol plasma binding protein.

hydrocarbon chains that make the molecule lipophilic and it therefore accumulates in plasma lipoproteins, fat deposits and cell membranes. It reacts readily with molecular oxygen and free radicals, protecting unsaturated fatty acids from peroxidation (Box 7.4). Laboratory rats require vitamin E for reproduction but evidence is not conclusive regarding a similar role for the vitamin in humans.

VITAMIN K illustrates the basic isoprene structure of the lipid vitamins most clearly (Fig. 7.28). The molecular mechanism for the involvement of this vitamin in the extremely complex cascade process of blood clotting is described in Fig. 7.29. The culmination of the cascade is the conversion of the soluble protein **fibrinogen** to the insoluble **fibrin**, forming a 'mesh' (the 'clot') that seals the wound. The conversion of fibrinogen to fibrin requires the action of the proteolytic enzyme **thrombin**, which occurs as the inactive **prothrombin**. During the formation of prothrombin a vitamin K-stimulated carboxylation of ten **glutamate** residues in the protein occurs, converting them to γ-**carboxylglutamate** residues. These are strong chelators of Ca^{2+}. During the clotting cascade process, the prothrombin–Ca^{2+} complex is converted to thrombin, which then catalyses clot formation.

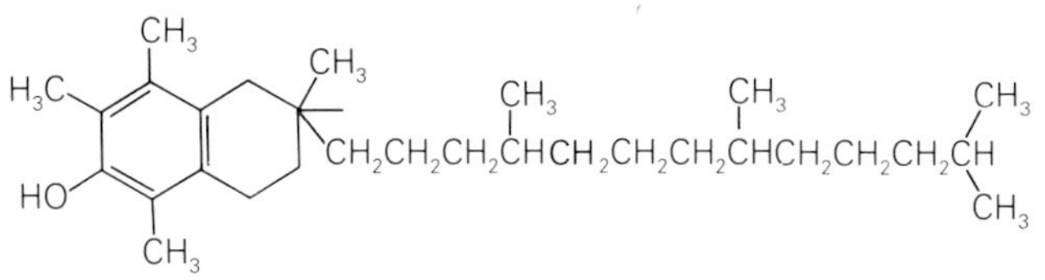

Fig. 7.27 α-Tocopherol.

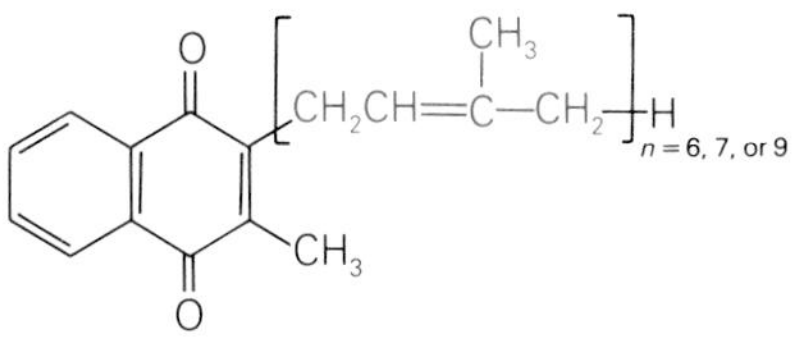

Fig. 7.28 Vitamin K_2: the isoprene units are highlighted.

Tocopherol: *the name alludes to the involvement of the vitamin in fertility. From the Greek* tokos, *child, and* pherein, *bear.*
Vitamin K: *the involvement of this vitamin in blood clotting was first observed by the Dane, Dam. K stands for* koagulations *vitamin.*

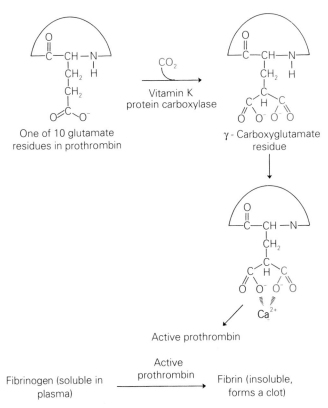

One of 10 glutamate residues in prothrombin

γ - Carboxyglutamate residue

Vitamin K protein carboxylase

CO_2

Ca^{2+}

Active prothrombin

Fibrinogen (soluble in plasma) → Fibrin (insoluble, forms a clot)

Active prothrombin

Fig. 7.29 Terminal steps in blood clotting.

7.6 Overview

The structures of lipids provide a complex picture, mirroring their many biochemical activities. The extremely hydrophobic nature of waxes makes them excellent waterproof coatings for free surfaces. The highly reduced, water-repellent triacylglycerols are ideal as compact energy-rich stores. This energy can be made available to the cell by a variety of oxidative processes.

The amphipathic phospholipids and glycolipids are important as structural elements of cellular membranes. Glycolipids play a major role in mediating a whole series of cell-surface recognition phenomena.

Polyisoprenoid lipids have the most complicated structures of all lipids. Some have specific roles (such as vitamin A in vision), while others (such as the steroid hormones) modulate a variety of biochemical and physiological processes.

Answers to Exercises

1. Example of alternative classification:
simple lipids (*cannot* be hydrolysed)
e.g. sterols, lipid vitamins, carotenoids
Complex lipids (*can* be hydrolysed)
e.g. acylglycerols, phosphoacyl-glycerols, glycolipids.

2.

H₂C—O—C—(CH₂)₁₄CH₃
HO—C—H
H₂C—O—C—(CH₂)₇CH=CH(CH₂)₇CH₃

3. 90 g.

4.

	M_r of fatty acid	Iodine number
Sardine oil	866	179
Beef fat	850	42
Coconut oil	654	9

Sardine oil: long fatty acids, but highly unsaturated, giving the low melting point necessary in an exothermic organism in a relatively cold environment.
Beef fat: relatively long fatty acids that are relatively unsaturated, giving high melting points. Organism endothermic.
Coconut oil: relatively unsaturated, but short fatty acid chains giving a low melting point. Again, exothermic organism.

5. Yes.

FILL IN THE BLANKS

1. _____ or neutral fats are composed of the alcohol _____ and up to _____ fatty

acids. Phospholipids always contain a _____ group and usually other components.

Glycolipids lack _____ but contain _____ .

_____ are important as _____ reserves and are stored in _____ _____ . Complex

lipids are important constituents of cell _____ . Isoprenoid compounds or _____

_____ are derived from the condensation of _____ units. Major compounds of these

types are the _____ and _____ vitamins. The former group includes the _____

of which cholesterol is the most abundant. Cholesterol is important as a constituent

of biomembranes, reducing their _____ , and as a precursor of _____ acids and

_____ _____ . Important isoprenoid vitamins are the _____, _____, _____ and

_____ vitamins.

Choose from: A, acylglycerols, adipose tissue, bile, carbohydrate (or sugars), D, E, energy, fluidity, glycerol, isoprene, K, lipid, membranes, phosphate (2 occurrences), polyprenyl compounds, steroid hormones, steroids, sterols, three, triacylglycerols.

MULTIPLE-CHOICE QUESTIONS

In all the following questions, select those options (any number may be correct) that correctly complete the initial statement, or correctly answer the question posed.

2. Sphingomyelins differ from other sphingolipids because they
A. are based on a ceramide core
B. are the only type to contain sialic acid residues
C. contain phosphate
D. are amphipathic
E. contain an amide link

3. State which of the following are correct.
A. All lipid-soluble vitamins are isoprenoid compounds.
B. Lipid-soluble vitamins are based on a perhydrocyclophenanthrene structure.
C. Lipid-soluble vitamins occur in a number of active forms.
D. Metabolites of vitamin A are involved in vision and in cell growth and differentiation.
E. The major function of vitamin K is to act as a biological anti-oxidant.

4. Which of the following is the odd one out?

A. Bile salts are polyisoprenoid compounds.
B. Bile salts are synthesized from cholesterol.
C. Bile salts are emulsifying agents.
D. Bile salts are synthesized in adipose tissue.
E. None of the above.

SHORT-ANSWER QUESTIONS

5. The diameter of a typical globular membrane protein is about 7.5 nm, while the average phospholipid molecule in a cell membrane occupies an area of about 0.07 nm^2. Calculate the total number of protein and lipid molecules in the outer layer of a cell membrane in a cell of 20 μm diameter. Assume that the cell is perfectly spherical and that there are 95 lipid molecules to every protein molecule.

6. How many isoprene units are there in β-carotene and in retinol?

8

Nucleic acids

Objectives

After reading this chapter you should be able to:

☐ describe the molecular composition of nucleic acids;

☐ relate the structure of nucleic acids to their sequences of nucleotides;

☐ relate the biological roles of nucleic acids to their structures;

☐ outline the basis of recombinant DNA technology.

8.1 Introduction

All cells contain a set of instructions that control the mechanisms by which the structure of the cell is elaborated and its activities maintained. These **genetic instructions**, or the *genome*, must also be capable of being copied faithfully and passed to subsequent generations of cells at reproduction. However, if variation between individuals is to occur, the instructions within the genome must also be capable of subtle modification. Furthermore, if new varieties or species are to appear (evolution), then these genetic materials must be malleable enough to acquire new or modified instructions, resulting in a change in the genome.

The genetic instructions are now known to reside in remarkable molecules called nucleic acids of which there are two types: **deoxyribonucleic acid** (DNA) and **ribonucleic acid** (RNA). DNA is known to be the repository of the genetic information in all cellular organisms. Acellular organisms (viruses) use either DNA or RNA for this purpose. However, RNA is an absolute requirement for the *expression* of the genome in all organisms.

The purpose of this chapter is to explain the structure and some of the fundamental properties of these molecules, whose functions are so central to life processes.

8.2 Discovery of nucleic acids

Nucleic acids were first isolated by the Swiss biochemist Miescher in 1869. During his investigations on the functions of the nucleus, Miescher extracted a material from a crude nuclear fraction of white blood cells present in pus obtained from used surgical dressings. The extracted material was found to have a high M_r, to be acidic and to contain appreciable amounts of phosphorus. Miescher called this material *nuclein*. Nuclein is now known to be a complex of nucleic acid and protein (or what would now be called **nucleoprotein**). Similar material was subsequently obtained from other animal tissues and also from yeast cells. Using the rather crude analytical

Genome: the full complement of genes of an organism (from the Greek word genea, *meaning breed).*
Nuclein: *meaning isolated from the nucleus.*

Reference Rosenfield, I., Ziff, E. and Van Loon, B. (1983) *DNA for Beginners.* Unwin Paperbacks, UK. An extremely funny 'comic' book, which nevertheless contains a large amount of biochemistry!

methods of that time the nucleic acid from yeast appeared to differ in molecular composition from that of animal tissues. In particular, while the animal nucleic acid contained the sugar **deoxyribose**, that from the yeast had ribose. As a result of this, for some time it was thought that animal cells contained deoxyribonucleic acid, with ribonucleic acid being characteristic of plant cells. However, this idea was soon questioned, and it is now known that all cellular organisms, both prokaryotic and eukaryotic, contain *both* DNA and RNA. In contrast, viruses may contain DNA *or* RNA, but not both.

Analyses of nucleic acid indicated that their basic 'building blocks' were nucleotides consisting of a base, a pentose sugar and phosphate (Fig. 8.1). By 1935, the general nature of nucleic acids had been established by the use of rather laborious chemical techniques. Since then methods of analysing and manipulating nucleic acids have advanced greatly.

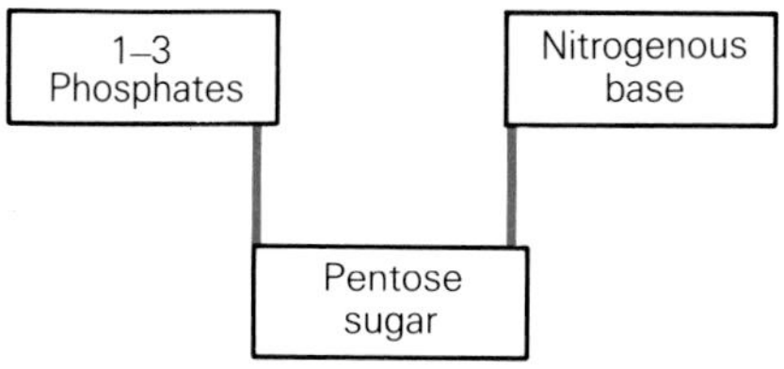

Fig. 8.1 A nucleotide consists of a base, a sugar, and one or more phosphate units.

See *Cell Biology*, Chapter 2

8.3 *Molecular composition of nucleic acids*

The complete hydrolysis of any nucleic acid yields three components:

- a pentose sugar
- a collection of nitrogenous bases
- phosphate

☐ Pentose sugars contain *five* carbon atoms. Examples: ribose, deoxyribose, xylose and xylulose.

The constituents of RNA and DNA are listed in Table 8.1. RNA contains the sugar D-ribose (Fig. 8.2a), and DNA 2-D-deoxyribose (Fig. 8.2b) (i.e. C-2 that lacks an oxygen). Both sugars occur predominantly in ring configurations.

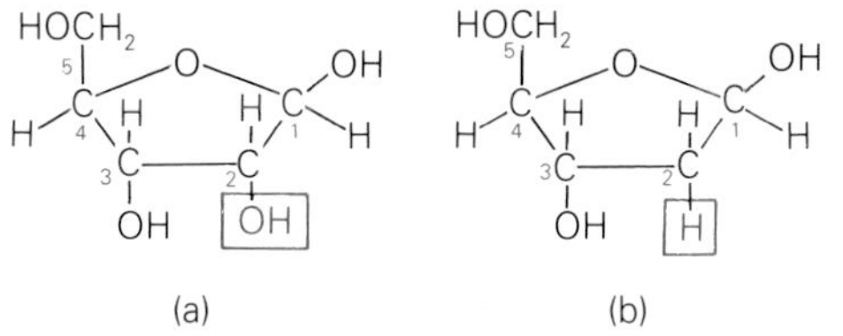

(a)   (b)

Fig. 8.2 (a) Ribose (ribofuranose), (b) Deoxyribose (2-deoxyribofuranose). These ring sugars are based on furan, a five-membered heterocyclic ring with one oxygen, therefore the sugars are referred to as furanose derivatives. The anomeric carbon (C–1) has its hydroxyl group above the plane of the ring, giving a β-anomer.

Table 8.1 *Components of nucleic acids*

	RNA	DNA
Acids	Phosphoric acid	Phosphoric acid
Sugars	D-ribofuranose	D-2-deoxyribofuranose
Purines	Adenine (A)	Adenine (A)
	Guanine (G)	Guanine (G)
Pyrimidines	Cytosine (C)	Cytosine (C)
	Uracil (U)	Thymine (T)

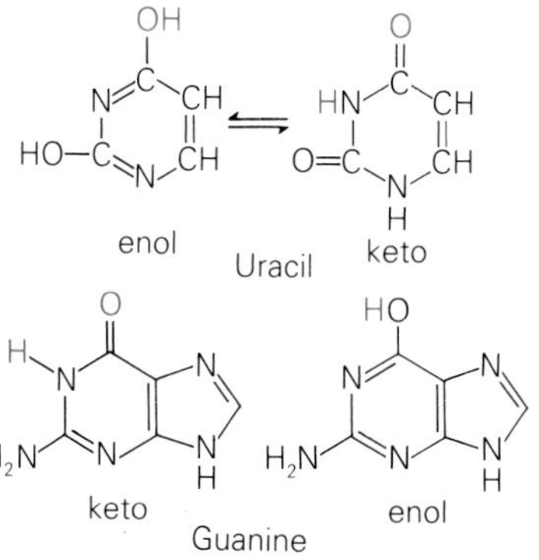

Nitrogenous bases

The nitrogenous bases of nucleic acids are related to either **purine** or **pyrimidine** (Fig. 8.3). Both RNA and DNA contain two purine bases called **adenine** (A) and **guanine** (G). RNA contains two pyrimidine bases: **cytosine** (C) and **uracil** (U).

DNA also contains cytosine but its second pyrimidine is **thymine** (T), which is 5-methyluracil. A possible biochemical reason for the difference is explained elsewhere. Minor amounts of other bases are also found in nucleic acids (see later). All the bases are flat or planar molecules and this allows them

☐ Tautomerism is shown by all the bases except adenine. Thus, uracil can oscillate between an *enol* (lactim) form and a *keto* (lactam) structure. Similarly *keto–enol* isomerization is shown by the purine-base guanine. The composition of the equilibrium mixture of the *keto–enol* forms depends largely upon the pH of the medium. At physiological pH values, the keto forms of the bases, as given in Fig. 8.3, predominate.

Reference Mainwaring, W.I.P., Parish, J.H., Pickering, J.D. and Mann, N.H. (1982) *Nucleic Acid Biochemistry and Molecular Biology*. Blackwell Scientific, Oxford, UK. A good solid account of nucleic acid biochemistry. A little old in a fast-moving field, but worth browsing through.

Exercise 1

Describe the differences in composition between DNA and RNA.

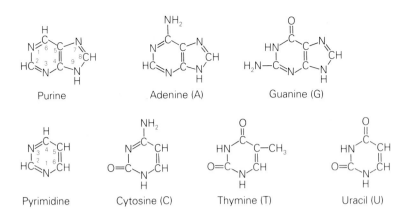

Fig. 8.3 Purine and pyrimidine and their derivatives, the bases commonly found in nucleic acids.

to stack rather like a tower of plates. In an aqueous environment, this stacking is enhanced by hydrophobic interactions between the rings of adjacent bases. These interactions play a major role in stabilizing the overall structure of nucleic acids.

Uncommon bases in nucleic acids

Nucleic acids contain small amounts of other bases, in addition to the five major ones listed above. Methylated bases are common. For example, the DNA of the bacterial virus, bacteriophage T_4 contains the base **hydroxymethylcytosine** (Fig. 8.4). Bacterial DNA is often methylated to give **5-methylcytosine, *N*^{4}-methylcytosine and *N*^{6}-methyladenine** (Fig. 8.4b–d). 5-Methylcytosine is also found in vertebrate and plant cells and the amount of methylation is variable. The DNA of animal viruses is usually non-methylated. In bacterial systems about one in every 200 adenines and cytosines may be methylated. Methylation appears to be extremely low or absent in yeast and insects. In contrast, 5-methylcytosine is common in vertebrates (3–5% of total cytosine), and extensive in higher plants (Table 8.2). Methylation reduces the susceptibility of DNA to certain hydrolytic enzymes and in eukaryotes is probably involved in the control of gene expression.

The small RNA molecules called **transfer ribonucleic acids** (tRNA) are notable for the wide variety of unusual bases they contain. A small selection of these bases is shown in Table 8.3. They are thought to confer resistance to hydrolysis on the tRNA, and assist in its binding to other structures in the cell, such as ribosomes.

(a) 5-Hydroxymethylcytosine (5HmCyt or 5HOMeCyt)

(b) 5-Methylcytosine (5MeCyt)

(c) N^4 Methylcytosine (4MeCyt)

(d) N^6-Methyladenine (6MeAde)

(e) 5-Hydroxymethyluracil (5HmUra or 5HOMeUra)

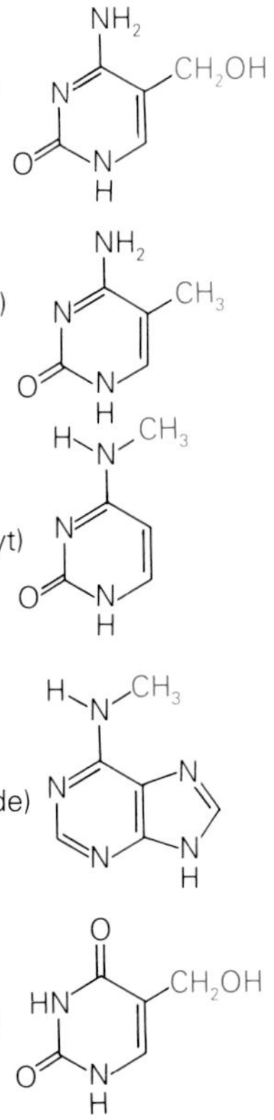

Fig. 8.4 The less common, or unusual bases: (a) 5-hydroxymethylcytosine, (b) 5-methylcytosine, (c) *N*^{4}-methylcytosine, (d) *N*^{6}-methyladenine, (e) 5-hydroxymethyluracil. Structural differences from parent bases are highlighted.

Table 8.2 *Base composition of DNA, expressed as a percentage of total, from different plant species (data from several sources)*

Species	A	G	C	*5MeCyt	T
Chlorella vulgaris	20.2	30.0	26.4	3.45	19.8
Lycopodium tristachyum	27.2	22.8	18.1	4.7	27.2
Dryopteris felix-mas (bracken fern)	28.3	19.8	18.2	5.6	28.4
Pinus sylvestris	30.1	19.3	16.5	3.6	30.5
Zea mays (sweetcorn)	25.6	24.5	17.4	7.2	25.3
Secale cereale (rye)	27.5	19.9	20.1	10.0	24.5

* 5MeCyt is 5-methylcytosine.

Reference Adams, R.L.P., Knowler, J.T. and Leader, D.P. (1986) *The Biochemistry of the Nucleic Acids*, 10th edn, Chapman and Hall, London, UK. A very readable and informative text. Covers virtually all aspects of nucleic acid biochemistry.

Table 8.3 *Some of the unusual bases isolated from tRNA molecules (structural differences from normal are highlighted)*

Parent base	Modified base*	Abbreviation	Structure
Adenine	1-methyladenosine	m^1A	
Cytosine	2-O-methylcytidine	Cm	
	5-methylcytidine	m^5C	
Guanidine	2'-O-methylguanosine	Gm	
	N^2-methylguanosine	m^2G	
	7-methylguanosine	m^7G	

Table 8.3 *Continued*

Parent base	Modified base*	Abbreviation	Structure
	N^2,N^2-dimethylguanosine	m$_2^2$G	
Uracil	pseudouridine	ψ	
	dihydrouridine	D	
	inosine	I	
	wybutosine	W	

* Shown as nucleosides.

Table 8.4 *Nomenclature of bases, nucleosides and nucleotides and nucleic acids*

Bases	Ribonucleoside	Ribonucleotide (5'-monophosphate)
Adenine	Adenosine	Adenylate (AMP)*†,†
Guanine	Guanosine	Guanylate (GMP)
Uracil	Uridine	Uridylate (UMP)
Cytosine	Cytidine	Cytidylate (CMP)

Bases	Deoxyribonucleoside	Deoxyribonucleotide (5'-monophosphate)
Adenine	Deoxyadenosine	Deoxyadenylate (dAMP)
Guanine	Deoxyguanosine	Deoxyguanylate (dGMP)
Thymine	Deoxythymidine	Deoxythymidylate (dTMP)
Cytosine	Deoxycytidine	Deoxycytidylate (dCMP)

* Any other ester must be specified, e.g. 2' UMP for Uridine 2'-phosphate.
† The di- and triphosphates are specified in same way, e.g. GTP for guanosine 5'-triphosphate.

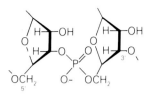

Fig. 8.5 A 5' → 3' phosphodiester bond between two ribose moieties.

The biosynthesis of bases is described in *Cell Biology*. The nitrogenous bases can be considered as part of a hierarchy of structures consisting of bases, **nucleosides, nucleotides** and ultimately **nucleic acids**. A nomenclature for bases, nucleosides and nucleotides is given in Table 8.4.

See *Cell Biology*, Chapter 6

8.4 Nucleic acids

Nucleic acids are polymers of nucleotides and may be described as either **polydeoxyribonucleotides** (DNA) or **polyribonucleotides** (RNA). In both cases the nucleotides are linked by **phosphodiester bonds** joining the 5' and 3' carbons of adjacent nucleotide sugars (Fig. 8.5). The sugar–phosphate linkages have a directional sense. The convention when writing them is to place the C-5' to the *left* and the C-3' to the *right* as indicated in Fig. 8.5. The

☐ Esters are formed by the reaction between an acid and alcohol:

$$R\text{–}OH + HO\text{–}\overset{\overset{\displaystyle O}{\|}}{C}\text{–}R' \rightarrow R\text{–}O\text{–}\overset{\overset{\displaystyle O}{\|}}{C}\text{–}R'$$

A phosphodiester bond is formed between the hydroxyls on two adjacent sugars and a phosphate group.

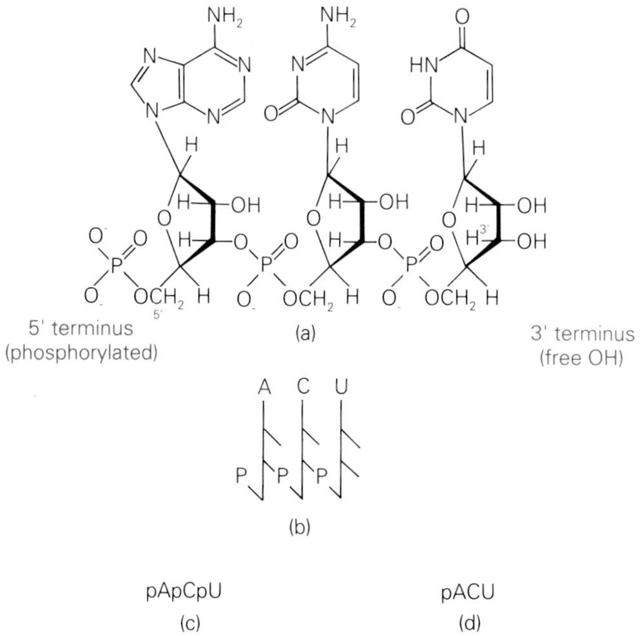

5' terminus (phosphorylated)   (a)   3' terminus (free OH)

pApCpU   pACU
(c)   (d)

Fig. 8.7 (a) Full structure of a triribonucleotide, with (b)–(d) increasingly abbreviated ways of representing the same structure.

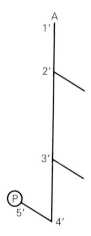

Fig. 8.6 A possible very simplied representation of AMP.

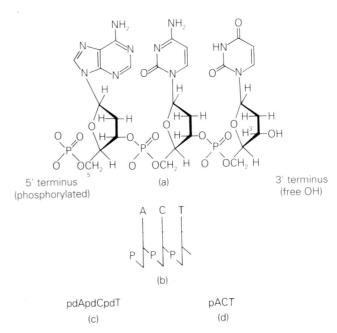

Fig. 8.8 (a) Structure of a trideoxyribonucleotide, with (b) and (c) abbreviated ways of representing the same structure.

pentose sugar can also be displayed diagrammatically as a straight line. Thus, AMP would be drawn as shown in Fig. 8.6. Figure 8.7 shows a structure of a triribonucleotide. The sequence of bases read in the 5' → 3' direction constitutes the **primary structure** of an oligonucleotide or nucleic acid. Figure 8.7 is a cumbersome, although fully detailed, way of expressing the sequence of bases of nucleic acids but, in fact, since the sugar–phosphate backbone simply repeats itself, a variety of shorthand forms are used. These abbreviated forms of expressing nucleotide sequences are shown in Fig. 8.7b–d. Similar methods are used to express the sequences of deoxyribose oligonucleotides but with the inclusion of a 'd' to denote deoxyribose (Fig. 8.8a–c). Note the importance of indicating the presence or absence of phosphate groups at the *termini* of the sequences. The positions of unknown bases are indicated by N ('nucleotide'): if it is at least known that to base is a purine or a pyrimidine, then Pu and Py are used, respectively.

8.5 Structure of DNA

The accepted structure of typical chromosomal DNA was first suggested in 1953 by Watson and Crick. This was the famous **double helix** (or **duplex**), the symbol of modern biology. Watson and Crick built an accurate molecular model representing the structure of DNA using experimental data of other workers. X-ray diffraction patterns obtained from fibres of DNA by workers at Kings College, London, showed DNA to have at least two possible **secondary structures**, dependent upon the relative humidity at which the data were collected. For example, the X-ray diffraction pattern obtained from DNA in low humidity (< 75% r.h.) arose from A-DNA (Fig. 8.9). This pattern was rather complicated and so difficult to interpret. In contrast, the X-ray diffraction pattern obtained for the so-called B-DNA, prevalent at high humidity (> 92% r.h.), was a much simpler (Fig. 8.9b), cross-like pattern known to be typical of helical structures. This pattern had a strong 0.34 nm reflection indicating the spacing between adjacent base pairs, and a 3.4 nm reflection corresponding to a complete turn of the helix. It could also be

Reference Watson, J.D., Hopkins, N.H., Roberts, J.W., Steitz, J.A. and Weiner, A.M. (1987) *Molecular Biology of the Gene*, Vols 1, 2, 4th edn, Benjamin/Cummings, Menlo Park, USA. A splendid, most comprehensive and up-to-date account of nucleic acid structure and replication.

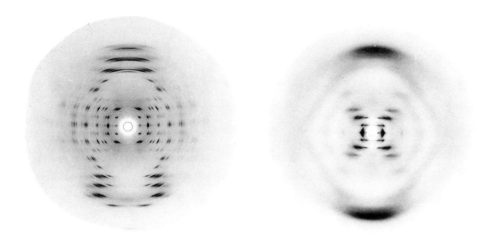

Fig. 8.9 X-ray diffraction patterns of (a) crystalline A-DNA, (b) semicrystalline B-DNA. Courtesy of Prof. M.H.F. Wilkins, Department of Biophysics, Kings College, London.

deduced from the data that the structure was a **dyad**, that is it had an asymmetry such that equivalent chains ran in opposite directions along the long axis.

A key observation that allowed the structure of DNA to be solved was based upon a knowledge of the base composition of DNA. Some years before, Chargaff had shown for many species of organisms that the molar amount of adenine in DNA was equivalent to the amount of thymine, while that of guanine was equivalent to that of cytosine, at least within experimental error (Tables 8.2 and 8.5). Therefore, it followed that the purine content equalled the pyrimidine content.

Figure 8.10 shows the structure of the B form of DNA proposed by Watson and Crick. The structure consists of two polydeoxyribonucleotide chains, with the sugar–phosphate groups on the outside. Both chains form **right-handed helixes**, each chain twisting around an imaginary central axis. These chains run in opposite directions, one in the $5' \rightarrow 3'$ direction, the other $3' \rightarrow 5'$. The chains are therefore described as **antiparallel**. The overall diameter of the double helix is about 2 nm, and the bases are on the inside, in a stacked configuration, extending between the two sugar–phosphate chains something like the steps of a spiral staircase.

☐ Right-handed helices are distinguished by the strands rising towards the right as they twist around an imaginary central axis.

Table 8.5 Base composition of the DNA from different organisms (data from a variety of sources)

Organism	A	G	C	5HmCyt*	T	Pu/Py	A/T	G/C†
Bacteriophage T₄r	32.2	18.0	–	16.3	33.5	1.01	0.96	1.10
φX174 (phage)	24.6	24.1	18	– 5MeCyt	0	–	–	–
Herpes virus	17.1	32.9	32.2	–	17.6	1.00	0.97	1.02
Adenenovirus, type 2	22.0	27.0	29.0	–	21.0	0.98	1.05	0.93
Anacystis nidulans (cyanobacterium)	23.5	32.5	25.5	–	18.5	1.27	1.27	1.27
Bacillus subtilis	28.9	21.0	21.4	–	28.7	1.00	1.01	0.98
Escherichia coli	23.8	26.0	26.4	–	23.8	0.99	1.00	0.91
Neurospora crassa (mould)	23	27.1	26.6	–	23.3	1.00	0.99	1.02
Saccharomyces cerevisiae (yeast)	30.1	19.3	21.7	–	28.9	0.98	1.04	0.89
Wheat	25.6	23.8	18.2	6.4	26.0	0.98	0.99	0.97
Onion	31.8	18.4	12.2	5.4	31.3	1.01	1.02	1.01
Pea	28.1	20.9	16.0	5.0	30.0	0.96	0.94	1.00
Frog	26.3	23.5	21.8	2.0	26.4	0.99	1.00	0.99
Rat (liver)	28.5	21.1	18.7	3.2	28.8	0.98	0.99	0.96
Human (liver)	30.3	19.5	19.9	–	30.3	0.99	1.00	0.98

* 5HmCyt and 5MeCyt are 5-hydroxymethylcytosine and 5-methylcytosine respectively.
† Pu/Py includes 5HmCyt and 5MeCyt where appropriate.

Reference Zimmerman, S.B. (1982) The three-dimensional structure of DNA. *Annual Review of Biochemistry*, **51**, 395–427. An informative overview of the different structural forms of DNA.

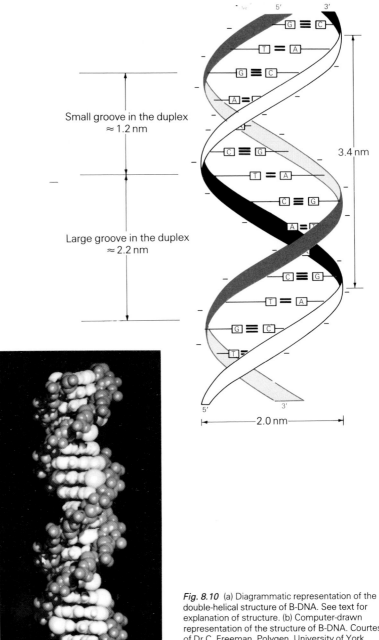

Fig. 8.10 (a) Diagrammatic representation of the double-helical structure of B-DNA. See text for explanation of structure. (b) Computer-drawn representation of the structure of B-DNA. Courtesy of Dr C. Freeman, Polygen, University of York.

Small groove in the duplex ≈ 1.2 nm

Large groove in the duplex ≈ 2.2 nm

3.4 nm

2.0 nm

Exercise 2

Given that the sequence of bases in one strand of DNA is GGTCATTCAC, predict the sequence of the complementary strand.

Each base pair occupies 0.34 nm of the length of the molecule and 10 base pairs (3.4 nm) give one complete turn of the helix. The two halves of the double helix are held together non-covalently by hydrogen bonds between **complementary bases**. Adenine always pairs with thymine and guanine with cytosine, thus generating Chargaff's ratios. It follows, therefore, that if the base sequence of one chain is known, then the other is simply its complement. The G : C pair is held together by *three* hydrogen bonds, while the A : T pair has only *two*, although the overall dimensions of both pairs are essentially the same (Fig. 8.11), allowing the formation of a regular double helix. The geometry of the B-DNA is characterized by having two grooves alternating along its length: the *major* groove is wider, while the other is called the *minor* groove.

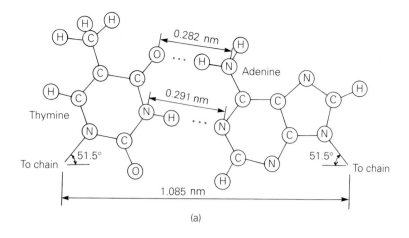

(a)

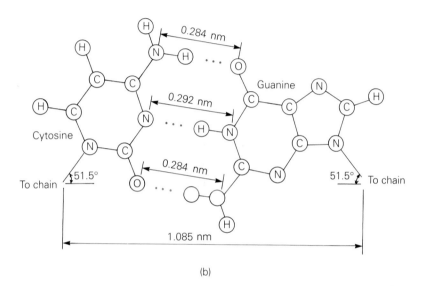

(b)

Fig. 8.11 Overall structure and dimensions of (a) A : T, and (b) G : C base pairs. Redrawn from Wilkins, M.H.F. and Arnott, S. (1965) *Journal of Molecular Biology*, **11**, 391.

A feature of the structure is that the central bases are not fully masked by the sugar–phosphate chains, but are observable within the helix. Naturally, each base shows the same orientation towards the grooves. The major groove is lined by the **C-6, N-7,** and **C-8** atoms of the purine bases, and the **C-4, C-5** and **C-6** atoms of the pyrimidine rings. The arrangement of these atoms can form recognition sites for proteins that interact with DNA.

See *Molecular Biology and Biotechnology*, Chapter 3

Factors stabilizing the double helix

Hydrogen bonding between bases plays a major role in stabilizing the double helix of DNA. Individual hydrogen bonds are weak, but the double helix is stabilized because there is an extremely large number of them occurring along its length. The hydrophobic interactions between adjacent bases referred to earlier also contributes to the great stability of the structure. The charged phosphate groups on the exterior of the molecule are able to react with the dipoles of water, further contributing to the stability of the DNA duplex. Mutual repulsion between the negative charges on adjacent phosphate groups would tend to favour a linear, rigid structure. However, in the cell these negative charges are masked by Mg^{2+} and, in eukaryotic cells, by basic proteins called **histones** forming a **nucleoprotein** complex. This allows the duplex to fold into higher ordered structural states.

□ Histones are highly basic proteins. They are designated by the letter 'H' and numbered H1 to H5. Histone H1 is rich in lysine residues. Two types of H2 occur, H2A and H2B, which contain smaller amounts of lysine than H1. Histones H3 and H4 are arginine rich (see *Cell Biology*, Chapter 3). These basic amino acids are found predominently at the amino termini of the proteins, and are responsible for mediating interactions with the negatively charged nucleic acids. The non-basic portions of the histones interact with other histones.

See *Molecular Biology and Biotechnology*

Reference Watson, J.D. (1968) *The Double Helix*. Weidenfeld & Nicholson, London. A highly personal history of the elucidation of the structure of DNA by one of its co-discoverers.

Reference Sayre, A. (1975) *Rosalind Franklin & DNA*. A fond biography of a scientist who took the first X-ray diffraction patterns of the A- and B-forms of DNA.

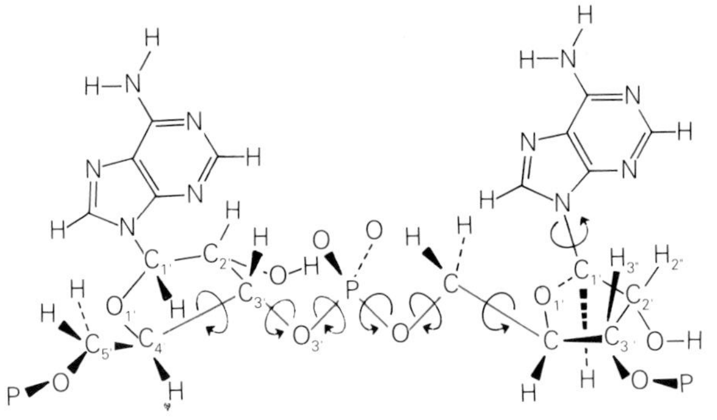

Fig. 8.12 Two adjacent nucleotides in a DNA molecule with the freely rotatable bonds highlighted. Redrawn from Olson, W.K. and Flory, P.J. (1972) Biopolymers, **11**, 1.

8.6 *Alternative structures of DNA*

The so-called B form of DNA is believed to be the major secondary structural type of DNA in solution and *in vivo*, although other forms occur. **Six bonds** between each nucleotide residue (Fig. 8.12) are freely rotatable and this allows for considerable variation in the conformations of DNA. One such conformation, recognized in the early studies, is the A-form already mentioned. A-DNA is less hydrated than B-DNA. This allows the polydeoxyribonucleotides to tilt extensively, shorter, fatter duplex, with about **11 base pairs** per complete helical turn. The bases are tilted to the extent of about 20° with respect to the extended helical axis, rather than the 6° or so in B-DNA. The X-ray diffraction studies mentioned earlier, used stretched fibres of DNA, containing many molecules of DNA orientated along the fibre. This is a relatively crude method of analysis, giving only general molecular spacings. More rigorous X-ray studies require crystals with highly ordered three-dimensional structures. It is not possible to crystallize DNA itself, although some small *synthetic* oligodeoxyribonucleotides will crystallize. For example, in 1980, X-ray diffraction analysis of the self-complementary dodecamer dCdGdCdGdAdAdTdTdCdGdCdG showed it to have a typical B-DNA structure. In contrast, the hexamer dCdGdCdGdCdG, and the tetramer dCdGdCdG have been shown to have completely different secondary structures. Deoxyribonucleic acids containing alternating G–C pairs form **left-handed double helical** structures called Z-DNA (Fig. 8.13). This structure is considerably slimmer than B-DNA, with **12 base pairs** per turn of the helix. The sugar–phosphate backbones form a zig-zag pattern along the length of the molecule, rather than showing the smooth rotation of B-DNA, hence the name Z-DNA.

Observations suggest that Z-DNA exists in specific regions of naturally

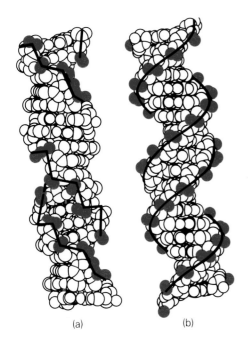

(a)　　　　　　　(b)

Fig. 8.13 Structures of (a) Z-DNA, (b) B-DNA. Redrawn from Wang *et al.* (1979) *Nature*, **282**, 680.

Table 8.6 *Some helical parameters associated with DNA*

Parameter	A-DNA	B-DNA	Z-DNA
Form	Fibre	Fibre	Crystal
Handedness	r.h.	r.h.	l.h.
Base pairs/turn	11	10	12
Helix length/base pair (nm)	0.29	0.34	G–C　0.35
			C–C　0.41
Rotation/base pair (range)	33°	36°	G–C −51°
	(16–44°)	(27–42°)	C–G −8.5°

Reference Rich, A.R., Nordheim, A. and Wang, H.-J. (1984) The chemistry and biology of left-handed Z-DNA. *Annual Review of Biochemistry*, **53**, 791–846. A splendid review of studies on Z-DNA.

occurring DNA. These regions apparently consist of about 12 to 24 sets of 12 alternating purine and pyrimidine residues in sequence: a feature which favours Z-conformation. The biological functions of Z-DNA are not known with certainty, although it is thought that it might have a role in regulating expression of the genetic information.

Some of the structural features of the A-, B- and Z-forms of DNA are summarized in Table 8.6.

CRUCIFORMS are another 'unconventional' group of structures, adopted by DNA, which are dependent upon specific base sequences. **Cruciform** structures are formed at **inverted repeats**. An inverted repeat occurs where a sequence of bases on one strand is followed, on the same strand, by the complementary sequence in reverse order (Fig. 8.14). Such an inverted repeat on one strand, must, of course, have an inverted repeat on the other complementary strand. If the strands of DNA separate (or 'denature'), then each strand will form a **hairpin loop** because of *intra*strand base pairing. Together, both loops give the cruciform structure (Fig. 8.14). Since cruciform formation is only possible when unwinding of the DNA occurs, any stress on DNA that causes unwinding will favour the formation of these unusual structures. It has, indeed, been suggested that cruciforms could serve as recognition sites for the binding of proteins to specific regions of DNA.

--- *Exercise 3* ---

A stretch of DNA bases in a strand of the double helix can base pair with itself, forming a cruciform structure. Which of the two structures, double helix, or cruciform, is likely to be the most stable?

See *Cell Biology*, Chapter 3

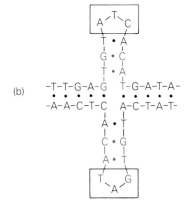

(a)
```
-T-T-G-A-G-T-G-T- | A-T-C | -A-C-A-T-G-A-T-A-
 • • • • • • • •   | • • • |  • • • • • • • •
-A-A-C-T-C-A-C-A- | T-A-G | -T-G-T-A-C-T-A-T-
```

(b)
```
                A ⌒ T
               /     \
              A       C
              T • A
              |   |
              G • C
              |   |
              T • A
              |   |
  -T-T-G-A-G  T   G-A-T-A-
  • • • • •   |   • • • •
  -A-A-C-T-C  A   A-C-T-A-T-
              |   |
              A • T
              |   |
              C • G
              |   |
              A • T
               \     /
              T       G
               \ A /
```

Fig. 8.14 (a) An inverted repeat sequence of bases in DNA that can give rise to (b) a cruciform structure.

(a)

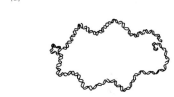

(b)

(c)

Fig. 8.15 Transition from (a) relaxed, to (c) negatively supercoiled DNA, via structure (b) where the main backbone has been hydrolysed, slightly unwound and the phosphodiester bonds rejoined. Redrawn from Felsenfeld, G. (1985) *Scientific American*, **253(4)**, 50.

SUPERCOILING is a tertiary structural form of DNA that is common in the **circular DNA** molecules of bacteria. **Supercoiling, superhelical** and **super-twisting** are all synonyms for the same phenomenon. Circular DNA, lacking supercoiled regions is called **relaxed** DNA, and will lie in a plane (Fig. 8.15). Supercoiling is perhaps best envisaged by imagining the closed circle of DNA to be broken, and the two strands of the duplex slightly unwound. The ends of the sugar–phosphate chains are then resealed (Fig. 8.15b). The molecule will now try to return to its original twist, but this can only be achieved if it becomes distorted such that it no longer lies in its original plane (Fig. 8.15c).

See *Cell Biology*, Chapter 3

Cruciform: *like a cross in shape.*

Reference Dickerson, R.E. (1983) The DNA helix and how it is read. *Scientific American*, **249(6)**, 86–102. A very readable account of how the structure of DNA varies and how its information content depends on this structure.

Reference Felsenfeld, G. (1985) DNA. *Scientific American*, **253(4)**, 44–53. An excellent synopsis of the structure and functions of DNA.

The DNA is now *negatively* supertwisted, that is the longitudinal axis of the double helix is twisted to give a superhelix. Negatively supercoiled DNA is the natural form of circular DNA in cells. In the cell this supercoiling is achieved by the action of enzymes called **gyrases** or **topoisomerases**. A considerable input of metabolic energy, in the form of ATP, is used in supercoiling.

$$\text{relaxed DNA} \xrightarrow[\text{ATP}]{\text{gyrase}} \text{negatively supercoiled DNA}$$

Several reasons have been suggested for the maintenance of negatively supercoiled DNA in cells. The packaging of the molecule is one possible reason: supercoiled DNA is more compact than relaxed DNA and occupies less space in the cell. Negative supercoiling favours an unwinding of the double helix, since this relieves **torsional stress** in the molecule. These unwound regions could thus interact more effectively with other molecules, including proteins. Thus, supercoiling is important in DNA (gene) replication and may also be important in the regulation of gene expression.

See *Molecular Biology and Biotechnology*, Chapters 1, 6 and 7

8.7 Dynamic structure of DNA

Detailed X-ray crystallographic studies of a variety of crystalline forms of DNA have indicated that there is a *large sequence-dependent variation* in the exact conformation adopted by DNA. The B-form of DNA is probably best regarded as an average approximate structure and this emphasizes the point that the double helix is not a fixed, rigid structure. Rather, the molecule is dynamic and is likely to be undergoing continuous internal rearrangements. For example, small sections of the duplex can move apart slightly, a phenomenon called **breathing**. The degree of breathing in any given region of double helix is dependent on the stability of the base pair, and the identity of its neighbours. Breathing is probably of importance in the interactions between DNA and proteins.

8.8 Base pairing in RNA

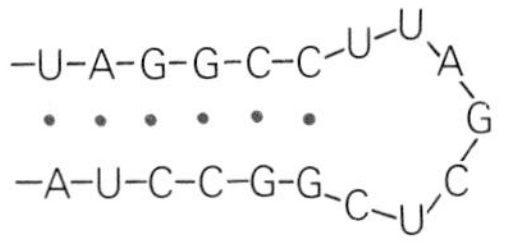

Fig. 8.16 Base pairing, an example of a hairpin loop in single-stranded RNA.

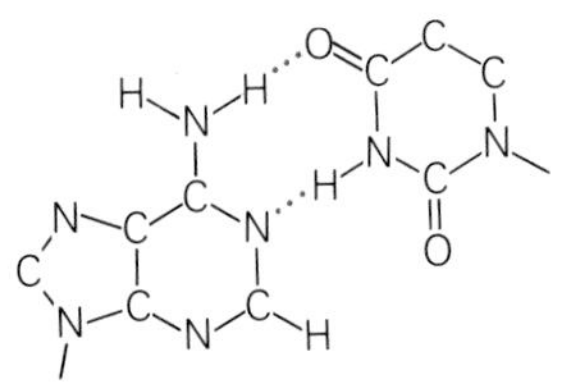

Fig. 8.17 Hydrogen bonding in an A : U base pair.

Unlike DNA, RNA molecules in cells usually consists of *single* polynucleotide chains since they are formed by copying the base sequence of *one* strand of the DNA. Single-strandedness does not preclude base pairing, however, since the molecules may double back on themselves forming hairpin bends in those portions that do not base pair (Fig. 8.16). Base pairing is of the usual complementary type for guanine and cytosine. However, uracil replaces thymine in RNA, giving an **A : U complementarity** that is hydrogen bond stabilized (Fig. 8.17). True double-stranded RNA is known in some viruses (see Section 8.10) where these are two complete, complementary molecules.

RNA double helixes can only assume the A-form. This is because the 2'-OH of ribose imposes a steric hindrance, such that the RNA double helices cannot adopt the B-structure.

DNA–RNA HYBRID duplexes are well known (Fig. 8.18). For example, such hybrids are intermediates in the synthesis of DNA in **retroviruses** replication. Sections of DNA–RNA hybrids about 5–10 nucleotides long are formed during the replication of cellular DNA. DNA–RNA *heteroduplexes* are artificially synthesized in many laboratories to further analytical studies on the structure and functions of nucleic acids.

See *Molecular Biology and Biotechnology*, Chapter 1

Gyrases/topoisomerases: enzymes that alter the rotational stress in DNA molecules. From the Greek gyrus, circle, topos, a place and isos, equal.

Heteroduplex: a double-helix in which one strand is composed of polydeoxyribonucleotide residues and the other of polyribonucleotide residues.

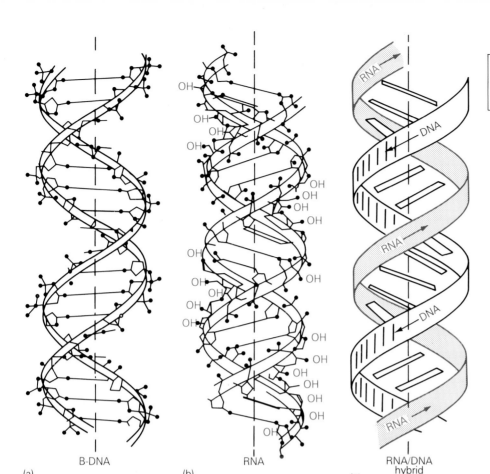

Exercise 4

Give the most likely RNA transcript of the DNA sequence: GGTCATTCAC.

(a) B-DNA   (b) RNA   (c) RNA/DNA hybrid

Fig. 8.18 Structures of (a) B-DNA, (b) RNA double helix and (c) DNA–RNA hybrid double helix (heteroduplex). Redrawn from Parish, J.H. (1972) *Theory and Practice of Experiments with Nucleic Acids*, Longman, London.

8.9 Biological functions of nucleic acids

The functions of nucleic acids are the storage, expression and replication of biological information. For most of the first half of this century the gene was generally regarded as being composed of protein, the DNA of the nucleus being ascribed the role of a mere scaffold, holding the protein genes in place. (DNA was thought to have too simple a structure to carry the genetic information.) However, overwhelming evidence now indicates that genes are not proteins, but are stretches of DNA. In all cellular organisms, DNA forms the genetic material, although in some viruses RNA carries the genetic information. The observation that somatic cells of organisms have twice the amount of DNA found in their gametes (Table 8.7) is consistent with the idea that DNA is the genetic molecule.

Importance of base sequences in DNA

It is apparent from the foregoing that, in general terms, all DNA molecules are alike in overall conformation. However, the DNA from any given species of organism is unique in its base sequence: this primary structure constitutes the genetic information relating to that organism. The information contained in this sequence has a variety of roles and functions.

Some genes, the structural genes, code for specific proteins, in other words the order of amino acid residues in a polypeptide that is characteristic for that

Reference Hunt, T., Prentis, S. and Tooze, J. (1983) *DNA makes RNA makes Protein*. Elsevier Biomedical Press. A series of readable articles, which appeared in the journal *Trends in Biochemical Sciences*. Topics include genome structure and organization, transcription, translation and post-translational modifications.

Reference Holzmüller, W. (1984) *Information in Biological Systems: the Role of Macromolecules*. Cambridge University Press, Cambridge, UK. A short text that introduces basic information theory, and how it applies to the functions of macromolecules in organisms.

Table 8.7 *Amounts of DNA associated with nuclei of gametes and somatic cells*

	Animals (DNA in pg/nucleus)					
	Carp	**Frog**	**Fowl**	**Rat**	**Ox**	**Human**
Thymus	–	–	–	6.3	6.6	–
Liver*	3.33	14.1	2.6	6–13	6.4–19.1	6.3–12.24
Kidney	–	–	2.3	6.7	5.9	6.34
Spleen	–	–	2.6	6.5	7.0	–
Lung	–	–	–	6.7	–	6.04
Pancreas	–	–	2.7	7.3	6.9	5.18
Leukocyte	–	–	–	6.6	7.0	6.50
Erythrocyte	3.49	14.6	–	–	–	–
Spermatozoa	1.64	6.9	1.3	3.3	3.3	3.4

Plants (DNA in pg/nucleus)					
Saccharomyces cerevisiae		*Aspergillus nidulans*		*Lilium longiflorum*	
Haploid	0.0245	Green haploid	4.38×10^{-8}	Microspore	53
Diploid	0.0495	Green diploid	10.17×10^{-8}	(haploid)	
Triploid	0.0670	White haploid	4.57×10^{-8}	Pollen grain	373
Tetraploid	0.1020	White diploid	8.40×10^{-8}	(two cell, haploid)	
				Apical root tip	313
				(tetraploid)	

* Liver cells may be multinucleate.

See *Molecular Biology and Biotechnology*, Chapters 2 and 3

type of protein is specified by the sequence of bases in the gene. The information in DNA is used indirectly, being copied to give an RNA molecule which specifies the primary structure of the polypeptide. Other types of genes code for RNA molecules that do not specify the primary structure of a polypeptide. In both cases the sequence of bases along the DNA is used to produce a complementary RNA sequence. This process is called **transcription** and enzymes that transcribe DNA to form RNA are called **RNA polymerases**. Yet another group of sequences in the DNA are regulatory genes and switch other genes on and off according to the needs of the organism. Some regions of the DNA have no known function.

There are four major types of RNA: the tRNA molecules mentioned earlier, **messenger RNA** (mRNA), **ribosomal RNA** (rRNA) and **small nuclear RNA** (snRNA). The first three are involved in protein synthesis, while the functions of snRNA molecules are more obscure.

snRNA molecules

See *Molecular Biology and Biotechnology*, Chapters 1–5

Small nuclear RNA molecules, as their name suggests, are only several hundred bases long, and are found in the nucleus. They are combined with proteins to give **ribonucleoprotein particles**. These are several types of snRNA, and all are extremely abundant with 10^4–10^6 copies of each being found per nucleus. The functions of snRNA are not known with certainty, although they have some roles in the processing of RNA.

Protein synthesis

The major roles of mRNA, tRNA and rRNA are understood in much more detail than those of snRNA. All have well-understood functions in the biosynthesis of proteins. For protein synthesis to occur, a copy of the structural gene is transcribed by RNA polymerase II to produce an mRNA molecule. The information specifying the sequence of amino acid residues in the protein is coded in the form of triplets of bases, called **codons**, along the

Reference Ptashne, M. (1986) *A Genetic Switch*. Cell Press and Blackwell Scientific, MA, USA. A short, profusely illustrated textbook. Good coverage of molecular interactions between proteins and nucleic acids, and their importance for gene regulation.

Table 8.8 *The 64 codons of the genetic code: the first base of the codon is given by the left-hand column, the second by one of the middle bases, and the third by the right-hand column (a codon for Proline is highlighted)*

5'-OH terminal base	Middle base				3'-OH terminal base
	U	C	A	G	
U	Phe	Ser	Tyr	Cys	U
	Phe	Ser	Tyr	Cys	C
	Leu	Ser	Term*	Term*	A
	Leu	Ser	Term*	Trp	G
C	Leu	Pro	His	Arg	U
	Leu	Pro	His	Arg	C
	Leu	Pro	Gln	Arg	A
	Leu	Pro	Gln	Arg	G
A	Ile	Thr	Asn	Ser	U
	Ile	Thr	Asn	Ser	C
	Ile	Thr	Lys	Arg	A
	Met	Thr	Lys	Arg	G
G	Val	Ala	Asp	Gly	U
	Val	Ala	Asp	Gly	C
	Val	Ala	Glu	Gly	A
	Val	Ala	Glu	Gly	G

* These three codons are chain terminators, and specify the end of the transcript of the structural gene.

Table 8.9 *Molecular composition of prokaryotic and eukaryotic ribosomes*

	Prokaryote	Eukaryote
Intact		
Dimensions (nm)	29×21	32×22
M_r	2.8×10^6	4.5×10^6
*s_0	70S	80S
Large subunit		
M_r	1.8×10^6	3×10^6
*s_0	50S	60S
Composition	2 RNA molecules	3 RNA molecules
	~ 3000 nucleotides	~ 5000 nucleotides
	~ 120 nucleotides	~ 160 nucleotides
		~ 121 nucleotides
	34 proteins	~ 45 proteins
Small subunit		
M_r	1.0×10^6	1.5×10^6
*s_0	30S	40S
Composition	1 RNA	1 RNA
	~ 1520 nucleotides	~ 2000 nucleotides
	21 proteins	~ 33 proteins

* s_0 is the sedimentation coefficient.

mRNA molecule. Since four bases can be arranged to give 4^3 (64) different triplet codons, it appears that most of the 20 standard amino acids found in proteins are specified by several codons (Table 8.8).

The mRNA is subsequently **translated** to give a polypeptide by small cytoplasmic particles called **ribosomes**. The ribosomes of eukaryotic cells are somewhat larger than those of prokaryotes, although both are composed of *two* subunits and both appear to function in essentially similar ways. Ribosomes are composed of proteins and rRNA molecules (Table 8.9). Ribosomal RNA is formed by the transcription of DNA by a different RNA polymerase, RNA polymerase I.

See *Molecular Biology and Biotechnology*, Chapter 4

During translation, metabolically activated amino acids are delivered to their appropriate codons in the mRNA–ribosome complex by tRNA molecules. Transfer RNA molecules are transcribed from DNA by yet another enzyme, RNA polymerase III. Cells contain at least one type of tRNA molecule for each type of amino acid. Each tRNA has a complementary sequence of three bases, or **anticodon**, to the codon for its amino acid, and is therefore able to recognize the correct codon on the mRNA. Thus, the amino acids in the polypeptide are arranged in the sequence specified by the order of codons in the mRNA, which was in turn dictated by the primary structure of the structural gene. These reactions are summarized in Fig. 8.19.

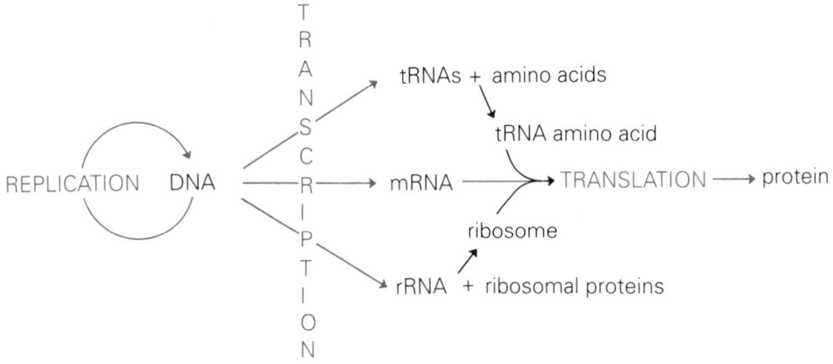

Fig. 8.19 Overview of molecules involved in replication, transcription and translation. The coloured lines represent a flow of information, not chemical transformations.

□ Holley was first to determine the primary structure of a nucleic acid. The sequence of tRNA^Ala from yeast was completed in 1965. The importance of this work was recognized with the award of a Nobel prize in 1968.

The process of protein synthesis has been pithily summarized by the phrase '*DNA makes RNA makes protein*'. The description of the processes involved in transcription and translation given here have been greatly simplified. For example, all the primary transcripts of mRNA, rRNA and tRNA are subjected to modification in eukaryotic cells, before their involvement in translation.

The structures of tRNA molecules are among the best understood of all nucleic acids. Since tRNA molecules are relatively small (M_r 26 000; about 80 nucleotides long), and contain many unusual bases (see earlier and Table 8.3), they were the first nucleic acids to have their primary structures elucidated. All tRNAs can be arranged schematically as a 'clover leaf'; this maximizes the base pairing, as shown for **tRNA^Phe** (the tRNA that delivers phenylalanine to the growing peptide chain during translation) in Fig. 8.20. However, X-ray diffraction studies reveal that tRNA molecules have a roughly L-shaped tertiary structure (Fig. 8.21).

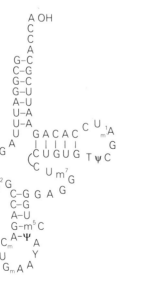

Fig. 8.20 Base sequence of tRNA^Phe arranged in the clover-leaf pattern. Modified bases are highlighted.

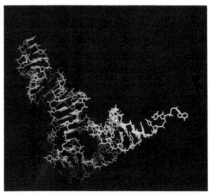

Fig. 8.21 Computer-drawn representation of yeast tRNA^Phe showing the 'L-structure'. Note the double-helical regions. Courtesy of Dr C. Freeman, Polygen, University of York.

Nucleic acids as the genetic material

In eukaryotes, DNA is confined to specific organelles: **nuclei, mitochondria** and **chloroplasts**. Although prokaryotes lack cellular organelles, there may be a degree of tertiary organization of the DNA. While the amount of DNA in the different cells of a given species of organism is constant, there are wide interspecific variations. Table 8.10 shows the haploid DNA content of a cell from a range of species.

In general there is an increase in the quantity of DNA per cell as the evolutionary tree is ascended, but there are also big overlaps (Fig. 8.22) between groups. The DNA is divided between chromosomes, each chromosome containing one double-helical DNA molecule. Accompanying the increased size of DNA in higher organisms there is an increase in the length of the DNA and the number of bases in each strand (Table 8.11). This increase reflects the requirement to carry more information in the DNA of complex organisms. However, large sections of DNA appears not to code for functional proteins, but rather to act as 'spacer regions' between genes, although these sequences may have as yet undiscovered functions.

Table 8.10 The DNA content in picograms (10^{-12} g) of a haploid cell from a range of species

Organism	DNA content of haploid cell (pg)
T$_2$ bacteriophage	0.0002
E. coli	0.01
Yeast	0.025
Coelenterate	0.30
Echinoderm	0.90
Insect	0.17
Fish (teleost)	0.5–1.5
Bird	1.0–2.0
Amphibian	7.5
Man	3.4
Mammals	2.4–3.2
Plant (corn)	7.8

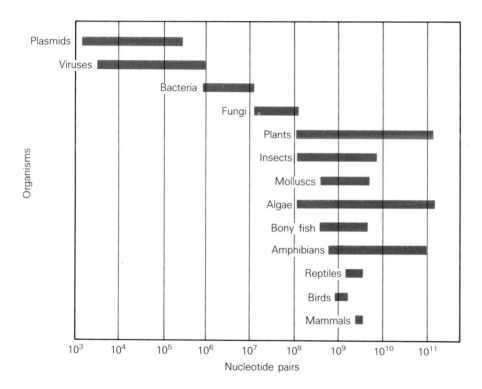

Fig. 8.22 Relative amounts of DNA in different types of cells, viruses and plasmids. Redrawn from Zubay, G. (1988) *Biochemistry*, 2nd edn, Macmillan, New York, p. 216.

Table 8.11 The length of nucleic acid molecules from a variety of species

Nucleic acid	No. nucleotides/ strand	Length (μm)
tRNA	70–80	0.003
5S RNA	120	0.04
16S RNA	1600	0.53
23S RNA	3100	1.00
Polyoma DNA	5000	1.7
E. coli DNA	4.5×10^6	1530 (1.5 mm)
Drosophilia (total haploid genome)	2×10^8	68 000 (68 mm)
Mammal (longest strand)	5×10^6	1800 (1.8 mm)
Human (total haploid genome)	3×10^9	10^6 (1 m)

Reference Judson, H.R. (1979) *The Eighth Day of Creation*. Jonathan Cape, London, UK. A splendidly researched, well-written history of some of the basic discoveries in molecular biology.

Viral nucleic acids

□ Virus coats are built from identical protein subunits in a limited number of arrangements. One kind of arrangement has icosahedral symmetry: this is a solid with 20 faces.

Viruses are comparatively simple organisms (if, indeed, they can be considered organisms). The simplest consist only of a protein capsule enclosing the nucleic acid, but many viral particles also have a membrane outer coat. Viruses may contain either RNA or DNA but not both. The shape of the protein capsule influences the structure of the nucleic acid. In **icosahedral** capsules up to 20–30% of the nucleic acid bases appear unstacked and are associated with the capsule protein. Figure 8.23 shows a typical arrangement of the viral RNA in relation to the protein coat.

Not only may viral nucleic acids be either DNA or RNA but they may also be either single- or double-stranded, circular or linear, depending upon the type of virus. Figure 8.24 shows the variety of structures of viral nucleic acids. Viruses usually contain a single copy of the nucleic acid but in some cases multiple copies are present. Table 8.12 shows the physical properties of some viral nucleic acids.

RNA, single-stranded, e.g. tobacco mosaic virus

RNA, double-stranded, e.g. reovirus (causes infantile gastro-enteritis)

DNA, single-stranded linear, e.g. parvovirus (causes bone marrow aplasia)

DNA, single-stranded circular, e.g. bacterio-phage ϕX174

DNA, double-stranded linear, e.g. bacterio-phage T_4 Herpes (causes cold sores)

DNA, double-stranded circular, e.g. polyoma (causes warts)

DNA, double-stranded with linked terminal proteins, e.g. adeno-virus (causes acute respiratory disease, pneumonia)

DNA, double-stranded with covalently sealed ends, e.g. poxvirus (causes small pox)

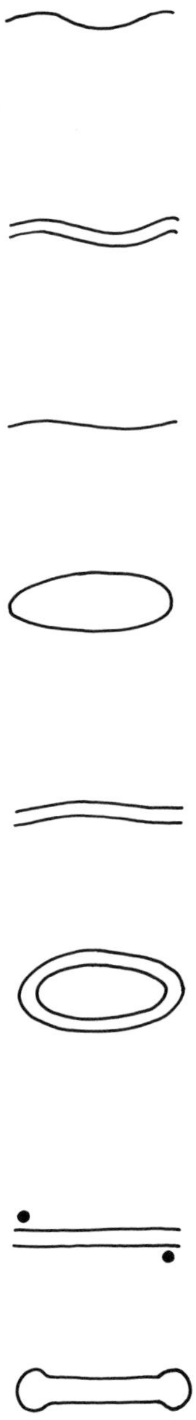

Fig. 8.24 The diversity of viral nucleic acid structures.

Table 8.12 *The molecular weights and lengths of viral nucleic acids*

Nucleic acid	Virus in which nucleic acid is found	Molecular weight of nucleic acid $\times 10^{-6}$	Length (μm)
Double-stranded DNA			
Adenovirus	Types 12, 18	21	70–80
Coliphage	T_3, T_7	25	–
Herpesvirus	Herpes simplex	100	180–200
Single-stranded DNA			
Coliphage	ϕX174	1.7	18–20
Double-stranded RNA			
Reovirus		15	75–80
Rice dwarf virus		15	
Single-stranded RNA			
Coliphage	R17	1.1	
Tobacco mosaic virus		2	
Picornavirus	Polio	2.5	21–30
Myxovirus	Influenza	4	

SINGLE-STRANDED VIRAL RNA has an M_r in the range 0.4–13×10^6. The RNA can be in one piece (picornaviruses) or fragmented (myxovirus RNA consists of at least seven pieces). Some viral RNA has a polyadenylic acid tail or a CCA 3' terminus. These viral RNA molecules, therefore, resemble eukaryotic mRNA molecules (Fig. 8.20) in these respects.

DOUBLE-STRANDED VIRAL RNA is found in a variety of icosahedral viruses of animals and plants and the RNA of these viruses is typically fragmented. Reovirus RNA is in ten pieces of M_r 0.63–2.6×10^6. These fragments are consistently isolated from viral preparations and are not artefacts of preparation.

SINGLE-STRANDED VIRAL DNA is found in small bacteriophages and can be circular (virus ϕX174) or linear (parvovirus).

DOUBLE-STRANDED LINEAR DNA is found in the majority of viruses so far investigated. The DNA of T-even bacteriophages is about 50 μm in length. Some viral DNA molecules contain abnormal bases. In T-even bacterio-

phages, 5-hydroxymethylcytosine (Fig. 8.4) replaces cytosine, and a bacterio-phage whose host is *Bacillus subtilis* contains 5-hydroxymethyluracil (Fig. 8.4e) instead of thymine. Circular double-stranded DNA is found in small animal viruses, such as the polyoma virus. Some linear DNA molecules have cohesive ends (see Box 8.3), which allows them to cyclize.

Bacterial DNA

The DNA of all prokaryotes is circular and double-stranded and each cell has a single copy, that is, one chromosome. In *Escherichia coli* the DNA has an M_r of 3×10^9 equivalent to 4×10^6 base pairs. Figure 8.25 is a gene map of the *E. coli* chromosome showing the location of all the bacterial genes. Prokaryotes often have additional, smaller pieces of circular DNA called plasmids. Bacterial chromosomal DNA is highly supercoiled. Fully extended *E. coli* DNA would form a circle of 1.5 mm circumference. Considering that the length of the bacterium is about 2 μm, there must be extensive folding of the DNA molecule to allow it to fit in the bacterial cell. The DNA of *E. coli* is complexed to large numbers of small basic proteins, which facilitate this folding.

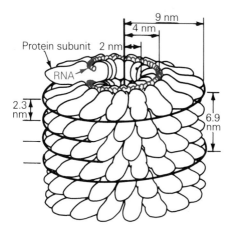

Fig. 8.23 The arrangement of RNA and protein subunits in Tobacco Mosaic Virus. The protein subunits assemble in a helix with 16 units to a turn. The RNA is arranged between the capsid proteins. Redrawn from Baron, S. (1986) *Medical Microbiology*, 2nd edn, Addison-Wesley.

☐ Bacteriophages or 'phages' are viruses that infect and replicate inside bacteria. In comparison with a number of other viruses they are quite large. The T-even bacteriophages (T_2, T_4 etc.) specifically infect *E. coli*, and are more virulent than the other phage types, causing lysis of the bacterium. The DNA is large (120 000 bases in T_4) and double-stranded.

☐ A number of bacteriophages have DNA in which the 5' end of each strand contains an unpaired section of 12 nucleotides. The sequences are complementary and allow a linear DNA to transform into a circular DNA with the aid of DNA ligase; they are called cohesive ends.

5'GGGCGGCGACCT   3'

         3'   CCCGCCGCTGCTGGA5'

linear DNA

circular DNA

See *Cell Biology*, Chapter 3

☐ Plasmids are small pieces of circular DNA of between 10^3 and 30×10^3 base pairs in length. Plasmids code for non-essential proteins and are often responsible for conferring properties like antibiotic resistance to bacteria. Plasmids often replicate independently of the bacterial chromosome. Plasmids are extensively employed in recombinant DNA technology (see *Molecular Biology and Biotechnology*, Chapter 9).

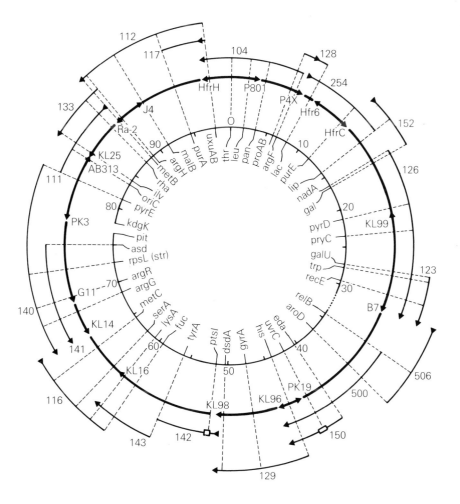

Fig. 8.25 A circular genetic map of *E. coli*. The position of genes is indicated on the inner circle. The relative position of genes is indicated in minutes as the time to transfer during bacterial conjugation. The position of the threonine locus is arbitrarily designated as 0 minutes. From Baron, S. (1986) *Medical Microbiology*, 2nd edn, Addison-Wesley.

Mitochondrial and chloroplast DNA

Both mitochondria and chloroplasts contain DNA as small double-stranded molecules. The mitochondrial DNA of plants is larger than that in animals; all plants seem to have similar-sized chloroplast DNA (Table 8.13). Mitochondria and chloroplasts contain multiple copies of DNA (Table 8.14) present in the **mitochondrial matrix** or in the **chloroplast stroma**. Mitochondrial and chloroplast DNA resembles bacterial DNA and unlike other eukaryotic DNA

See *Cell Biology*, Chapter 3

Exercise 5

Given that the chromosome of *E. coli* is some 4×10^6 base pairs long, calculate the percentage volume of the cell (2 μm long $\times$ 0.75 μm diameter) occupied by the DNA molecule. (Assume the bacterium is a perfect cylinder in shape.)

Table 8.13 The relative sizes of mitochondrial and chloroplast DNA from a range of species

Mitochondria from	Type	Molecular weight $\times 10^{-6}$	
Animals	Circular	9–12	Single size in any species
Higher plants	Circular	Varies	Several sizes of DNA
Fungi	Circular	22–50	
Protozoa (*Paramecium*)	Circular	18–27	
Chloroplast Algae	Circular	90–120	
Higher plants	Circular	85–97	Single size of DNA in any one species

---

Box 8.1
Mitochondrial DNA and human evolution

Mitochondrial DNA (mtDNA) is unusual in that it is inherited in a strict maternal fashion, in other words from the female gamete or oocyte. Spermatozoa contribute only *nuclear* DNA to the zygote. Mitochondria lack DNA repair enzymes, and as a consequence mtDNA is subject to a mutation rate about ten times higher than that experienced by nuclear DNA. This high mutation rate means there is great variation in the base sequences of mtDNA and these variations can be estimated by determining restriction fragment length polymorphisms (RFLPs). Restriction endonucleases are used to digest the mtDNA producing a series of fragments. These fragments can be separated by electrophoresis, resulting in a banding pattern on the electrophoresis gel characteristic of the sequence of the mtDNA and the endonuclease used. The more similar the banding patterns the greater the similarity between different mtDNA molecules and the more closely the individuals are related.

Cann and co-workers have analysed the mtDNA from 147 individuals representing African, Asian, Australian, Caucasian and New Guinean populations by RFLP using 12 different restriction endonucleases. Their results are consistent with an evolutionary tree with two main roots. The first root leads exclusively to Africans, the other to some Africans plus the other four populations. The sequence diversity was greatest with African mtDNA, suggesting that they are the oldest group, since older mtDNA would have accumulated more mutations. Cann and associates have also suggested that contemporary mtDNA must have arisen from a single (female) individual. For present-day mtDNA to have descended from a large number of ancestors would suggest that heterogeneity among ancestral mtDNA, was much less than in modern DNA.

This study implies that modern man (the gender-led proper noun seems particularly inappropriate in this case) arose from a single woman (Eve?), who lived in Africa (Eden?), some 140 000–290 000 years ago (18 000 generations), assuming a 2–4% divergence in DNA sequences per million years.

Naturally these findings have been challenged, and other studies indicate that many 'Eves' and, indeed, males may have contributed to our nuclear DNA. Of clear interest is the work, currently in progress, investigating *paternal* ancestry using Y chromosome-specific probes. Will a common paternal ancestor, Adam, be identified?

Cann, R.L., Stoneking, B. and Wilson, A.C. (1987) *Nature*, **325**, 31–6.

Reference Beale, G. and Knowles, J. (1978) *Extranuclear Genetics*. Edward Arnold, London. Rather old now, but still an excellent introduction to mitochondrial and chloroplast DNA.

Reference Gorvell, L.A. (1983) Mitochondrial DNA. *Scientific American*, **248(3),** 60–73. A very readable account of the structure and replication of mitochondrial DNA.

molecules, is not associated with **histone** proteins. Mitochondrial DNA codes for some mitochondrial proteins (others are derived from nuclear DNA), and for the rRNA and tRNA molecules needed for protein synthesis *in* the mitochondrion. The genetic code used by human mitochondria differ slightly from that of prokaryotic DNA and also that of eukaryotic chromosomes (Table 8.15). Mitochondrial and chloroplast ribosomes are similar in structure to bacterial ribosomes.

□ Kinetoplasts form part of a highly specialized mitochondrion found in certain flagellated protozoa, such as *Trypanosoma*, the causative agent of sleeping sickness. Kinetoplast DNA (kDNA) consists of many small circular (*minicircles*) DNA molecules ($< 2.4 \times 10^3$ base pairs) interlinked with fewer larger circular (*maxicircles*) DNA molecules ($> 30 \times 10^3$ base pairs). The maxicircles are believed to function as conventional mtDNA. Seemingly, the minicircles are never transcribed, but may have a structural role.

Table 8.14 *The relative amounts of organelle DNA in cells from a variety of species*

Mitochondrial DNA from	Cell type	No. of DNA molecules/cell	No. of organelles/cell
Rat	Liver	5–10	1000
Mouse	L-cell line	5–10	100
Frog	Egg	5–10	10^7
Yeast	Vegetative	2–50	2–50
Chloroplasts			
Chlamydomonas	Vegetative diploid	80	2
Maize	Leaves	20–40	20–40

Table 8.15 *Differences in the genetic code between mitochondrial and non-mitochondrial DNA*

Codon	Normal meaning in eukaryotes and prokaryotes	Meaning in mitochondria from		
		Humans	Yeast	*Neurospora*
UGA	STOP	Trp	Trp	Trp
CUU				
CUC	Leu	Leu	Thr	Leu
CUA				
CUG				

Eukaryotic chromosomes

The term ***chromosome*** was originally used to describe the microscopic structures that become visible in cell nuclei before cell division (Fig. 8.26). The word chromosome has also come to be applied to the corresponding genetic structure of bacteria and viruses (see earlier). However, the chromosomes of eukaryotic cells are much more complex than those found in bacteria and viruses. Each eukaryotic chromosome contains a *single* DNA molecule (that is a double helix), which may be twenty or more times larger than that found in bacterial chromosomes (Table 8.11). The DNA is tightly complexed with proteins called histones, which are thought to have structural and protective functions. There is only a limited number of types of histone proteins, but they are present in many copies and constitute most of the protein content of the chromosome. Additionally, there is a group of diverse proteins, loosely bound to the DNA called, somewhat unimaginatively, **non-histone proteins**. These are present in much smaller quantities than histones. The functions of many non-histone proteins are not known with certainty: some have roles in transcription (see earlier), while others are important in the replication of DNA.

Fig. 8.26 Scanning electron micrograph of a human chromosome in metaphase (× 23 000). The chromosome consists of two identical chromatids joined at their centromere. Courtesy Dr C.J. Harrison, Paterson Laboratories, Christie Hospital, Manchester.

Chromosome: a discrete nucleoprotein complex. In eukaryotic cells chromosomes are most clearly seen at metaphase. From the Latin chroma, colour, and some, body, referring to the intense staining that clearly reveals mitotic figures during cell division.

8.10 Replication of DNA

DNA carries the genetic information in its sequence of bases and this must be transmitted from one generation to the next. During reproduction there must be accurate copying of the base sequence of the DNA so that each of the two new daughter cells receives a copy of the original parental genome. The mechanisms by which this is achieved are described in *Molecular Biology and Biotechnology*, Chapter 1.

8.11 Enzymic hydrolysis of nucleic acids

The use of enzymes allows nucleic acids to be selectively hydrolysed. Enzymes that hydrolyse nucleic acids are called **nucleases** or **phosphodiesterases**. A large number of different types of nucleases are known, which differ in their mode of action (specificity). Thus, different phosphodiesterases are specific for RNA, DNA or both; for double- or single-stranded nucleic acids and for *a*- or *b*-type cleavage. There are two major types of nucleases: those that cleave internal bonds, **endonucleases**, and those that hydrolyse terminal nucleotides, **exonucleases**. Most effective hydrolysis occurs with a combination of both, since the endonuclease generate more 'ends', effectively increasing the concentration of substrate for exonuclease activity. The specificities of some general nucleases are summarized in Table 8.16. The selective use of such enzymes and analysis of the products can allow the complete base sequence of small oligonucleotides to be determined.

Following complete hydrolysis the base composition of a sample can be determined by, for example, ion-exchange chromatography. Chargaff used paper chromatography in the determination of his famous rules concerning the base ratios of DNA.

The measurements of the base composition of the DNA from a diversity of species have indicated that although the ratios of $A:T$ and $G:C$ are invariant, the proportions of $(A+T):(G+C)$ show enormous variation. The data in Table 8.17 show 'lower' organisms having the widest variations in $G+C$ content (19–70%), while 'higher' organisms have a $G+C$ content nearer to 50%.

□ A phosphodiester bond may be hydrolysed on its C-3' or C-5' side. The former is called *a* hydrolysis, the latter *b* hydrolysis:

RNA is susceptible to alkaline hydrolysis (*b* hydrolysis) which releases nucleoside 3'-phosphates. However, the absence of a 2'-OH renders DNA resistant to alkaline hydrolysis.

Table 8.16 *Substrate and bond specificities of some common, general endo- and exonucleases*

Enzyme	Substrate	Mode of attack	Type of cleavage	Example
Pancreatic ribonuclease	RNA	Endo (*b*-type)	At C and U to give fragments with py-3'-P	↓ …py-p-N…
Takadiastase T_1	RNA	Endo (*b*-type)	Between G and N on 3' (of G)	↓ …G-p-N…
Micrococcal	RNA, DNA	Endo (*b*-type)	Between A and N on 5' (of A)	↓ …N-p-A…
Snake venom nuclease	RNA, DNA	Exo (*a*-type)	Starts 3'-OH, terminus; cleavage at 3'; no base specificity	↓ …N^1-p-N^2-p-N (3'-OH) …
Spleen nuclease	RNA, DNA	Exo (*b*-type)	Needs, and attacks at 5'-OH terminus; bond cleaved at 5' position; no base specificity	↓ …(5'-OH)N^1-p-N^2-p-N^3…

Table 8.17 *The guanine and cytosine (including 5-hydroxymethylcytosine and 5-methylcytosine) composition of a variety of organisms (data from several sources)*

Group	Organism	G + C (%)
Virus	Bacteriophage AR9	27.1
	Bacteriophage T$_4$	34.3
	ϕX174	42.1
	Adenovirus, type 2	56.0
	Herpesvirus	65.0
Bacteria	*Clostridium perfringens*	26.8
	Bacillus subtilis	42.4
	Escherichia coli	52.4
	Mycobacterium	73.0
Fungi	Yeast	41.0
	Aspergillus niger	50.1
	Neurospora crassa	53.7
'Lower' plants	*Ginkgo biloba*	34.9
	Dryopteris felix-mas	43.6
	Nitella sp.	48.6
	Chlorella vulgaris	59.9
'Higher' plants	Wheat	48.4
	Pea	37.4
	Onion	36.0
Protozoa	*Plasmodium falciparum*	19.0
	Tetrahymena pyriformis	29.2
	Euglena gracilis	57.1
Higher animals	Human	39.4
	Rat	43.0
	Horse	43.0
	Rhesus monkey	45.2

Restriction endonucleases

Restriction endonucleases are a group of nucleases from microorganisms that recognize specific, short base sequences of double-stranded DNA, and then hydrolyse both strands. Restriction endonucleases are named by giving them the first three letters of the microorganism from which they were isolated (e.g. *Eco* for *E. coli*). A fourth letter is used, if necessary, to indicate the strain of bacterial species (*EcoR* for *E. coli*, R strain). Where more than one restriction enzyme is found in a particular strain, then these are differentiated by using Roman numerals, e.g. *Eco*RI, *Eco*RII. Restriction endonucleases are classified into types I, II and III.

Types I and III restriction endonucleases hydrolyse the nucleic acid in an unpredictable manner at positions other than their recognition sites. Their products are therefore heterogeneous in size. Type I enzymes consist of three subunits. In addition to nuclease activity, the enzymes methylate the DNA (see earlier), using **S-adenosylmethionine** (SAM) as the methyl group donor. Following the cleavage of the DNA, the enzymes then show extensive ATPase activity.

Type III restriction endonucleases comprise two subunits. They catalyse two self-competing reactions: methylation of the DNA and nuclease action. They do not have an absolute requirement for SAM, but are stimulated by it. Nuclease cleavage usually occurs some 20 to 30 bases from the 3' end of the recognition site.

Reference Malcolm, A.D.B. (1981) The use of restriction enzymes in genetic engineering. In *Genetic Engineering,* Vol. 2 (ed. R. Williamson), Academic Press, London. A clearly written, well-balanced introduction to restriction endonucleases and some of their uses.

Box 8.2
Methylation of DNA

Methylases (or methyltransferases) are a group of enzymes that also recognize palindromic sites. Thus, type I restriction endonucleases have corresponding methylase enzymes. Once methylated, the palindromic site on the host DNA is no longer recognized by the endonuclease and is therefore protected against hydrolysis. In contrast, invasive DNA, for example that present as a result of a viral attack *is* susceptible, and can be degraded.

In eukaryotic cells, methylation of some bases of DNA (see earlier) seems to be important in the control of gene expression. The bases that form targets for methylation in animals are usually cytosines that are immediately followed by a guanine (CG).

$$...CG... \quad \text{methylase} \quad ...^mC \ G...$$
$$...GC... \quad \rightarrow \quad ...G^mC...$$
$$\textit{S-adenosylmethionine}$$

In plants, methylation is of cytosines, separated from a guanine by another base (CNG). Methylation occurs at the cytosines on both strands.

$$...CNG... \quad \text{methylase} \quad ...^mCN \ G...$$
$$...GNC... \quad \rightarrow \quad ...GN^mC...$$
$$\textit{S-adenosylmethionine}$$

However, not all susceptible cytosines are methylated. Further, the pattern of methylation within the DNA genome is cell-specific, and this pattern is maintained from generation to generation. There is increasing evidence that functionally active genes are less methylated than inactive genes, i.e. methylation of a gene stops, or depresses, its level of expression. The mechanisms for altering methylation in such a tissue-specific manner are not understood, nor is the manner in which methylation reduces the activity of the gene. However, it is possible that methylation of bases will alter the interactions between DNA and regulatory proteins, effecting levels of transcription.

Doerfler, N. (1983) DNA methylation and gene activity. *Annual Review of Biochemistry*, **52**, 93–124. An excellent account of the methylation of bases of DNA and its possible roles in regulating gene expression.

Exercise 6

You are provided with the following oligodeoxyribonucleotide:

CCGGATCCAAGCTT
GGCCTAGGTTCGAA

which of the restriction endonucleases in Table 8.18 would produce the largest number of fragments?

The most widely used of these enzymes are the type II. Type II restriction endonucleases differ from the other types in requiring only Mg^{2+} for activity. They are always dimeric with M_r 30–40 000. Type II restriction endonucleases recognize sequences 4 to 6 nucleotides long, called **palindromes**: the order of bases being the same in both strands but running in opposite directions. Type II enzymes hydrolyse at fixed sites *within* the recognition sequence (Table 8.18 and Fig. 5.7), generating a unique set of fragments for any given DNA molecule. These fragments may be separated by electrophoresis to give a specific and characteristic banding pattern. Cleavage by a different enzyme would give a different pattern (Fig. 8.27). A bacterial chromosome, containing about 3×10^6 base pairs would be hydrolysed by any single type II enzyme to several hundred fragments. A small viral or plasmid DNA may, however, have only a handful of recognition sites, and so generate only a limited number of fragments.

8.12 Overview

Organisms are composed of a remarkable variety of molecules. A typical bacterium may contain about 2000 proteins, eukaryotic cells about 50 000. In addition, many other molecules contribute to the diversity of organisms. This enormous amount of information is coded for by giant molecules called nucleic acids. DNA is the genomic material of cellular organisms, but RNA is used by some viruses for this purpose.

DNA shows a variety of detailed molecular structures, however, all DNA

Palindrome: a word or phrase that reads the same backwards or forward e.g. madam, refer consequently sequences of DNA that read the same in either direction. From the Greek palin, *again, and* dromos, *a running.*

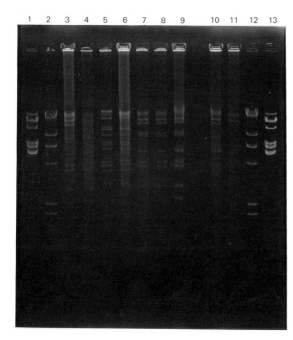

Fig. 8.27 Electrophoretic separations on a slab gel of digests of λ-phage and plasmid nucleic acids by restriction endonucleases. Lanes 1 and 2 show digestion products of DNA by *Bam*HI and *Hind*III respectively. These digests are repeated in lanes 13 and 14. Lanes 3 and 4 shows effects of *Hind*III and lanes 5 and 6 *Bam*III digestion products on the plasmid R39 isolated from *Salmonella johannersburg*. Lanes 7 and 9 show digest of R39, from *E. coli*, with *Hind*III. Lanes 11 and 12 show products of digestion of a deletion mutant of R39 with *Hind*III. Note the different banding patterns obtained from the action of different nucleic acids with the same restriction endonuclease. Courtesy of Drs P. and P.C. Gowland, Department of Biological Sciences, Manchester Polytechnic.

molecules basically have a double-helical structure. This consists of two sugar–phosphate chains on the outside of the molecule while internally nitrogenous bases are hydrogen-bonded together in pairs of near-identical size. Thus, although the bases are of irregular shape, the molecule can adapt a regular helical conformation. Further, the sequence of bases constitutes the genetic information. Variations in the sequence between different individuals gives rise to the bewildering complexity of organisms. The bases hydrogen-bond together in specific (complementary) pairs. Hence, separation of the duplex allows the base sequences to be copied forming two daughter helices, essentially identical to the original, parental, DNA molecule and this is the basis of replication in biological systems.

The information in the base sequence of DNA can also be copied to give complementary RNA molecules. The RNA molecule can then be translated to give the protein products of the structural genes. Thus, RNA allows the genome to be expressed giving rise to individual organisms; without RNA the genetic message would be inert.

The DNA double helix is not a static molecule. Conformational changes throughout the molecule aid in its interactions with other molecules. In eukaryotes, basic proteins called histones stabilize, protect and condense the extended structure of the DNA molecule, allowing it to pack within the confined space of the cell. Other interactive proteins are important in the regulation of expression of the genomes. This enables the organism to develop and respond to changes in the environment. Thus, the marvellous diversity of living systems, apparent at the macroscopic level, has its origins in the informational content of genomic nucleic acids.

Box 8.3
*Recombinant DNA
technology*

Type II restriction endonucleases are proving to be invaluable tools in many biochemical studies. They allow large molecules of DNA to be reproducibly and specifically cleaved into smaller, manageable sections, each of which can then be studied more easily.

Type II enzymes can cut both strands at the symmetrical centre of the palindrome site or asymmetrically. The enzyme *Hae*IV leaves products with flush or blunt ends, whereas *Eco*RI generates overhangs that constitute **cohesive or sticky ends**. If a second DNA molecule is hydrolysed with the *same* restriction enzyme, then fragments with complementary sticky ends are generated. These can recognize the fragments of the first molecule and bind to them. The non-covalent join can be made permanent (covalent) by using the enzyme **DNA ligase** to produce a new DNA molecule that is a recombination product of the two original DNA molecules.

These types of investigations form the basis of **recombinant DNA technology**. The figure shows how a recombinant molecule can be formed. A bacterial plasmid is cleaved with a restriction endonuclease to produce pieces with cohesive ends. A piece of DNA, from the genome perhaps of a completely different species, containing the gene of interest, may be generated using the same enzyme. The plasmid and DNA recognize each other by virtue of their complementary sticky ends and are ligated to give a recombinant plasmid.

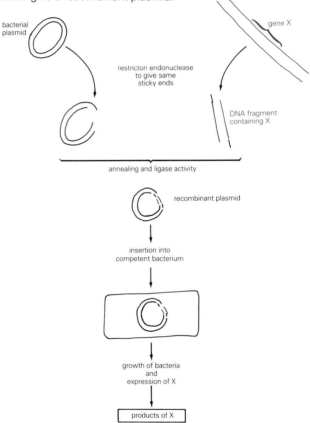

bacterial plasmid

gene X

restricton endonuclease to give same sticky ends

DNA fragment containing X

annealing and ligase activity

recombinant plasmid

insertion into competent bacterium

growth of bacteria and expression of X

products of X

Hardy, K. (1986) *Bacterial Plasmids*, 2nd edn. Van Nostrand Reinhold, UK. A short up-to-date text dealing with various aspects of the molecular biology and biology of plasmids.
Watson, J.D., Tooze, J. and Kurtz, D.T. (1983) *Recombinant DNA: A Short Course*. W.H. Freeman, New York. An extremely good short introduction to recombinant DNA technology. Written and illustrated in the style of the journal *Scientific American*.

The recombinant plasmid is then inserted into the host bacterial cell, where the foreign gene is expressed, in other words the recipient bacterium synthesizes the product of another species. Recombinant DNA technology is a powerful technique. It is already being used to synthesize large amounts of biomolecules that were previously difficult to obtain in sufficiently large quantities for research or therapeutic uses. Examples are insulin ('humulin' is recombinant human insulin) and interferons. *Molecular Biology and Biotechnology*, Chapter 9, gives a more extended explanation of recombinant DNA technology.

Table 8.18 *Examples of some restriction endonucleases, the microorganisms from which they are obtained and the palindromes recognized (sites of cleavage are highlighted)*

Restriction endonuclease	Microorganism	Palindrome recognized
*Bam*HI	*Bacillus amylolquifaciens* H	$\downarrow$ ...GGATCC... ...CCTAGG... $\uparrow$
*Eco*RI	*E. coli* R	$\downarrow$ ...GAATTC... ...CTTAAG... $\uparrow$
*Hae*II	*Haemophilus aegyptus*	$\downarrow$ ...PuGCGCPy... ...PyCGCGPu... $\uparrow$
*Hind*III	*H. influenzae* Rd	$\downarrow$ ...AAGCTT... ...TTCGAA... $\uparrow$
*Hae*III	*H. aegyptus*	$\downarrow$ ...CCGG... ...GGCC... $\uparrow$
*Sma*I	*Serratia marcescens* Sb	$\downarrow$ ...CCCGGG... ...GGGCCC... $\uparrow$

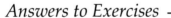

Answers to Exercises

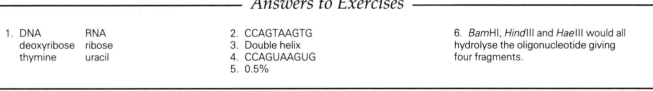

1. DNA   RNA
   deoxyribose   ribose
   thymine   uracil

2. CCAGTAAGTG
3. Double helix
4. CCAGUAAGUG
5. 0.5%

6. *Bam*HI, *Hind*III and *Hae*III would all hydrolyse the oligonucleotide giving four fragments.

QUESTIONS

FILL IN THE BLANKS

1. Nucleic acids are polymers of _____ linked by _____ bonds through the _____ and _____ carbon atoms of adjacent _____ residues. Each base pair of DNA occupies _____ of the length of the molecule and _____ base pairs form one complete turn of the _____ . A molecule of DNA consists of a double helix in which the chains of _____ run _____ to each other. The duplex forms a _____ _____ helix.

The two halves of the double helix are held together by _____ _____ between _____ bases. In _____ _____ DNA, the molar amount of _____ is equivalent to that of _____ , while that of _____ is equivalent to _____ .

_____ of the chemical bonds in each _____ are _____ allowing DNA to assume one of a number of _____ structures. The commonest is _____-DNA, while _____ is a _____ handed helix.

Choose from: B, Z, adenine, antiparallel, complementary, cytosine, deoxyribo-nucleotide, deoxyribonucleotides, double stranded, guanine, helix, hydrogen bonds, left nucleotides, phosphodiester, right handed, rotatable, secondary, six, sugar, ten, thymine, 0.34 nm, 3',5'.

MULTIPLE-CHOICE QUESTIONS

2. Identify the odd one out:
A. A – T
B. G – C
C. A – U
D. T – C

3. Which of the following best summarizes Chargaff's ratios?
A. A/T = G/C
B. (A+ T) = (G + C)
C. (A + C) = (G + T)
D. (A+ T)/(G + C) = 1

4. Which of the following sequences are complementary?
A. 5'pdApdTpdGp3' : 5'pdTpdApdCp3'
B. 5'pdGpdApdC 3' : 5'pdGpdTpdC 3'
C. 5'pdCpdCpdT 3' : 3'pdGpdGpdA 5'
D. GCCGT : CGGCA

5. Which of the following are true?
A. Methylation of guanine and cytosine is common in plants.
B. Complementary base pairing is only possible in DNA.
C. Alternating G,C bases favour B-DNA.
D. A high humidity favours Z-DNA.

SHORT-ANSWER QUESTIONS

6. (a) The DNA of the virus ϕX174 is single stranded. If the cytosine content is 18%, what can be deduced about the proportion of the other bases?
(b) Tobacco mosaic virus (TMV) contains double-stranded RNA. If the adenine content is 28% what is the most probable composition of the genome?

7. The following diagram represents an electron micrograph of a disrupted viral particle (VP) with its circular DNA released. If the viral particle has a diameter of 40 nm, what is the maximum possible number of codons in a transcript of the DNA?

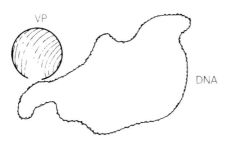

8. The molar guanine and cytosine content of samples of DNA isolated from several species, together with the melting temperatures associated with these samples, are given in the table:

(a) Plot a graph to illustrate these results.
(b) Using the graph formulate an expression that relates the melting temperature of a sample of DNA to its purine content.
(c) During an investigation of two further samples of DNA, the labels of the samples listing the results became mixed up. The investigator decided that sample A (G + G = 39.6%) had a T_m of 85.5°C, while sample B (G + C = 41.8%) had a T_m of 86.5°C. Was the investigator justified in this assumption? Explain your answer.

Species	G + C (%)	T_m (°C)
Yeast spp.	35.8	84.0
Human (spermatozoa)	37.5	84.7
Rat liver	40.0	85.7
Bacillus subtilis	44.0	87.5
Escherichia coli	51.0	90.0
Pseudomonas aeruginosa	68.0	97.0
Mycobacterium tuberculosis	70.3	98.00

9. The following oligonucleotide was treated with bovine pancreatic RNase I. Four products were isolated from the reaction mixture:

A–G–U–G–G–C–A–G–C–A–G–G–A
RNase I   ↓
A–G–Up + G–G–Cp + A–G–Cp + A–G–G–A

Using the above results, explain the specificity of action of RNase I as fully as possible.

10. A biochemist recently isolated a nuclease enzyme from his favourite bacterium, *Runfasttoloo*. When the oligonucleotide, X, was incubated with extracts of the nuclease, the following products were obtained.

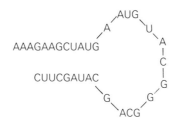

AAAGAAGp + CUAUGAAUGUACGGGCAGCAUAGp + CUUC

Explain the substrate specificity of this enzyme.

11. An oligoribonucleotide was hydrolysed with the aid of some of the nucleases listed in Table 9.18.:
(a) Pancreatic ribonuclease produced two fragments, a trinucleotide and a dinucleotide.
(b) Treatment with sodium hydroxide produced pAp, uracil, and 3′ phosphates of A, C and G.
(c) Treatment with takadiastase T_1 produced uracil and a tetranucleotide.
What is the sequence of the nucleic acid?

ESSAY QUESTION

12. Write an essay on the structure of DNA, emphasizing those features that make it suitable for storing and transmitting genetic information.

Answers to Questions

Chapter 1

1. Biological membranes surround all cells as well as most cell *organelles*. They are constructed from a *bilayer* made of *lipid*, in which the molecules point their *hydrophobic* ends towards each other. However, membranes constructed solely in this way would be *impermeable* to both hydrophilic and hydrophobic molecules. Therefore, the other major component of membranes is *protein* which can regulate the *movement* of materials into and out of cells. These molecules characteristically have *hydrophobic* regions on their surfaces whereby they interact with the membrane *lipid*. This is in contrast to the majority of globular *proteins* whose *hydrophobic* regions tend to be hidden inside the molecules.

2. Biological macromolecules are typically formed from smaller molecules by linking them together in *condensation* reactions. Such reactions require an input of *energy* to drive them, and the overall process involves removal of *water*. Biological macromolecules can *interact* with other molecules, both large and small, in highly *specific* ways. This is possible because macromolecules can form *complementary* shapes that recognize each other and *bind* tightly together. The bonds involved in such *interactions* are almost always *weak*, *non-covalent* bonds. Although such bonds are individually much *weaker* than C–C bonds, the fact that there are very *many* of them involved in an interaction makes the union *strong*.

3. The protein haemoglobin is made up of *four* subunits, identical in *pairs*. Each subunit contains one *haem* group which has a central *iron* atom that functions in *oxygen* transport. The subunits fit very tightly together and are joined by *non-covalent* bonds. The haem groups fit into *hydrophobic* clefts in the polypeptides and are linked to the protein *non-covalently*. Because of the way in which the subunits fit tightly together, a change in *shape* of one subunit leads to a *conformational* change in the other subunits. A consequence of this behaviour is that the *oxygen* binding curve for haemoglobin is *sigmoidal* in shape.

4. A, D, E.
5. E.
6. A, B, C.
7. A, B, C, D.
8. E.
9. B, D.
10. A, B, D.
11. A, B, D, E.
12. B.
13. • Protein: enzyme catalysis, antibody activity, structural.
• Nucleic acids: carrying genetic information (DNA), acting as an adaptor during protein synthesis (tRNA), carrying information from nuclear DNA to site of protein synthesis on the ribosomes (RNA).
• Polysaccharides: store of energy (starch, glycogen), mechanical structure (cellulose), gums and mucilages in plants (protection).
14. Weak bonds form easily but reversibly. Biological structures therefore come together only when shapes are complementary; units are self-assembling. Structures can easily be taken apart at the appropriate times with little or no expenditure of energy.
15. (a) DNA carries information in the genes on the chromosomes.
(b) Phospholipids form the foundation of the biological membranes.
(c) Enzymes catalyse reactions with high specificity enabling the reactions to proceed at low (biological) temperatures

and ensuring minimal by-product formation.

16. A membrane consisting only of a bilayer of phospholipid would be impermeable to both hydrophobic and hydrophilic molecules. Compounds need to get into and out of cells. Therefore proteins are present in the liquid bilayer to allow this to occur.

17. Proteins: amino acids link via peptide bonds. Each amino acid has a side-chain with distinctive properties. The sequence of different side-chains in a protein gives it its distinctive properties and enables it to carry out its biological function.

- Polysaccharides: the monosaccharide sugars have numerous hydroxyl groups by which the units may be linked together enabling a variety of different properties and structures to be generated in the polysaccharides. The range of possibilities is expanded by there being a range of different monosaccharides which, in addition, may be further modified by adding sulphate or amino groups.
- Nucleic acids: the four different bases of the nucleosides, in a linear sequence enable information to be carried in the form of a code. Because the bases 'pair', DNA may be duplicated at cell division.

18. Heat denatures proteins (and also disrupts structures such as membranes) and so large temperature differences could not be achieved. Heat engines require heat to go from a region of high temperature to a region of low temperature. The greater the temperature difference the higher the efficiency possible.

Chapter 2

1. The forces that maintain the three-dimensional structure of a protein are mainly non-*covalent*. They include *hydrophobic* interactions, *van der Waals* forces, *hydrogen* bonding and *salt bridges*.

The structure of a protein molecule comprises a *hierarchy* of structural *levels*. Secondary structures are *regularly* shaped conformations stabilized by *hydrogen bonds*. Folding of secondary structures generates *supersecondary* structures, which associate to form

domains. It is often difficult to distinguish between these last two levels of structure. The *tertiary* structure of a *globular* protein is the completely folded polypeptide chain and includes one or more *domains*. Protein molecules are often composed of several polypeptides, each referred to as a *subunit*; the overall assembly is called the *quaternary* structure of the protein.

2. (a) surface
   (b) interior
   (c) surface
   (d) surface
3. Two.
4. Look for extended, continuous sequences of predominantly hydrophobic amino acid residues.
5. A and B.
6. A. True.
   B. False.
   C. False.
   D. False.
   E. True.
7. α,β,α connected together by two β-turns (the second Type II).
8. (i) a, 3; b, 2; c, 4; d, 1.
   (ii) No.
   (iii) Surface.

Chapter 3

1. Keratins are a *family* of *fibrous* proteins found in *hair, nail* and *horn*. Their main functions are the *protection* of the external surfaces of the animal but they also have an important intracellular role in the *cytoskeleton*. Their amino acid composition shows a high content of *cysteine* residues and amino acid residues with *hydrophobic* side-chains as well as polar side-chains. Their primary structure shows a *seven* amino acid repeat in which the first and fourth residues are *hydrophobic*, the fifth and seventh *polar*. The secondary structure is *α-helical* and *three or four* helices interact to form a *microfilament*. Individual *α-helices* are cross-linked both within and between *microfilaments* by *disulphide bonds*.

2. The predominant secondary structure in silk is the *β-pleated sheet*. This structure is resistant to *stretching* because the tension is borne by *covalent bonds* in a backbone that is nearly fully *extended*. On one side of the sheet

glycine is the exposed amino acid side-chain, and on the other *alanine* and *serine* residues. Adjacent sheets lie with *similar* amino acid residues together.

3. The amino acid composition of collagen is unusual in that it contains the residues *hydroxyproline* and *hydroxylysine*. Assembly of collagen polypeptides to form a *triple* helix requires the presence of *extension* peptides, later cleaved to form *tropocollagen*. Molecules form a *staggered* structure called a *microfibril* and several of these form a mature *collagen fibre*. The strength of fibres is increased by *cross-links* involving the amino acids *lysine* and *allysine*. Collagen is also glycosylated by addition of *glucose* and *galactose* sugar residues to *hydroxylysine* in the polypeptide chain.

4. The other major protein of connective tissue is *elastin*. Like collagen, it contains a high proportion of *glycine* and *proline* residues, but has less *hydroxyproline*. It forms a three-dimensional network of monomers cross-linked through *lysine* and *allysine* residues, forming the unusual amino acid structures of *desmosine* and *isodesmosine*.

5. B.
6. A, B and E.
7. C.
8. A, B, C, D and E.
9. A, B, C and E.
10. B.
11. A, B and C.
12. A and C.
13. C, D and E.
14. Proteins with a high content of α-helices; elastic to a certain amount of tension, then snap; intertwine to form long, strong fibres, which are cross-linked to increase rigidity.
15. Amino acid composition; modified amino acids; lysine cross-links; no disulphide bonds (important structures in secreted globular proteins); triple helix; shape; polymerization; insolubility; repetitive sequence.
16. See Figure 3.9.
17. See Box 3.8.
18. Fibrous proteins tend to have an unbalanced ratio of amino acid residues, being rich in a few and poor in most. They are often deficient in some or all of those amino acids that cannot be synthesized and are therefore essential dietary components (arginine, histidine, isoleucine, leucine, lysine, methionine,

phenylalanine, threonine, tryptophan and valine).

19. See chapter Overview for properties. Functions depend on the predominant secondary structure, e.g. extensibility of α-helix allows keratins to stretch; inelasticity of β-pleated sheet allows silk to be very strong and supple; rigidity of triple helix makes collagen a strong basis for skeletal structures and connective tissue; three-dimensional cross-linking of random coils in elastin allows the stretch and elasticity needed for example in ligaments and skin.

20. Structure and properties are largely dependent on the major secondary structures – keratin in hair and fibroin in silk. In keratin the predominant secondary structure is the α-helix and an amino acid composition and sequence which allows for the interaction and cross-linking of several helices. The helices can be stretched because hydrogen bonds holding them together can be broken relatively easily; disulphide cross-links help them to relax. The more disulphide bonds, the greater the rigidity. In silk the predominant secondary structure is the β-pleated sheet. It is nearly fully extended so will not stretch and is very strong in the direction of the sheet because this is covalently bonded. However, relatively weak forces hold adjacent sheets together so they will move relative to one another providing suppleness.

21. Both are insoluble, fibrous, highly cross-linked structures and have an amino acid composition in which a few residues predominate. They have repeated sequences, modified amino acids and cross-links through allysine. Collagen is a rigid rod-like molecule, is not extensible because of the nature of the triple helix, and contains a high proportion of hydroxyproline. Tropocollagen monomers are organized into cross-linked fibres. Collagen provides strength to a tissue, e.g. bone, cartilage. Elastin monomers form random coils, which can be greatly extended. Monomers are cross-linked through desmosine and isodesmosine so that tissues with a lot of elastin can be stretched in several directions and relax to their original shape and size. Elastin is particularly important in ligaments and skin.

Chapter 4

1. Enzyme catalysis does not alter the *equilibrium* constant of a reaction but does reduce the *activation energy* barrier. The Michaelis constant has dimensions of *substrate concentration* and represents the concentration of *substrate* at which the velocity is *half* V_{max}. The constant k_{cat} is derived from the maximum velocity by accounting for the total concentration of *enzyme*. The value of k_{cat}/K_m represents the *second* order rate constant for the reaction
$$E + S \rightarrow E + P$$
and is a measure of the *specificity* of an enzyme.
2. C, D and E.
3. E.
4. Specificity constant with A is $25/1.5 = 16.7$ dm^3 mol^{-1} s^{-1}.

Specificity constant with B is $90/5.5 = 16.4$ dm^3 mol^{-1} s^{-1}. Therefore both A and B are hydrolysed with essentially the same effectiveness.
5. $K_m = 5.3$ mmol dm^{-3}; $V_{max} = 9.7$ μmol min^{-1}.
6. A. $K_m = 8.3$ mmol dm^{-3}; $V_{max} = 34.5$ μmol min^{-1}.
      B. K_m increased to 13.2 mmol dm^{-3}, while V_{max} was unaffected, i.e. competitive inhibition.
      C. $K_i = 15$ mmol dm^{-3}.

Chapter 5

1. The original theory to explain the binding of substrate to enzyme based on the *complementarity* of structure between these two components was called the *lock and key* model. The concept of complementarity in the transition state is called the *induced fit* model. Enzyme catalysis occurs by a *lowering* of the *activation* energy of the *transition* state.

Hen egg white lysozyme can catalyse the hydrolysis of *chitin* which is a polysaccharide comprised of $\beta1$–4-linked N-acetylglucosamine residues. The hexasaccharide is cleaved to produce the *tetrasaccharide* and *disaccharide* products. The two amino acid residues directly involved in the hydrolysis are *Glu-35* and *Asp-52*.
2. A, E and F.

3. A, C, D and E.
4. Replace the hydrophobic amino acid at the base of the binding pocket with one that is negatively charged.
5. Thiol proteinases (–SH group at active site), e.g. ficin, papain, bromelain; acid proteinases (–COOH at active site), e.g. pepsin, chymosin; dehydrogenases (common nucleotide binding regions).
6. This is to restrict the access of water to the active site and therefore produce a chemical environment different from that of the solution in which the enzyme and substrate are dissolved. Also enables the binding between substrate and enzyme to be maximized.
7. Monoclonal antibodies should be raised to an appropriate analogue of the proposed substrate. Ideally, this substrate analogue should represent the transition-state intermediate in the hydrolysis of the ester. Therefore, on binding of the substrate itself to the monoclonal antibody it will assume a strained conformation that will be conducive to hydrolysis of the ester bond.

8. (a) Analysis of structure by X-ray crystallography.
(b) Identification of active-site amino acids by chemical modification studies.
(c) Use of competitive inhibitors and model substrates.
(d) Kinetic studies and pH profiles.
(e) Site-directed mutagenesis of codons for appropriate amino acid residues.

Chapter 6

1. *Amylose* differs from *amylopectin* in having less branching. However, *glycogen* has even more $\alpha1$–6 branch points than *amylopectin*, although all three polysaccharides are composed of α-D-*glucose* residues. *Cellulose* consists of β-D-glucose residues, while the similar carbohydrate, *chitin* contains N-acetylglucosamine.

The configuration of differing polysaccharides vary. Amylose forms a *helical* structure. Cellulose and chitin both consist of extended *ribbons*. Pectate has a *buckled ribbon* conformation which is stabilized by Ca^{2+} binding to adjacent *galactouronate* residues giving an *egg-box* structure.

Peptidoglycans of bacteria occur in a number of types although all are cross-linked in some manner forming a single 'bag-shaped' molecule which gives extracellular support. Proteoglycans of animal cells also occur extracellularly forming part of the cell matrix. A proteoglycan complex consists of polysaccharides called glycosaminoglycans linked to proteins all bound to a single molecule of hyaluronate.

Complex oligosaccharides usually consist of branched structures which lack repeating sequences. These are therefore suitable for participating in recognition phenomena. Such oligosaccharides generally occur as part of glycoproteins and glycolipids.
2. A2,3,4; B1,3,4; C2,3,4; D1,5,6.
3. A3; B1; C2; D2; E1; F1.
4. A1,2,3; B1,4; C3; D2,3; E2.
5. Gelatinize starch in boiling water. Separate amylose using an alcohol such as butanol to 'guest' complex with amylose, thus precipitating it. Purity could be checked by periodate oxidation (one IO_4^- reduced per glucose residue and no formate released).
6. Break H-bonding between polymers (the probable cause of insolubility). This could be done by treatment with alkali or dimethyl sulphoxide.
7. By examining sizes of the products using chromatography (gel, paper; HPLC). A range of sizes released during hydrolysis would indicate random attack.
8. Periodate oxidation would not act on 1–3-glucan. Methylation analysis would also give different products. For example, 2,3,6 or 2,4,6-tri-O-methyl derivatives respectively. Enzymes with specificity for hydrolysing the two types of bond could also distinguish between the glucans, degrading the one and not the other. ^{13}C NMR would produce different spectra.
9. By incorporating labelled (^2H or ^3H)

water into the gel and studying isotope release.
10. Assuming they were soluble, by examining molecular sizes using gel filtration chromatography in the absence and presence of dissociating agents, for example alkali, dimethyl sulphoxide.
11. By looking at fragments produced by chemical or enzymic hydrolysis: (B–B) or (A–B–B) or (B–B–A) would indicate the former.

Chapter 7

1. Acylglycerols or neutral fats are composed of the alcohol glycerol and up to three fatty acids. Phospholipids always contain a phosphate group and usually other components. Glycolipids lack phosphate but contain carbohydrate. Triacylglycerols are important as energy reserves and are stored in adipose tissue. Complex lipids are important constituents of cell membranes. Isoprenoid compounds or polyprenyl compounds are derived from the condensation of isoprene units. Major compounds of these types are the steroids and lipid vitamins. The former group includes the sterols of which cholesterol is the most abundant. Cholesterol is important as a constituent of biomembranes, reducing their fluidity, and as a precursor of bile acids and steroid hormones. Important isoprenoid vitamins are the A, D, E and K vitamins.
2. C.
3. A, C and D.
4. D.
5. Number of protein molecules = 2.47×10^7.
Number of lipid molecules = 2.35×10^9.
6. β-carotene, 8; retinol, 4.

Chapter 8

1. Nucleic acids are polymers of nucleotides linked by phosphodiester bonds through the 5′ and 3′ carbon atoms of adjacent sugar residues. Each base pair of DNA occupies 0.34 nm of the length of the molecule and ten base pairs form one complete turn of the helix. A molecule of DNA consists of a double helix in which the chains of deoxyribonucleotides run antiparallel to each other. The duplex forms a right handed helix.

The two halves of the double helix are held together by hydrogen bonds between complementary bases. In double stranded DNA, the molar amount of adenine is equivalent to that of thymine, while that of guanine is equivalent to cytosine.

Six of the chemical bonds in each deoxyribonucleotide are rotatable allowing DNA to assume one of a number of secondary structures. The commonest is B-DNA, while Z-DNA is a left handed helix.
2. D.
3. D.
4. B, C and E.
5. A. True. B. False. C. False. D. False.
6. (a) A + T + G = 82%.
(b) U = 28%; G = C = 22%.
7. 1764.
8. (b) $T_m = 0.4 (G + C) + 69.6$.
(c) Yes.
9. Endonuclease, 5′ or b, pyrimidine–purine bonds.
10. Endonuclease, 5′ or b, restriction site G–C.
11. pAACGU
12. Stable: phosphodiester bonds; interchain hydrogen bonding; lack of 2′-OH groups.
Information: capacity to store huge amounts of information because of ability to vary sequence of large numbers of bases.

Glossary

A

Active site: *the region of an enzyme that is responsible for catalysis as well as specific recognition of a substrate. It was realized many years ago that large enzyme molecules with a single active site could be in physical contact with small substrate molecules over a very small proportion of their total surface areas.*

Allosteric: *from the Greek allos, other, ie. the binding of a ligand to a site on the protein molecule not associated with its normal biological activity changes the shape and therefore activity of the protein.*

B

Basement membrane: *layer of connective tissue to which cells are attached.*

Biomolecules: *molecules, large and small, found in organisms and manufactured by cells.*

Bond energy: *the energy required to break a chemical bond. Hydrophobic effects and hydrogen bonds are weak, with bond energies of about 4–8 and less then 20 kJ mol^{-1} respectively. Covalent bonds are much stronger. Disulphide bonds have energies of about 200 kJ mol^{-1}.*

C

Carboxonium ion: *a carbon atom that posseses a transiently stable positive charge. Carbonium ions are often produced as reaction intermediates.*

Catabolism: *the breaking down, usually by a process of oxidation, of food materials or stored molecules to release biologically useful energy to drive life processes.*

Catalysts: *these speed up a chemical reaction by lowering the activation energy. Many chemical catalysts are known and used (e.g. platinized asbestos) but biological catalysts, enzymes, are infinitely more specific in their reactions. In principle a catalyst should be unchanged at the end of the reaction.*

Chaotropic agents: *those that increase disorder (chaos), e.g. some salts which destroy the highly ordered state that water molecules normally adopt.*

Cholesterol: *a sterol lipid found in most cell membranes, serum and bile. From the Greek chole, bile and stereos, solid.*

Chromosome: *a discrete nucleoprotein complex. In eukaryotic cells chromosomes are most clearly seen as metaphase. From the Latin chroma, colour, and some, body, refering to the intense staining that clearly reveals mitotic figures during cell division.*

Coenzymes: *organic molecules, usually vitamin derivatives, that participate in, and are essential for, enzyme catalysed-reactions.*

Cofactors: *non-protein compounds that are essential for the catalytic activity of an enzyme. Cofactors that are tightly bound to enzymes are called prosthetic groups. Cofactors may be small organic molecules (that is coenzymes) or metal ions, e.g. Mg^{2+}, Ca^{2+}.*

Collagens: *a family of glycoproteins found in connective tissue, composed of cross-linked triple helices.*

Collagenases: *enzymes that degrade collagen. Notable examples occur during the development of gas gangrene (causative organism Clostridium histolyteum) and in the metamorphosis of tadpole tails.*

Configurational change: *rearrangement of covalent bond structure, e.g. mannose and glucose have different configurations of –H and –OH about C-2.*

Conformation: *the general structure or shape.*

Covalent modification: *adding of an additional group to the enzyme molecule by forming a covalent chemical bond. Alternatively a covalent bond may be broken to remove part of the molecule.*

Cruciform: *like a cross in shape.*

Cytokeratins: *keratin polypeptides found as part of the cytoskeleton, the intracellular scaffold.*

D

Desmosine and isodesmosine: *derivatives of lysine, which occur in elastin.*

Dielectric constant: *a measure of the relative effect a substrate has on the force with which two oppositely charged groups attract one another. The larger the dielectric constant, the weaker the force.*

E

Elastase: *an enzyme that degrades elastin; specific for neutral aliphatic amino acids.*

Elastin: *connective tissue protein cross-linked in such a way as to provide elasticity.*

Enzymes: *macromolecular biological catalysts. The majority of enzymes are proteins, but recently certain RNA molecules have been discovered to possess catalytic activity. From the Greek en zyme, ferment.*

Executive: *the agency with the skill or abilities to perform functions, i.e. the 'doers'.*

Extracellular matrix: *the substance in which connective tissue cells, like all cells, are distributed, being largely a mixture of proteins and complex carbohydrates secreted by the cells.*

F

Fibroin: *the main protein of silk, largely composed of β-pleated sheets, making the structure strong and inelastic.*

G

Gelatin: *the soluble product formed when collagen is boiled in water; the triple helices are disrupted by the heat and do not reform in the absence of the extension peptides.*

Genome: *the full complement of genes of an organism. From the Greek genea, breed.*

Greek key: *a supersecondary structure named after its resemblance to a pattern common on ancient Greek pottery.*

Gyrases/topoisomerases: *enzymes that alter the rotational stress in DNA molecules. From the Greek gyrus, circle, and topos, place and isos, equal.*

H

Haemoglobin and myoglobin: *derived from the Greek haimo and myos, blood and muscle respectively.*

Hepatocyte: *parenchymal liver cell. From the Greek hepar, liver and kytos, cell.*

Heteroduplex: *double helix in which one strand is composed of polydeoxyribonucleotide residues and the other of polyribonucleotide residues.*

Heteropolymers: *combinations of more than one type of monomer. Structures may be simple as in -(A.B)$_n$)- or highly complicated as on -AC × BABDEG-.*

Homologue: *one member of a family of structurally related molecules. This can be illustrated simply by reference to the aliphatic alcohols which form a homologous series n-methanol (CH$_3$OH), n-ethanol (CH$_3$CH$_2$OH), m-propanol (CH$_3$CH$_2$CH$_2$OH), etc. Homologous proteins are ones with a substantial amount of amino acid sequence in common.*

Homopolymers: *formed by combining identical monomer building blocks e.g. AAAAA…*

Hydrophobic: *means, literally, water-hating, in contrast to hydrophilic, which means water-loving.*

These terms may be used to describe whole molecules or groups that form part of molecules. Thus, a molecule may be said to have hydrophobic regions.

K

Keratins: a family of proteins whose secondary structure is predominantly an α-helix, particularly found in external protections such as hair, and in epithelial cells.

L

Lactone: formed by the intramolecular condensation of hydroxyl groups to give a cyclic ester with the resultant liberation of a molecule of water.

Legislative: the branch or agency with the power to make laws and set limits.

Lipids: the oily, greasy and waxy materials that can be extracted from organisms. From the Greek lipos, fat.

M

Macromolecules: means 'large molecule', but this is not very helpful as it does not define what 'large' is. In biological systems it means molecules with M_r in the thousands to millions, and includes proteins, nucleic acids and polysaccharides. Vitamin B_{12} (cobalamin), M_r 1355, would not be regarded as macromolecule while insulin, M_r 5780, would.

Metabolism: the sum of reactions going on in a living cell. Anabolic reactions are the building-up reactions, and the catabolic reactions are the breaking-down (to provide energy) reactions.

Micelle: this was first used in 1879 to describe a colloidal particle with a double layer of ions round it, with detergents and phospholipids in water it refers to the small vesicles which form in which hydrophobic tails point inwards and hydrophilic heads point onwards.

Monosaccharides: a very large range of carbohydrates of general formula $(CH_2O)_n$, where n is commonly 3–7. Such units link together to form disaccharides, oligosaccharides and polysaccharides.

N

N, $C^α$, C': the atoms of an amino acid residue that contribute to the backbone of the polypeptide, i.e. the amino nitrogen, the α-carbon and carboxyl carbon atoms respectively.

Neokink: a localized distortion in the normal structure of B-DNA caused by an alteration in the orientation of the phosphodiester bonds.

Nuclein: original name for an extract of DNA, meaning isolated from the nucleus.

O

Oils and fats: these are triglcerides or triacyl-glycerols. Fats are solid at room temperature whereas oils are liquid.

Oligosaccharides: carbohydrates comprising 3–20 sugar residues. A Greek affix denotes the number of residues; tetrasaccharide (4), octasaccharide (8), etc. From the Greek, oligos, small.

Organic chemistry: the chemistry of carbon compounds (excepting carbonates and CO_2, CO, etc.); all the rest is inorganic chemistry.

P

P_{50}: the partial pressure of oxygen at which the haemoglobin or myoblobin is 50% saturated with oxygen.

Palindrome: word or phrase that reads the same backwards or forward e.g. madam, refer: consequently sequences of DNA that read the same in either direction. From the Greek palin, again, and dromos, running.

Photosynthesis: the process by which plants and some bacteria take in CO_2 and use light energy to convert it to highly reduced carbon compounds such as carbohydrate, $(CH_2O)_6$.

Polyhedron: a regular shaped solid bounded by plane faces. From the Greek polys, many, and hedra, sides.

Polyunsaturated: usually used in the context of fat or fatty acid, having multiple double bonds. From the Greek polys, much or many.

Post-translational modification: changes made to the structure of a protein after the polypeptide has been synthesized e.g. chemical changes to amino acids or cross-linking residues.

Protein: from the Greek protos, first. The Dutch chemist, Mulder, introduced this word in 1835 implying 'first rank', on the most important substance in living matter.

Pyrophosphate: formed by the combination of two (or more) phosphates with the removal of a molecule of water.

R

Recessive: characteristic that is only expressed if there is no dominant gene to mask its effects.

Residue: a monomer linked to others in a polymer.

S

Sialic acids: a number of related acidic sugars: N-acetyl, N-glycolyl, 4-7-8-9-O-acetyl neuraminic acid.

Sigmoidal: S-shaped.

Squames: the scale-like structures of which the stratum corneum of skin is composed.

Substrate: the compound that an enzyme converts into product during catalysis.

Substrate analogue: a compound with a similar chemical structure to the true substrate.

T

Teichoic acids: material found in the cell walls of bacteria, from the Greek teichos, wall.

Tocopherol: a lipid-soluble vitamin. The name alludes to the involvement of the vitamin in fertility. From the Greek tokos, child, and pherein, bear.

Tropocollagen: the triple helical structure of collagen remaining after the terminal prepeptides have been removed.

Tropoelastin: monomer of elastin.

V

Vitamin K: a lipid-soluble vitamin involved in blood clotting. The role in clotting was first observed by the Dane, Dam (K for koagulations vitamin).

Index

Page references to Tables are in *italic* and those to Figures are in **bold**. References to Boxes and Side-notes are indicated by (B) and (S) after the page numbers respectively.

Methylases (methyltransferases) 190 (B)

Methylated nitrogenous bases 168, *169–70*, 190 (B)

Methyltransferases (methylases) 190 (B)

Micelles 23, 24, 153

Michaelis-Menten equation 87–94

Microfibrils 75, 135

Microorganisms, elemental/ chemical composition *1*

Mitochondrial DNA 186–7

Molecular complementarity 6–7, 27, 33–4, 49, 111

Molecular interactions 6–7, 25, 34

Molecular recognition 33–4, 133, 156, 175

Monosaccharides 4–6, *125*

Mucins 127 (B)

Multifunctionality (macromolecules) 136 (B)

Myoglobin 54–5

Neokink (DNA) 111

Neural peptides 41

Neutral fats (triacylglycerols) 22, 147–8, 150 (B), 151

Nitrogenous bases 167–70

Non-histone proteins 187

Nuclear magnetic resonance, of enzymes 106

Nucleases (phosphodiesterases) 188–92

Nucleic acids 7 (B), 166–72

Nucleophiles 118 (S)

Nucleoprotein 166, 175

Nucleosides 7 (B), 171

Nucleotides 7 (B), 171

Oligosaccharides, complex 131–6, 156

Oligosaccharins 129 (B)

Opsin 162

Optical rotations 8 (B)

Organelles 19

Organic chemistry 2 nomenclature 4 (B)

Oxidation reactions 16, 148

Oxidoreductases *107*

Oxygen transport 54–60

Palindromes 190, *191*

Pectins *128*, 130

Pentose sugars 167

Peptide bonds 35, 42–3, *44*

Peptides 6, 35, 41

Peptidoglycan 129

Peptidohydrolases (proteases) 117 (S)

Perhydrocyclopentano- phenanthrene 158

Permeability, selective (membranes) 21

Peroxidases 153 (B)

Peroxidation 163

pH 11 (B), 116–17

Phages 184–5

Phenylalanine 38

Phosphatidyl choline (lecithin) 153

Phosphatidyl ethanolamine (cephalin) 153

Phosphatidyl serine 153

Phosphoacylglycerols (glycerol phospholipids) 153–4

Phosphoanhydride bonds 124

Phosphodiester bonds 171

Phosphodiesterases (nucleases) 188–92

Phosphofructokinase 101, 102 (S)

Phospholipids 22–3, 153–4

Phosphorylation 17, 101

Photosynthesis 14, 136, 161

Plants
cells 20 (B)
elemental composition *1*
polysaccharides 129, 131

Plasmids 185, 192 (B)

Polarity (molecular)
amino acids *36–8*
and solubility in water 12

Polydeoxyribonucleotides 171–2

Polymers 4–6

Polypeptides 6, 35

Polyprenyl lipids (isoprenoids) 156, 158–63

Polyribonucleotides 171

Polysaccharides
complex oligosaccharides 131–6
general structure 5 (B), 6, 124–8, 134–6
nomenclature 131 (B)

Polysaccharides, capsular 130

Polysaccharides, storage (reserve) 130–1

Polysaccharides, structural (fibrous/matrix) 128–30

Polyunsaturated fats 22 (B), 149

Post-translational modifications (proteins) 73

Proline *37*, 38

Prosthetic groups 108 (B)

Proteases (peptidohydro- lases) 117 (S)

Proteinases 117 (S)

Proteins
chemical composition 6 (B), 35–41
denaturation 60–2
general functions 7, 33–4
interactions with other molecules 25–7
post-translational modifications 73
synthesis 180–2
see also Fibrous proteins; Globular proteins

Proteoglycans 130

Pseudoplastic solutions 130, 131 (B)

Purines 167–8

Pyrimidines 167–8

Pyrophosphates 17

Pyruvate dehydrogenase complex (PDC) 50

Rancidity 153 (B)

Recombinant DNA 192 (B)

Reduction reactions 16

Replication (DNA) 28, 178, 182

Resilin 70

Restriction endonucleases 111–12, 189–92

Retinal 162

all-*trans* Retinol (vitamin A) 161–2

Retroviruses 178

Rhodopsin 51 (S), 162

Ribonucleic acid, *see* RNA

Ribonucleoproteins 180

Ribosomal RNA, *see* rRNA

Ribosomes 50, 181–2, 187

Ribozyme (RNA) 95 (B), 126 (S)

RNA
base pairing 178
as catalyst 95 (B), 126 (S)
chemical composition 7 (B), 166–72

mRNA 180–2

rRNA 180–2

snRNA 180

tRNA 168, 180–2

RNA, viral 184–5

RNA polymerases 180

Salinity 11 (B)

Salt bridges, *see* Ionic bonds (salt bridges)

Saponification 150 (B), 151–2

Saturated fatty acids 22 (B), 149

Scurvy 73 (B)

Selective permeability (membranes) 21

Serine *36*, 39

Serine proteinases 117–21

Sialic acid 155–6

Silks 69–70, 71 (B)

Site-directed mutagenesis, enzyme studies 106, 120

Skin 161–2
see also Keratins

Small nuclear RNA, *see* snRNA

Sphingolipidoses 157 (B)

Sphingolipids, *see* Glycolipids; Sphingomyelins

Sphingomyelins 154

Starch 130–1, 138–41

Steroids 158–9

Substrate analogues 97

Substrate specificity (enzymes) 27, 33, 111–14, 133

Subtilisin 121

Sugar-phosphates 17

Sulphatides 155

Superoxide dismutases 153 (B)

Synchrotron X-radiation, of enzymes 105 (B)

Taurine 161

Tautomerism, *see* Isomerism

Teichoic acids 130

Temperature, effect on enzymes 11 (B), 27, 96

Thalassaemias 56–7 (B)

Thermodynamics 15
and proteins folding 52–4

Threonine *36*, 39

Thrombin 163

Thymine 167–8

Tocopherols (vitamin E) 153 (B), 162–3

Tonofilaments 69

Topoisomerases 178

Transcription 180–2

Transfer ribonucleic acid, *see* tRNA

Transferases *107*

Transferrin 34

Translation 180–2

Transmembrane proteins 24, 34

Transport proteins 34

Triacylglycerols (neutral fats) 22, 147–8, 150 (B), 151

Tropocollagen 74

Trypsin 101 (S)

Tryptophan *36*, 38